HOMOGENEOUS HYDROGENATION

Catalysis by Metal Complexes

VOLUME 15

The titles published in this series are listed at the end of this volume.

HOMOGENEOUS HYDROGENATION

PENNY A. CHALONER

The School of Chemistry and Molecular Sciences, University of Sussex, Falmer, Brighton, United Kingdom

MIGUEL A. ESTERUELAS

Department of Inorganic Chemistry, University of Zaragoza, Zaragoza, Spain

FERENC JOÓ

Institute of Physical Chemistry, Kossuth Lajos University, Debrecen, Hungary

and

LUIS A. ORO

Department of Inorganic Chemistry, University of Zaragoza, Zaragoza, Spain

KLUWER ACADEMIC PUBLISHERS

DORDRECHT / BOSTON / LONDON

Library of Congress Cataloging-in-Publication Data

Homogeneous hydrogenation / by Penny A. Chaloner ... [et al.].
p. cm. -- (Catalysis by metal complexes ; v. 15)
Includes index.
ISBN 0-7923-2474-9
1. Hydrogenation. 2. Catalysts. I. Chaloner, Penny A.
II. Series.
QD281.H8H66 1993
574'.23--dc20 93-20972

ISBN 0-7923-2474-9

Published by Kluwer Academic Publishers,
P.O. Box 17, 3300 AA Dordrecht, The Netherlands.

Kluwer Academic Publishers incorporates
the publishing programmes of
Martinus Nijhoff, Dr W. Junk, D. Reidel and MTP Press.

Sold and distributed in the U.S.A. and Canada
by Kluwer Academic Publishers,
101 Philip Drive, Norwell, MA 02061, U.S.A.

In all other countries, sold and distributed
by Kluwer Academic Publishers Group,
P.O. Box 322, 3300 AH Dordrecht, The Netherlands.

Printed on acid-free paper

Printed in the Netherlands

TABLE OF CONTENTS

FOREWORD

During the past two decades several excellent reviews have been published on the topic of homogeneous hydrogenation, but the most relevant contribution has been the book by B.R. James (Homogeneous Hydrogenation, Wiley, 1973). This tremendously successful book gives a description of all the important results which served as solid basis for the extensive and dramatically fast later development of the field. Today it is hopeless to attempt a compilation of all the important results into one volume. Therefore the aim of the present work is not to supplement earlier reviews but to give a general account of the main principles and applications of homogeneous hydrogenation by transition metal complexes.

Special attention is devoted to the mechanisms by which homogeneous hydrogenation may occur, describing also some views about the role of the recently discovered complexes of molecular hydrogen (Esteruelas and Oro, Chapter 2). Sources of hydrogen, other than H_2 are also considered (Esteruelas and Oro, Chapter 3). The latest achievements in highly stereoselective hydrogenations made many new applications in organic synthesis possible. These applications are discussed by presenting examples of the reduction of important unsaturated substrates (Chaloner, Chapter 4). As a recent way of the execution of these reactions the hydrogenations in aqueous or biphasic/phase transfer catalysed systems is also described (Joó, Chapter 5). New developments with supported hydrogenation catalysts are included (Chaloner, Chapter 6). Finally, the biochemical way of H_2-activation is reviewed, highlighting the differences in realization of hydrogenation in natural and synthetic systems (Joó, Chapter 7).

There are many important reactions involving hydrogen such as hydroformylation or the Fischer-Tropsch synthesis (to mention only two) which – despite many mechanistic similarities – are better treated separately in their own context. Topics such as the photochemically assisted hydrogenation are not covered in separate sections, but examples are mentioned throughout the text; others, such as ionic hydrogenation are not covered at all.

We adhered to the IUPAC systematic nomenclature as much as possible. In case of compounds with well established trivial names, however, this may hinder the comparison with the original references. For this reason, the index contains the trivial, as well as the systematic names.

Despite the mentioned division of labor, all the chapters incorporate comments and suggestions from all the co-authors. This cooperation was a constant source of inspiration and advice, as was the support of our colleagues and co-workers. We are grateful to all of them. Special thanks are due to the Editor for his encouragement and understanding throughout the process of writing and also for his care exercised during the process of publishing. The financial support of the following organizations is gratefully acknowledged: the Hungarian National Research Foundation, OTKA (J.F.), the Spanish Ministry of Education and Science, DGICYT (M.A.E. and L.A.O.), and the Universitas Foundation (J.F.).

Brighton – Debrecen – Zaragoza
The Authors

CHAPTER 1

INTRODUCTION

The scope and importance of homogeneous hydrogenation

The term “homogeneous hydrogenation” refers to a chemical reaction in which one or more hydrogen atoms (and only those) are incorporated into the product(s) of the reaction, on the action of a catalyst dissolved in the same phase as the reactants. In the everyday use this term is often used in a more limited meaning, *i.e.* as the homogeneously catalysed addition of H_2 to unsaturated organic substrates. Without doubt, this latter class of reactions certainly comprises the major field of homogeneous hydrogenation. However, because of similarities of the catalysts and the mechanisms, two other processes must also be considered. First, H_2 is not the sole possible source of hydrogen, and other molecules (so called hydrogen donors, DH_2) are often found as reactants in hydrogenations. These are the catalytic *hydrogen transfer* reactions, or with a simpler phrase: transfer hydrogenations. Second, during or subsequent to the addition of hydrogen, either from H_2 or from DH_2 , fragments may be reductively split off the original substrate leading to products of *hydrogenolysis*.

Under ambient conditions H_2 is a rather unreactive molecule, therefore catalytic hydrogenations – either homogeneous or heterogeneous – always involve steps of H_2-activation. Activation results in a substantial transformation of molecular hydrogen, either to H^- (hydride), $H^\cdot$ (hydrogen radical) or H^+ (proton). In many reactions the subsequent transfer of these forms of hydrogen to the substrate was found less demanding; therefore the term homogeneous hydrogen activation is often used synonymously with homogeneous hydrogenation.

Reductions with molecular hydrogen (such as *e.g.* that of Fe(III) to Fe(II) or Ru(IV) to Ru(III)) do not always lead to H-containing products. Since the activation of H_2 implies the same mechanistic prerequisites as the more general hydrogen additions these simple reductions are also

P.A. Chaloner, M.A. Esteruelas, F. Joó and L.A. Oro, Homogeneous Hydrogenation 1–3.

considered in this book. However, despite their enormous importance in the early years of research into hydrogen activation, at present these reactions are of only marginal significance.

Although on the basis of dollar value of industrial production homogeneous hydrogenation methods are far behind such established industries as hydroformylation or polymerization, the past and present importance of homogeneous hydrogenation can hardly be overestimated. It played a key role in the fundamental understanding of catalytic reactions. It is tempting to speculate the origin of this central role but two important points certainly should be considered. First, while being rather unreactive, H_2 is still more reactive than some other important small molecules (CO, N_2, *etc.*) and its activation can be studied under relatively mild conditions. Second, and not completely independent of the first, the product distribution in a simple reaction such as alkene hydrogenation, generally poses less severe difficulties with regard to analysis and rationalization than the case of reactions under high pressure at elevated temperature. Consequently, hydrogenation offered the route from the very simple to the very complex, allowing accumulation of the necessary knowledge before the next leap in advancement. The impressive developments of organometallic chemistry in the last three decades gave further impetus to research into hydrogenation by expanding our understanding of the structure and reactivity of compounds serving as homogeneous catalysts. Since many of the homogeneous hydrogenations take place under mild conditions where powerful spectroscopic techniques can be easily used both to assign structures and to follow reaction kinetics, catalytic mechanisms are easier to study than in heterogeneous catalysis.

As a result of all these efforts the basic energetic and mechanistic questions of the activation of molecular hydrogen are well known, although the discovery of complexes of molecular hydrogen in 1984 may caution against such statements. Many hydrogenation reactions could be examined almost step by step along the reaction coordinate and clear experimental evidence could obtained for the composition and structure of reaction intermediates, in many cases supported by further detailed studies of their less reactive, often isolated analogues. In addition to the contribution of research into homogeneous hydrogenation on the development of *homogeneous catalysis* in general, the mechanistic and structural ideas developed during this research are very important for *heterogeneous catalysis*, as well.

In accordance with the spirit of the series, this book is devoted to homogeneous hydrogenations *catalyzed by metal complexes*. It is well known that many metals having partially filled *d* or *f* electron shells are able to function as hydrogenation catalysts, and the vast majority of homogeneous hydrogenation catalysts are comprised of *transition metal complexes*, due to the fact that they present several interesting features, such as

i) the ability to form strong bonds, in a variety of oxidation states, with compounds containing π-electron systems; this ability is particularly relevant for low valent platinum metal complexes allowing them to enter into redox cycles,
ii) the ability to accommodate several different ligands in their coordination sphere with a variability of the coordination number, and
iii) the possibility of modifying the electronic and/or steric environment at the active site by an appropriate selection of the participative as well as the non-participative ligands.

At this stage it is convenient to remark that a very important point to emerge from the study of homogeneous hydrogenation by metal complexes, and especially by those with tertiary phosphine ligands, is the opportunity to use tailored ligands in order to enhance the reactivity and promote the selectivity of the transition metal centered catalyst. Other factors, such as solvent, temperature, pressure, *etc.* are also very important and can be varied in a controlled manner in order to control selectivity.

CHAPTER 2

THE MECHANISMS OF HOMOGENEOUS HYDROGENATION

A wide variety of soluble transition metal complexes are known to act as hydrogenation catalysts under very mild conditions. This field has been surveyed comprehensively in recent reference books [1] and review articles [2]. Since 1965, most reports on homogeneous hydrogenation have involved complexes of the platinum group metals with phosphine ligands, and there has been intense activity in this area.

Very detailed mechanistic studies have now firmly established reaction pathways for some hydrogenation catalysts of the 1960s, while other recent developments have focused on mechanistic aspects of new hydrogenation catalysts. Another contribution to the field is the recent discovery of dihydrogen complexes, which provides a deeper insight into the oxidative addition of molecular hydrogen to transition metal catalysts.

It is not the authors' intention to review in detail the very extensive literature here. Instead we intend to focus on the most representative catalysts, with special emphasis on the mechanistic aspects of homogeneous hydrogenation. The discussion is organized according to catalyst type. For each major catalyst system, consideration is given to the nature of the metal catalyst (mainly Rh, Ir, Ru and Os), typical conditions for homogeneous hydrogenation, the types of substrate that can be reduced, the selectivity of the reaction, and the mechanism of the process. Separate sections are devoted to binuclear catalysts, clusters of platinum group metals and transition metal catalysts derived from non platinum group metals. On the other hand, some relevant palladium and platinum catalysts are treated in Chapters 4 and 5.

Table *2.1* shows some representative examples of homogeneous hydrogenation catalysts for the hydrogenation of organic substrates under very mild conditions. Scheme *2.1* presents an idealised picture of the two possibilities for the hydrogenation of alkenes by metal complexes which do not contain an M-H bond. According to James [1a], the first

P.A. Chaloner, M.A. Esteruelas, F. Joó and L.A. Oro, Homogeneous Hydrogenation 5–85.

TABLE 2.1
Some representative examples of homogeneous catalysts for the reduction of organic substrates

Fe	Co	Ni	Others
$[Fe(CO)_5]$	$[CoH(CN)_5]^{3-}$	$[Ni_2(CN)_6]^{4-}$	$[H(CO)(PPh_3)_2Ru(\mu\text{-bim})Ir(COD)]$
$[FeH(\eta^2\text{-}H_2)(PP_3)]$	$[Co_2(CO)_8]$	$[NiI_2(PPh_3)_2]$	$[\{(\eta^5\text{-}C_5Me_5)_2MH\}_2]$ (M=La,Nd)
$[Fe(CO)_3(diene)]$	$[Co(\eta^3\text{-}C_3H_5)\{P(OMe)_3\}_3]$	$[Ni(CO)_4]$	$[\{Cr(\eta^5\text{-}C_5H_5)(CO)_3\}_2]$
	$[CoH(CO)_2(P\underline{n}Bu_3)_2]$		$[Cr(CO)_6]$
Ru	**Rh**	**Pd**	
$[RuHCl(PPh_3)_3]$	$[RhCl(PPh_3)_3]$	$[PdCl_2]$/diamine/$NaBH_4$	
$[RuHCl(CO)(PPh_3)_3]$	$[RhH(CO)(PPh_3)_3]$	$[PdCl_2]$/DMF	
$[RuH_2(PPh_3)_4]$	$[Rh(NBD)(PR_3)_2]^+$	$[PdCl_2(py)_2]$	
$[RuH_2(\eta^2\text{-}H_2)(PPh_3)_3]$	$[\{RhH(P\{O\underline{i}Pr\}_3)_2\}_2]$		
$[(\mu\text{-}H)_2Ru_3(\mu_3\text{-}O)\ (DPPM)_2]$	$[\{(\eta^5\text{-}C_5Me_5)RhCl_2\}_2]$		
$[Ru_4H_4(CO)_{12}]$			
Os	**Ir**	**Pt**	
$[OsHCl(CO)(P\underline{i}Pr_3)_2]$	$[IrH_3(PR_3)_3]$	$[PtH(SnCl_3)(PPh_3)_2]$	
$[OsH_2Cl_2(P\underline{i}Pr_3)_2]$	$[IrH_5(PR_3)_2]$		
$[Os_3H_2(CO)_{10}]$	$[Ir(COD)(PR_3)_2]^+$		
$[Os_4H_4(CO)_{12}]$	$[Ir(COD)(py)(PCy_3)]^+$		
	$[\{(\eta^5\text{-}C_5Me_5)IrCl_2\}_2]$		

possibility is referred to as the unsaturated (or alkene) route involving initial coordination of the substrate followed by activation of molecular hydrogen by the resulting complex; the second possibility, the so called hydride route, which seems to be more general, implies initial reaction with molecular hydrogen followed by coordination of the substrate. The second common mechanism presented in Scheme *2.2* is usually adopted by hydrogenation catalysts which contain an M-H bond [3].

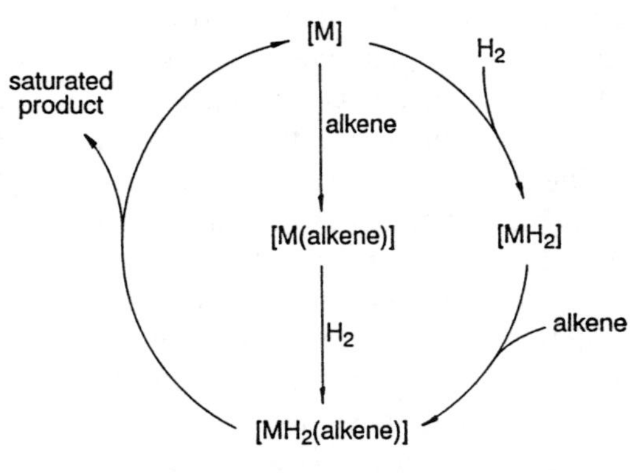

Scheme 2.1.

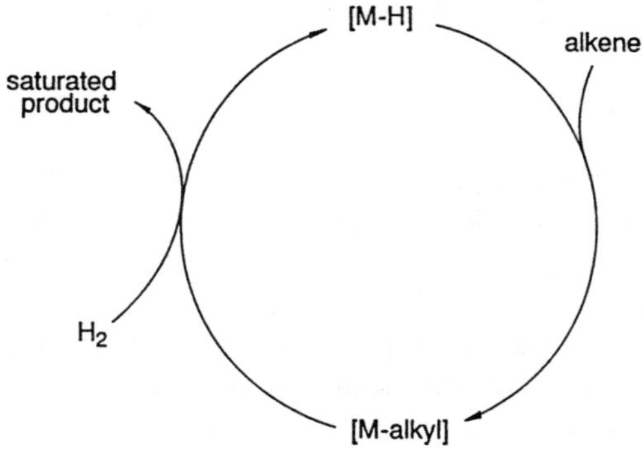

Scheme 2.2.

Scheme *2.1* and *2.2* show idealised proposals for the homogeneously catalysed hydrogenation of alkenes, but the actual mechanisms are, in general, more complex, involving multistep reactions. Thus, although rate data and derived rate laws have been determined for many catalytic cycles, a complete knowledge of the mechanism requires the rate law of the individual steps to be determined independently. Unfortunately, few catalytic cycles have been studied at the required level of detail, mainly because of experimental difficulties in studying the individual steps in isolation.

On the other hand, extensive studies on rhodium and iridium catalysts have shown that more than one active species can be present during the reaction and these species can have their *own* separate catalytic cycles (see sections *2.1* and *2.2*). Furthermore, it has been observed for some enantioselective reactions (see chapter *4*) that the major hydrogenation product arises from a minor intermediate, which cannot always be detected. Therefore, it is very difficult to be certain of the real mechanisms of homogeneous hydrogenation, although reasonable hypotheses consistent with experimental observations can be used as a practical approach. Therefore, it is prudent to be open to new mechanistic approaches which may be suggested by new experimental findings.

2.1. Wilkinson-type catalysts

2.1.1. [RhCl(PR_3)$_3$] COMPLEXES

The [RhCl(PPh_3)$_3$] complex, usually known as Wilkinson's catalyst, is probably the most widely studied of all the known homogeneous hydrogenation catalysts [4]. It was discovered independently by Osborn and Wilkinson [5] and by Coffey [6], and is most easily prepared by treating $RhCl_3.3H_2O$ with triphenylphosphine in ethanol [7].

This catalyst is widely and routinely used to hydrogenate a range of unsaturated organic molecules [8]. The reaction generally occurs under mild conditions (room temperature and atmospheric pressure of H_2). Aromatic organic compounds are inert, so benzene and toluene are useful solvents. The addition of a polar co-solvent such as ethanol or acetone produces a significant increase in the hydrogenation rate. The catalyst is slowly deactivated by chlorinated hydrocarbons such as $CHCl_3$ or CCl_4 due to oxidation to [$RhCl_3$(PPh_3)$_3$]. However, small amounts of O_2 or H_2O_2 activate the system; the enhancement of the hydrogenation rate by O_2 is often accompanied by double-bond migration. This isomerisation is inhibited if PPh_3 is present when O_2 is added to [RhCl(PPh_3)$_3$] [9].

[MXL_3] complexes other than [RhCl(PPh_3)$_3$] have been used as catalysts for the hydrogenation of alkenes and alkynes. Studies of the hydrogenation of cyclohexene with [RhX(PPh_3)$_3$] as catalysts show an increase in the rate of X = Cl<Br<I [10]. The effect of varying L in [$RhClL_3$] show that [RhCl($AsPh_3$)$_3$] and [RhCl($SbPh_3$)$_3$] are less effective as hydrogenation catalysts than the triphenylphosphine complex [11]. With [RhCl(PR_3)$_3$] complexes, the rate of reaction depends on the phosphine. The following data [12] illustrate this point: the activity of the catalyst increases as PR_3 varies in the order P(C_6H_4-4-Cl)$_3$ (1.8) $<$ P(C_6H_5)$_3$ (41) $<$ *P(C_6H_4*-4-CH_3)$_3$ (86) $<$ P(C_6H_4-4-OCH_3)$_3$ (100) (the figures in parentheses are the relative rates of hydrogenation of cyclohexene). The complexes with more basic tertiary phosphines, such as PEt_3 or $PPhEt_2$, are catalytically inactive. The complex [IrCl(PPh_3)$_3$] is also catalytically inactive, although it reacts readily with molecular hydrogen to form the stable dihydride [IrH_2Cl(PPh_3)$_3$] [13a]. Under normal conditions phosphine is not dissociated from the dihydride and coordination of the substrate is thus impeded. Recent studies have shown that the related complex [IrH_2Cl($PiPr_3$)$_2$] which unlike [IrH_2Cl(PPh_3)$_3$] is coordinatively unsaturated, catalyses the hydrogenation of benzylideneacetone to 4-phenylbutan-2-one via the intermediate [IrH_3($PiPr_3$)$_2$]. Kinetic studies carried out in 2-propanol at 60°C suggest that the dihydrogen $[IrH_2Cl(\eta^2\text{-}H_2)(PiPr_3)_2]$ is an intermediate in the heterolytical activation of hydrogen in changing from $[IrH_2Cl(PiPr_3)_2]$ to $[IrH_3(PiPr_3)_2]$ [13b].

The complex $[RhCl(PPh_3)_3]$ catalyses the chemospecific hydrogenation of alkenes in the presence of other easily reduced groups such as -CHO or $-NO_2$, terminal alkenes in the presence of internal alkenes, and less substituted alkenes in the presence of more substituted alkenes. The widely accepted mechanism for these reactions is shown in Scheme *2.3*. The problem of elucidating the intimate details of this general mechanism is compounded by the fact that it is a multistep process, and by the fact that several stereoisomers are possible for each of the intermediates.

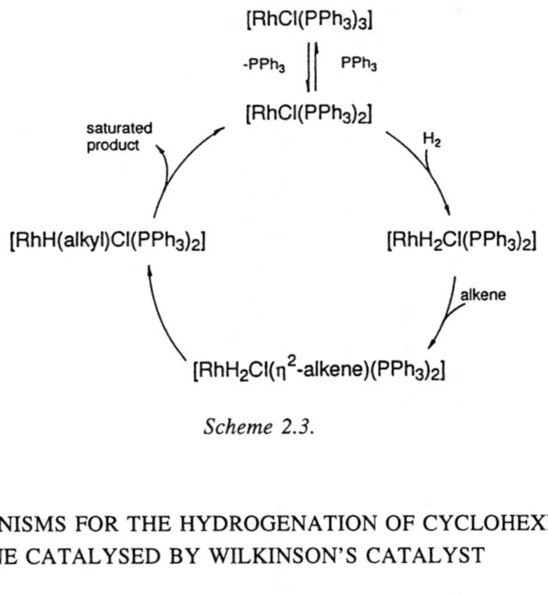

Scheme 2.3.

2.1.2. MECHANISMS FOR THE HYDROGENATION OF CYCLOHEXENE AND STYRENE CATALYSED BY WILKINSON'S CATALYST

Depending on the alkene being hydrogenated, kinetic studies of the overall catalytic reactions give a variety of different rate laws which can be fitted to different variants of the mechanisms shown in Scheme *2.3*. Consequently, and due to the complexity of such systems, it was necessary to study the individual steps of the catalytic cycles separately and directly in order to elucidate the mechanistic features more rigorously. Halpern accomplished this in the case of cyclohexene by ^{31}P NMR spectroscopy [14]. These studies led to the elucidation of the mechanism shown in Scheme *2.4*, in which the dotted outline shows

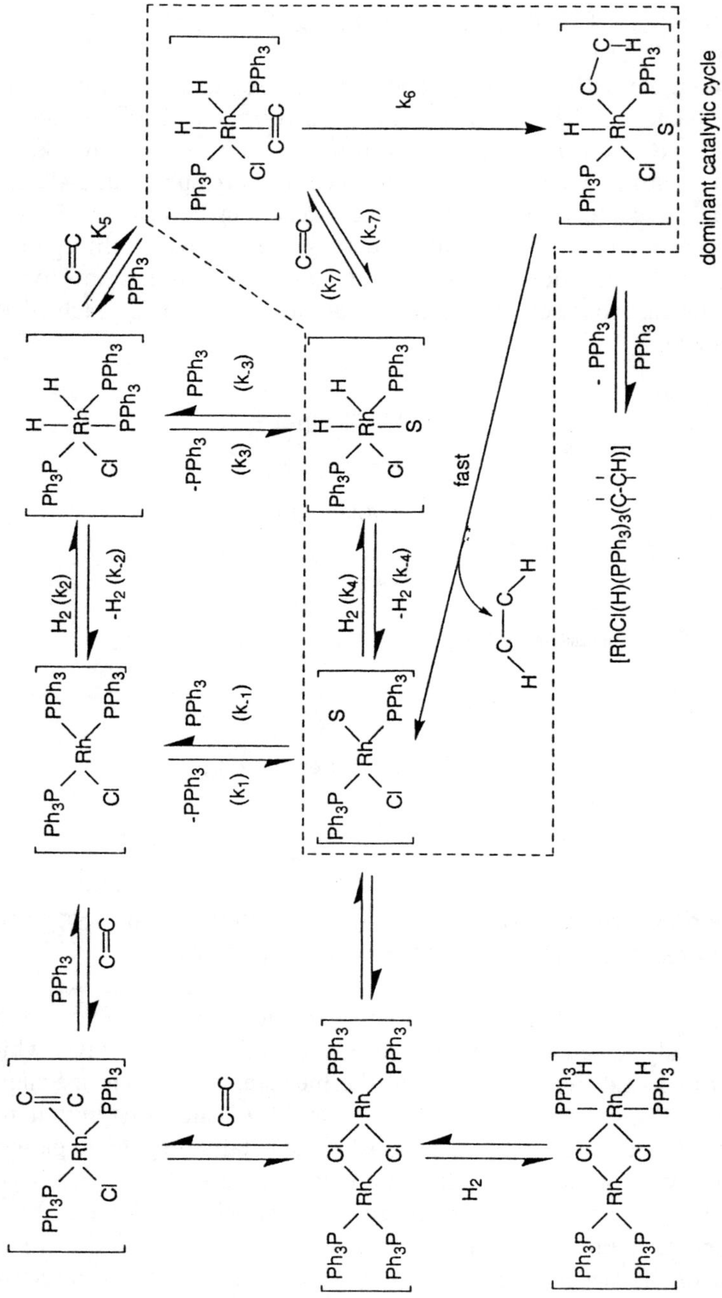

Scheme 2.4.

the dominant route. The rate-law derived for this mechanism is given by equation *2.1*.

$$\frac{[Rh]_{Tot}}{rate} = \frac{1}{k_6} + \frac{[PPh_3]}{K_5\, k_6\, [cyclohexene]} + \frac{[PPh_3]}{K_1\, k_4\, [H_2]} \qquad (2.1)$$

where the rate and equilibrium constants (Table *2.2*) have been evaluated from direct measurements on the individual stages of the reaction sequence. The excellent agreement between the rates calculated from equation *2.1* using the values shown in Table *2.2* and the measured catalytic rates provides a reliable test of the mechanism at the given level of detail.

TABLE 2.2
Rates and equilibrium constants for the hydrogenation of cyclohexene and styrene catalysed by $[RhCl(PPh_3)_3]$[a]

Cyclohexene	Styrene
$k_1 = 0.68\ s^{-1}$	$K_5 = 1.7 \times 10^{-3}$
$k_2 = 4.8\ M^{-1}\ s^{-1}$	$k_6 = 0.11\ s^{-1}$
$k_{-2} = 2.8 \times 10^{-4}\ s^{-1}$	$K_9 / k_4 = 5 \times 10^{-4}\ s$
$k_{-1} / k_4 = 1.0$	$k_{10}\, K_8 = 1.0 \times 10^{-4}\ s^{-1}$
$K_5 = 3.0 \times 10^{-4}$	
$k_6 = 0.22\ s^{-1}$	

a) In benzene at 25ºC.

According to Scheme *2.4*, the predominant hydride route consists of oxidative addition of a hydrogen molecule prior to alkene coordination. Both the associative pathway through $[RhCl(PPh_3)_3]$ and the dissociative pathway through $[RhCl(PPh_3)_2]$ could operate for hydrogenation, depending on the concentration of free PPh_3. But $[RhCl(PPh_3)_2]$ reacts with H_2 at least 10^4 times faster than $[RhCl(PPh_3)_3]$ [15], and is likely to be the active intermediate. The octahedral dihydride alkene complex $[RhH_2Cl(alkene)(PPh_3)_2]$ is the key intermediate in the catalytic cycle. The intramolecular alkene insertion step to give the alkyl hydride intermediate, $[RhH(alkyl)Cl(PPh_3)_2]$, is generally believed to be the rate-determining step for the whole process. It is suggested that the reductive elimination of alkane from this alkyl hydride intermediate to regenerate [$RhCl(PPh_3)_2$] is fast.

At least five rhodium-containing species have been observed directly in this reaction system. These include: [$RhCl(PPh_3)_3$], [$RhH_2Cl(PPh_3)_3$] [16], [{$Rh(\mu\text{-}Cl)(PPh_3)_2$}$_2$], [$(PPh_3)_2Rh(\mu\text{-}Cl)_2RhH_2(PPh_3)_2$] and, in the case of ethene, [$RhCl(C_2H_4)(PPh_3)_2$] [15b]. None of these actually lies within the dominant catalytic cycle shown in Scheme *2.4* and, to the extent that these species accumulate in the system, they contribute to reducing the catalytic rate.

For the hydrogenation of styrene, similar studies reveal an expanded mechanism, depicted schematically in Scheme *2.5* and related to the mechanism of Scheme *2.4* through the superposition of a parallel path involving the intermediate [$RhH_2Cl(styrene)_2(PPh_3)$]. This path accounts for most of the rate of catalytic hydrogenation of styrene under the conditions commonly used for this catalyst. The rate-law derived for this mechanism is given by equation *2.2* and the independently determined rate and equilibrium constants for styrene are given in Table *2.2.*

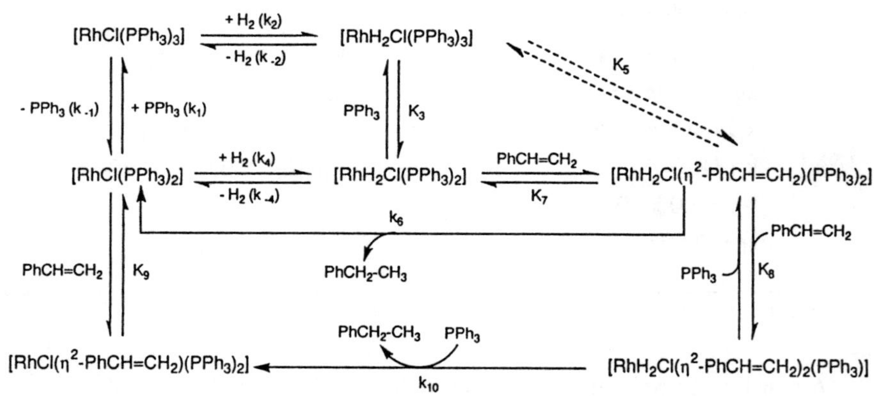

Scheme 2.5.

$$\frac{[Rh]_{Tot}}{rate} = \frac{([styrene] + K_5^{-1}[PPh_3])[PPh_3]}{(k_{10}\, K_8\, [styrene] + k_6[PPh_3])[styrene]} + \frac{K_1\, K_9\, [styrene] + [PPh_3]}{K_1\, k_4\, [H_2]} \qquad (2.2)$$

2.1.3. POTENTIAL ENERGY PROFILE FOR THE HYDROGENATION OF ETHENE CATALYSED BY $[RhCl(PH_3)_2]$ [17]

The potential energy profile (Figure *2.1*) for the dominant route of the catalytic cycle shown in Scheme *2.4* has been studied by the *ab initio* MO method using PH_3 in place of the ligand PPh_3 and C_2H_4 and C_2H_5 as models for the alkene and alkyl group, respectively [17]. The geometries of the transition states and the intermediates are shown in Scheme *2.6*.

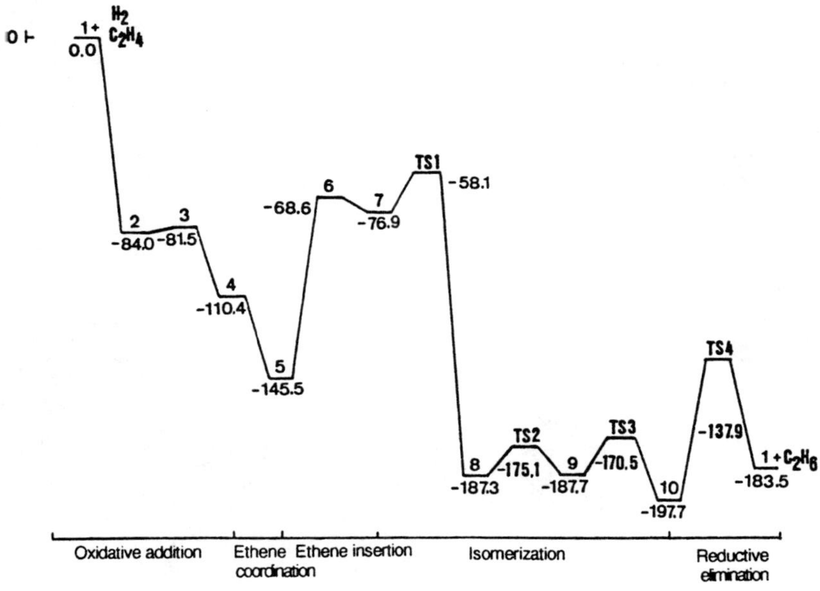

Fig. 2.1.

The oxidative addition of H_2 to $[RhCl(PH_3)_2]$ (*1*→*4*) is exothermic by 110.3 kJ/mol and the activation barrier from *2* to the dihydride *4* is 2.5 kJ/mol. In the complex *2* the Rh-H bond lengths are only 0.13 Å longer than those in *4*, whereas the H-H distance (0.863Å) is substantially longer than that in molecular hydrogen (0.735Å). This shows the existence of a strong three-centred interaction between H_2 and the Rh atom. The structure of *2* is similar to that determined experimentally for the dihydrogen complex $[W(CO)_3(PR_3)_2(\eta^2\text{-}H_2)]$ in which the H-H distance is 0.75Å (X-ray) and 0.82Å (neutron) and the W-H distance is 1.95Å (X-ray) and 1.89Å (neutron) [18]. Theoretical calculations [19]

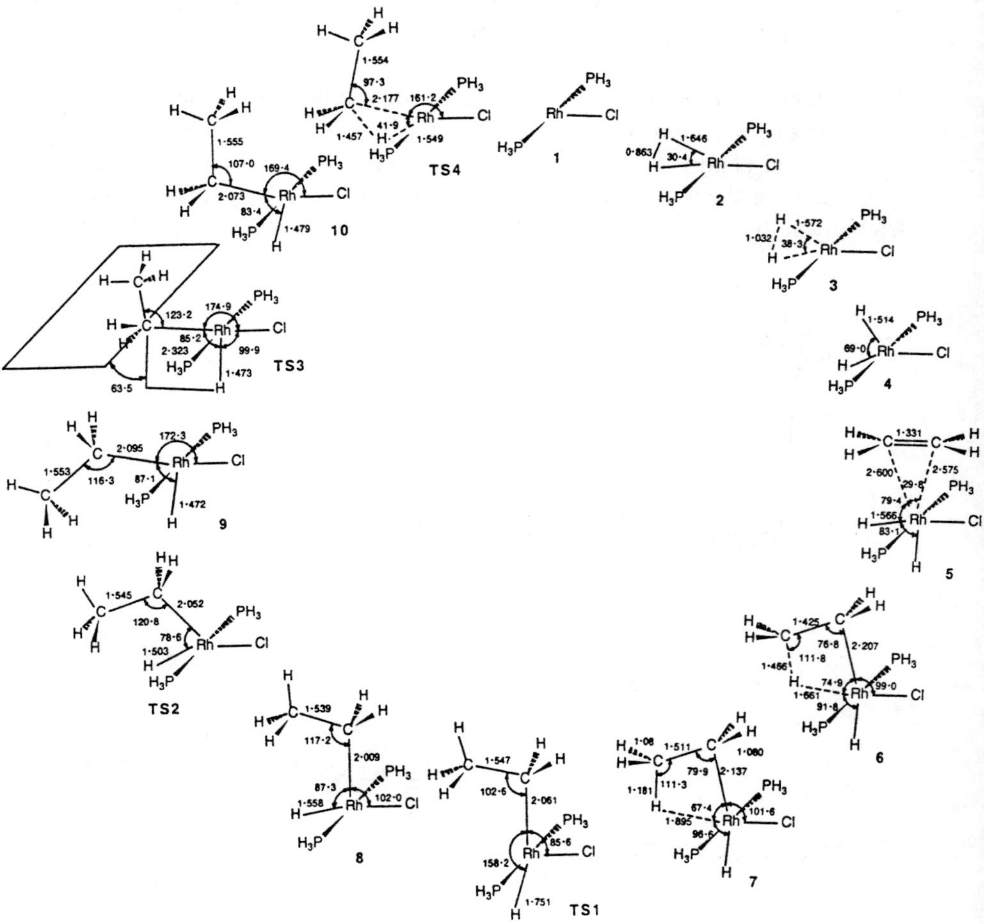

Scheme 2.6.

show that the intramolecular oxidative addition from the tungsten dihydrogen complex to give a dihydride is endothermic by 71.1 kJ/mol, and attributed the stability of this compound to the lowering of the *d* orbital energy in the metallic fragment by the π-acceptor CO ligands. This lowering prevents π-back-donation from the occupied orbitals of the metallic fragment to the σ* orbital of H_2, which would promote the dissociation of H_2 to give the dihydride. In addition, the dihydride $[WH_2(CO)_3(PR_3)_2]$ is 7-coordinate and therefore unstable. The rhodium complex *4* has no such strong acceptor ligands and is 5-coordinate, so the oxidative addition takes place easily.

The alkene coordination reaction takes place without a barrier. The stabilisation energy due to this coordination has been calculated to be 35.1 kJ/mol. The intramolecular migratory alkene insertion reaction to give $[RhH(C_2H_5)Cl(PH_3)_2]$ [20], combined with the first step of the isomerization of the *trans*-ethyl hydride complex to a *cis*-complex, is exothermic overall, and appears to take place as one combined step. This combined step has the highest activation barrier (about 84–104 kJ/mol) and is found to be the rate-determining step for the catalytic cycle. The theoretical calculation suggests that the rate-determining step includes part of the *trans-cis* isomerization process. The overall *trans-cis* isomerization is highly exothermic and takes place with a small activation energy, preferably *via* hydride and chloride migration rather than *via* alkyl migration.

The structural changes during the insertion process are such that the symmetrically η^2-coordinated alkene at first shifts its position to the left to give an η^1-coordinated species and then picks up the hydride from the metal at its uncoordinated carbon (follow the process *5*→*6*→*7* in Scheme *2.6*). The reaction is thus best described as a hydrogen migration.

The final reductive elimination of ethane to regenerate *1* is nearly thermoneutral and has a substantial activation barrier which is nevertheless smaller than that of the rate-determining step. Therefore, there can be no H/D scrambling for a deuterated alkene.

Similar calculations show that the complex $[RhCl(C_2H_4)(PH_3)_2]$ is stable, the C_2H_4 binding energy being 146 kJ/mol, and that the oxidative addition of H_2 is much more difficult than for *1*.

Steric hindrance due to bulky phosphines and substituents on the alkene would make both the alkene complex and the alkyl complex less stable. This would lower the energy gain in the alkene coordination step and decrease the activation barrier for reductive alkane elimination. Steric hindrance would also disfavour the alkene coordination prior to oxidative addition of H_2.

2.2. Cationic rhodium and iridium catalysts

2.2.1. THE CATALYST SYSTEMS

From examining the mechanistic features of the catalytic cycle for $[RhCl(PPh_3)_3]$ it is confirmed that homogeneous hydrogenation catalysts must, in general, have several free coordination sites, different potential coordination numbers for a given oxidation state, and different oxidation states which are readily accessible. Osborn and co-workers [21] made the important discovery of the existence of the cationic rhodium and iridium complexes with the general formula $[M(diene)L_a]^+$ (a = 2 or 3) which meet the criteria mentioned above. Under H_2 the diene is hydrogenated, generating the reactive $[ML_2]^+$ fragment. As expected, this is an excellent catalyst and it has several unusual properties which depend partly on its ionic character [22]. For example, the intermediates $[MH_2(solvent)_2L_2]^+$ can be isolated relatively easily from coordinating solvents such as acetone or ethanol. As might be expected in a system that binds solvent well, rates of reduction of various substrates are very solvent dependent [23, 24], as the substrate competes with the solvent for the metal.

Interestingly, this cationic system is active for a large number of donor ligands L. Wilkinson's catalyst requires PPh_3 or a ligand with a very similar cone angle, because the steric compression of the relatively bulky phosphine encourages ligand dissociation, which is required to generate a site at the metal. In Osborn's system, no dissociation of L is required, and thus the steric requirement is relaxed and L can be varied more widely. For example, simple amines or nitriles can be used in certain circumstances [25–27]. Even $[Rh(NBD)L_3]^+$ salts were shown to be active [21e]. Unlike Wilkinson's system, the side-reaction leading to isomerisation of the alkene is a more serious competitor with hydrogenation itself.

Interestingly, selectivity effects have been found for various dienes [28, 29] and alkynes [29a, 30], the reduction of both of which could be stopped at the monoene stage. Almost exclusive *cis*-addition of H_2 to the alkyne was observed. Ketones can be also reduced using the rhodium system [21b].

Along with exceptional rates for the hydrogenation process, an interesting feature of the iridium system, studied extensively by Crabtree and co-workers [2b], is that some alkene dihydride intermediates can be observed directly. However, a disadvantage of this system is that it is rather readily deactivated under certain conditions to give the dimer complexes $[L_2HIr(\mu\text{-}H)_3IrHL_2]^+$. We propose to highlight the similarity between the rhodium and iridium systems.

Halpern also made some significant observations in the chemistry of cationic rhodium catalysts with bidentate ligands [14]. Because much of this relates to asymmetric induction, it is largely discussed in Chapter *4*. However, we consider that some features of the $[Rh(COD)(DPPE)]^+$ complex (DPPE = bis (1,2-diphenylphosphino ethane) should also be discussed in this section.

2.2.2. METHODS OF SYNTHESIS [31, 32]

A very general and convenient method for preparing cationic complexes of the type $[M(diene)L_a]^+$ is to treat the dimeric compounds $[\{M(\mu\text{-}Cl)(diene)\}_2]$ (M = Rh, diene = 1,5-cyclo-octadiene (COD), bicyclo[2,2,1] heptadiene (NBD), tetrafluorobenzobicyclo[2,2,2]octatriene (TFB). M = Ir, diene = COD) with an equimolar amount of a silver salt and the corresponding ligands *L* (equation *2.3*).

$$\tfrac{1}{2}[\{M(\mu\text{-}Cl)(diene)\}_2] + aL + Ag^+ \longrightarrow [M(diene)L_a]^+ + AgCl \qquad (2.3)$$

M = Rh, diene = COD, NBD, TFB
M = Ir, diene = COD

Other alternative methods used by some authors are shown in the following Scheme.

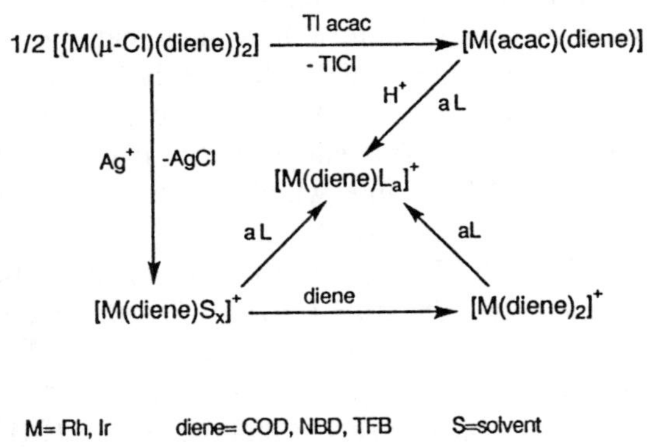

Scheme 2.7.

2.2.3. REACTION WITH HYDROGEN

The elucidation of the interaction between hydrogen and transition metal complexes is an important aspect of the study of hydrogenation reactions. Schrock and Osborn made an important early contribution to this field [21]. In solvents such as acetone, ethanol or acetonitrile, the $[Rh(NBD)L_a]^+$ complexes react quantitatively with molecular hydrogen to give the dihydride complexes $[RhH_2S_xL_a]^+$ and norbornane. The ^1H

NMR spectra of these species are consistent with the structure shown below (Figure *2.2*).

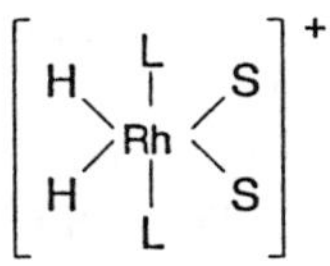

S= solvent

Fig. 2.2.

The $[Rh(COD)L_a]^+$ and $[Rh(TFB)L_a]^+$ complexes also yield the dihydride $[RhH_2S_xL_a]^+$. However, the rate of reduction of the diene varies markedly. For example, $[Rh(NBD)(PPh_3)_2]^+$ reacts with hydrogen some 10^2 times faster than $[Rh(COD)(PPh_3)_2]^+$.

It is unusual that $[Rh(NBD)L_3]^+$ reacts with hydrogen since it is formally five-coordinate and thus coordinatively saturated. Presumably NBD or L must dissociate before hydrogen can be added. NMR Spectroscopic evidence indicates that added $PPhMe_2$ exchanges rapidly with coordinated $PPhMe_2$ in $[Rh(NBD)(PPhMe_2)_3]^+$ while added NBD does not alter the spectrum. This suggests that it is L, rather than one arm of the chelating NBD ligand, which dissociates prior to addition of hydrogen to what is then $[Rh(NBD)L_2]^+$.

The reaction of $[Rh(NBD)L_2]^+$ with hydrogen most likely yields a short-lived intermediate, *cis*-$[RhH_2(NBD)L_2]^+$, which must yield the dihydride $[RhH_2S_xL_2]^+$ under hydrogen in the presence of solvents such as acetone or ethanol (Scheme *2.8*). In agreement with this proposal, Crabtree and co-workers have observed that the cations $[Ir(COD)(PR_3)_2]^+$ (PR_3 = PPh_3, PPh_2Me) react with hydrogen at –80°C in CD_2Cl_2 to give the dihydrides *cis*-$[IrH_2(COD)(PR_3)_2]^+$, which are stable at least for several hours at –80°C. When the solutions are warmed to –20°C in the absence of an excess of H_2, hydrogen is partly lost to give $[Ir(COD)(PR_3)_2]^+$. In acetone, the reactions lead to the cations $[IrH_2(acetone)_2(PR_3)_2]^+$ [33].

All these observations can be rationalized in terms of the reactions shown in Scheme *2.8*. Cationic dihydrogen complexes have not been observed under the above conditions but the related compound $[IrH_2(\eta^2\text{-}H_2)_2(PCy_3)_2]^+$ has recently been observed after protonation of $[IrH_5(PCy_3)_2]$ in CD_2Cl_2 at –80°C [34]; addition of acetonitrile gives $[IrH_2(MeCN)_2(PCy_3)_2]^+$.

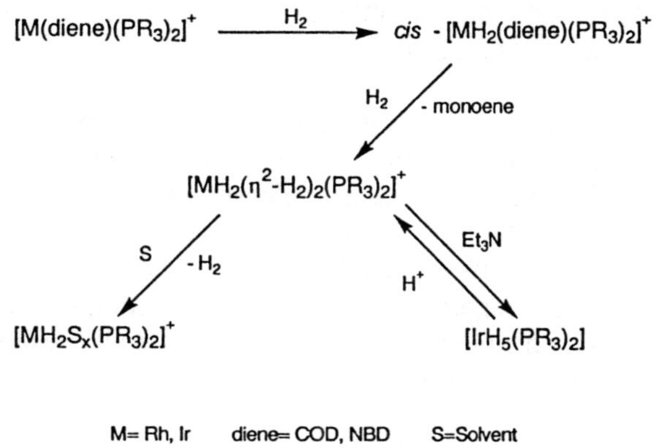

Scheme 2.8.

The reaction of the $[Ir(COD)L_2]^+$ complexes with hydrogen is inhibited by relatively strong electron-donating ligands such as pyridine (py), Cl^-, Br^- or I^- (e.g. $[Ir(COD)(py)_2]^+$) but occurs readily with π-acceptor ligands such as PR_3 or alkene (e.g. $[Ir(COD)_2]^+$). According to Crabtree, this is due to the tendency of these cationic iridium complexes to show Lewis acid behaviour in their reactions. Thus, the addition of hydrogen is a reductive rather than an oxidative process. The fact that the Pauling electronegativities of Ir(2.2) and H(2.1) are comparable is consistent with equation *2.4* [35].

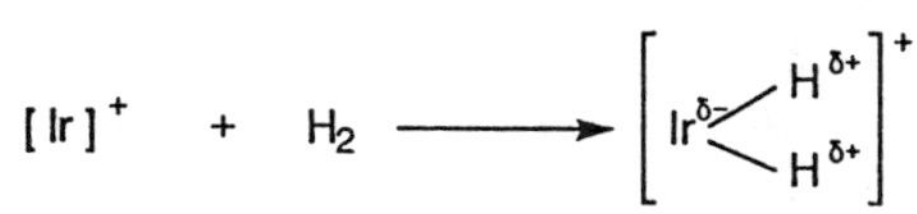

(2.4)

Other features of the chemistry of these iridium cations can also be understood in terms of their Lewis acidity. For example, some $[Ir(COD)(PR_3)_2]^+$ complexes add HX (X = Cl, Br, I) by an ionic mechanism; X^- attack followed by protonation. The positive charge on the complex is not wholly responsible for this behaviour since $[IrCl(COD)(PPh_2Et)]$ shows the same order of addition [36]. It has also been observed that in the presence of 3,3-dimethyl-1-butene, the $[IrH_2(acetone)(PPh_3)_2]^+$complex dehydrogenates cycloalkanes in refluxing 1,2-dichloroethane [37].

2.2.4. MECHANISM FOR THE HYDROGENATION OF ALKENES CATALYSED BY $[M(diene)L_a]^+$ COMPLEXES

Table *2.3* lists some representative results for the hydrogenation of 1-hexene and *cis*-2-hexene in acetone catalysed by the complexes $[Rh(NBD)L_a]^+$. Though the system is clearly complicated by large and variable amounts of alkene isomerization in addition to hydrogenation, it is possible to discern two general trends; i) 1-hexene is hydrogenated and isomerized more rapidly than *cis*-2-hexene; and ii) catalysts which contain more basic phosphine appear to hydrogenate alkenes more rapidly and the rate of isomerization is also increased.

TABLE 2.3 [21e]
Representative results for the hydrogenation of 1-hexene and *cis*-2-hexene in acetone catalysed by the complexes $[Rh(NBD)L_a]^+$

Complex	Substrate	Concen. (mM)	K_h[a]	K_i[b]	K_h+K_i[c]
				($\times 10^4$ s^{-1})	
$[Rh(NBD)(PPh_3)_2]^+$	1-Hexene	5.3	0.1	d	-
$[Rh(NBD)(PPh_2Me)_2]^+$	1-Hexene	3.7	3.0	4.5	7.5
	cis-2-Hexene	5.3	-	-	2.0
$[Rh(NBD)(PPhMe_2)_2]^+$	cis-2-Hexene	5.3	-	-	5.0
$[Rh(NBD)(PPhMe_2)_3]^+$	1-Hexene	3.5	6.0	6.0	12.0
	cis-2-Hexene	5.3	-	-	5.5
$[Rh(NBD)(PPh_2OMe)_2]^+$	cis-2-Hexene	5.3	-	-	3.5

a) K_h is the initial rate constant for the appearance of hexane. b) K_i is the rate constant for the appearance of *cis* and *trans*-2-hexene. c) Separation of the two rates was not feasible due to concomitant hydrogenation of the trans-2-hexene and trans-3-hexene. d) Not measured.

There have been no rigorous kinetic studies to establish the intimate features of the catalytic cycle of these reactions. However, some clues can be obtained from deuterium labelling experiments. Thus, Schrock and Osborn [21e] observed that hydrogen and deuterium scramble in the absence of alkene, and the hydrogen from water or 1-hexene exchanges with molecular deuterium. This behaviour is characteristic of group 8 monohydride complexes [38]. Monohydride species can arise in these catalytic systems by deprotonation of the cationic dihydride species (equation *2.5*).

$$[RhH_2S_xL_n]^+ \rightleftharpoons RhHS_yL_n + H^+ \quad (2.5)$$

The following observations support this idea: i) if the $[Rh(NBD)(PPhMe_2)_3]^+$ complex is treated with molecular hydrogen and $H[ClO_4]$ after flushing with N_2, and 1-hexene is then injected, after 1h, less than 1% isomerization has occurred. In an identical system in the absence of acid, isomerization is 50% complete in ca. 30 min.; ii) when the same complex is treated as in (i) substituting NEt_3 for $H[ClO_4]$, the isomerization of 1-hexene is extremely rapid and, iii) when the $[Rh(NBD)\{P(OPh)_3\}_2]^+$ complex is treated with H_2 in the presence of $H[ClO_4]$, one hour after injecting 1-hexene the solution contains 96% 1-hexene, while in the absence of $H[ClO_4]$, hydrogenation occurs very rapidly.

Scheme *2.9* presents a rough mechanistic Scheme which qualitatively accounts for these observations. There are three possible paths by which an alkene can be hydrogenated. Path A involves a monohydride catalyst which both extensively isomerizes and also hydrogenates alkenes and is favoured in the presence of NEt_3 or when the catalyst contains basic phosphine ligands. In path B the cationic dihydride is the active catalyst,

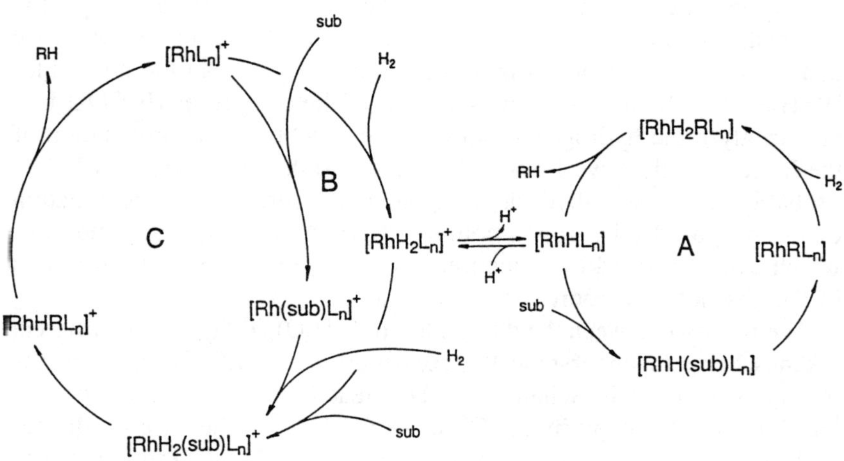

Scheme 2.9.

and, in general, hydrogenates alkenes less efficiently; there may only be limited isomerization. $[RhH_2S_xL_n]^+$ is favoured in the presence of $H[ClO_4]$. Path B is, of course, strictly analogous to the mechanism shown in Scheme *2.3*. Path C involves direct hydrogenation of $[RhS_x(alkene)L_2]^+$. It may operate to some extent, since the catalysts are produced *via* direct hydrogenation of a diene complex $[Rh(diene)L_n]^+$. Note that the proposed intermediate alkyl-hydride complex $[RhH(R)S_xL_n]^+$, is common to both pathways B and C.

Hydrogenation with minimal concomitant isomerization is best accomplished under acidic conditions. Under these conditions hydrogenation proceeds *via* path B. Thus, it is suggested that the alkene enters the coordination sphere by displacing a solvent molecule. The alkene would then be *cis* to one hydride but *trans* to the other. If the *cis* hydride migrates to the alkene and the stereochemistry does not change further, the resulting intermediate alkyl-hydride would have *trans* stereochemistry. This stereochemistry would not facilitate reductive elimination of alkene in the next step. Consequently, either the first hydride migration is accompanied by stereochemical rearrangement, or rapid rearrangement must occur after this migration so that the β-elimination process, which is generally facile, does not compete with the reductive elimination reaction. Therefore, the fact that catalyst solutions isomerize alkenes after removal of hydrogen without addition of acid seems to imply that the rate of isomerization by $[RhHS_yL_n]$ must be much faster than that of hydrogenation by $[RhH_2S_xL_n]^+$.

The uncharged species $[RhHS_yL_n]$ (generated according to equation *2.5* and, therefore, present under neutral or basic conditions) are considerably more efficient hydrogenation catalysts than the dihydrides $[RhH_2S_xL_n]^+$. Dichloromethane solutions of the complexes $[Ir(COD)L_2]^+$ also catalyse the hydrogenation of alkenes. Deliberate deprotonation of the iridium catalyst system with NEt_3 leads to the formation of $[IrH_5L_2]$. Probably, these penta-hydride compounds are formed by deprotonation of the proposed dihydrogen complex $[IrH_2(\eta^2\text{-}H_2)_2L_2]^+$ (Scheme *2.8*), which can be formed by reduction of the diene in excess hydrogen and in the absence of a coordinating solvent.

The reaction between the dihydride, $[IrH_2(H_2O)_2(PPh_3)_2]^+$, and different alkenes helps to understand the contribution of paths B and C to the mechanism shown in Scheme *2.9*. This reaction has been examined by ^1H NMR spectroscopy in CD_2Cl_2 at –80°C [24]; in the case of alkenes such as ethene (C_2H_4) or propene (C_3H_6), complex *13* (equation *2.6*) was formed. Styrene also gave some *13* (ca. 50%) at –80°C, but on warming to –50°C, no *13* was left and the equilibrium was shifted toward *12*. Bulkier alkenes, such as cyclooctene, cyclopentene and 3,3-dimethyl-1-butene gave only species *12* in equilibrium with *11*.

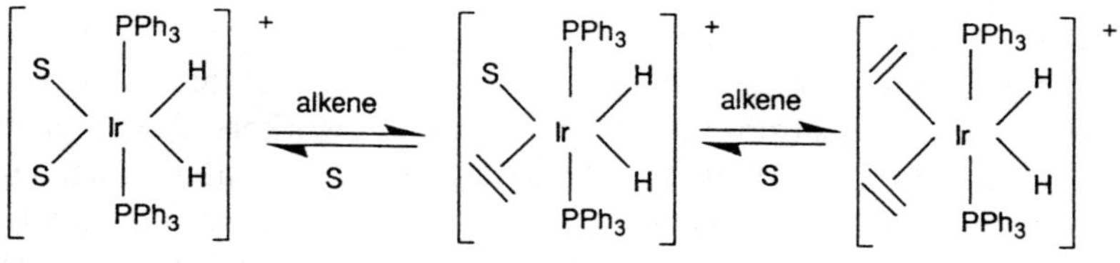

11 12 13

S= H_2O (2.6)

For *12* containing cyclopentene or cyclooctene, two *CH* alkenyl resonances were observed in the ^1H NMR spectra. Alkene rotation is probably frozen out at –80°C and the molecule adopts the electronically preferred configuration shown in equation *2.6*, for which the two termini of the C=C group become inequivalent. This configuration is probably adopted so that the C=C group can avoid sharing *d* orbitals with the good π-acceptor PPh_3 ligands [24, 39]. This configurational preference leads to the complexes *12* and *13* tending to have coplanar M(C=C)H groups, an arrangement which should lead to the most rapid hydrogenation.

On warming reaction mixtures of *11* and alkenes to room temperature, alkene hydrogenation takes place. The organometallic product detected in solution depends on the alkene. For C_2H_4 and C_3H_6, $[Ir(alkene)_2(PPh_3)_2]^+$ seems to be formed; this reacts with H_2 at –80°C to give *13*. For styrene, the product $[Ir(\eta^6\text{-}C_6H_5Et)(PPh_3)_2]^+$ is obtained. The reaction mixtures formed with cyclopentene or cyclooctene also result in hydrogenation on warming. The organometallic products are of the type $[Ir(alkene)S(PPh_3)_2]^+$ at –80°C. At this temperature, they do not add H_2, perhaps because of the increased electron density present at the metal, which would inhibit reductive addition.

It is plausible that the alkene dihydride species *13* can be an intermediate in the catalytic hydrogenation of alkenes by $[Ir(COD)L_2]^+$. The main difference between the conditions used to observe *12* and *13* and the catalytic conditions is the presence of the coordinating solvent S. It has been already mentioned that the formation of the dihydrides $[MH_2S_xL_n]^+$ most probably proceeds *via* the dihydrogen complexes $[MH_2(\eta^2\text{-}H_2)_2L_2]^+$. Thus, it is likely that under catalytic conditions, the intermediate *13* could be formed by hydrogen addition to the species

$[Ir(alkene)_2L_2]^+$ or by displacement of the two dihydrogen ligands of complexes of type $[IrH_2(\eta^2\text{-}H_2)_2L_2]^+$. So, paths B (substituting $[IrH_2(\eta^2\text{-}H_2)_2L_2]^+$ for $[IrH_2S_2L_2]^+$) and C (Scheme *2.9*) are also plausible routes leading to the hydrogenation of alkenes catalysed by dichloromethane solutions of the cationic iridium systems. In this context, it is interesting to mention that Caulton and coworkers have recently demonstrated that $[IrH_2(\eta^2\text{-}H_2)(PPhMe_2)_3]^+$ catalyses the hydrogenation of ethene [40]. $[IrH(alkene)_xL_2]$ could be present at undetectable concentrations and could also catalyse the hydrogenation and isomerization of alkenes by path A.

2.2.5. SELECTIVE HYDROGENATION OF DIENES TO MONOENES

The reduction of norbornadiene catalysed by $[RhCl(PPh_3)_3]$ in benzene is extremely slow compared with the rate of reduction of monoenes [1a]. Since $[RhCl(NBD)(PPh_3)]$ can be isolated from the catalyst solution, it is presumably the energetically favoured species under catalytic conditions, and evidently does not react readily with molecular hydrogen. The method for preparing the catalytic species discussed here gives an example where the hydrogen reacts rapidly with a diene complex. Therefore, it can be assured that this catalyst system will not be deactivated in a similar way.

These cationic systems catalyse the selective, sequential hydrogenation of norbornadiene, butadienes or cyclohexadienes [28, 29]. During the reduction, both double bonds of the dienes are in fact coordinated. This is suggested, in the case of NBD, by the fact that the norbornene-d_2 reaction product selectively formed under D_2 is solely *endo*-d_2 and, after that, the norbornane-d_4 is solely *endo*-2,3-*exo*-5,6 (equation *2.7*)

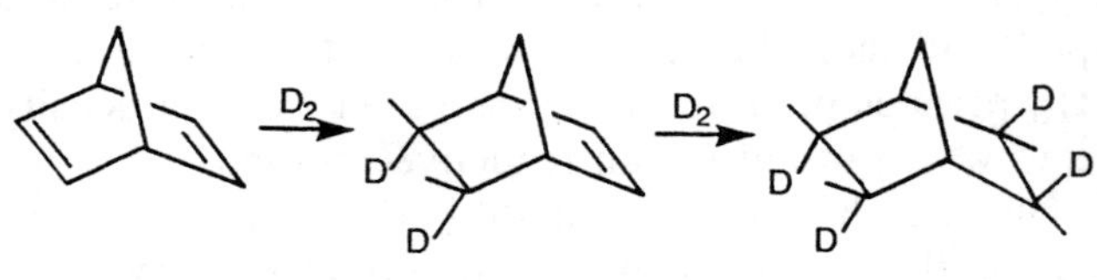

(2.7)

Metals are known to coordinate in the *exo* position in norbornene complexes. Therefore, hydrogenation of 2,3-disubstituted norbornenes gives *endo*-2,3-disubstituted-norbornane and of course D_2 adds *exo* to norbornene. However, the metal is also *exo* in the known complexes containing monodentate NBD. Therefore, only if both double bonds are coordinated can D_2 add first specifically *endo* to NBD.

The rate of hydrogenation of NBD is relatively dependent on the basicity of the ligands L. This is analogous to the reduction of monoenes by catalyst systems of this type and there is no reason to suspect that the overall mechanism of norbornadiene hydrogenation is not essentially identical to that shown in Scheme *2.9*. However, it should be noted that the rate of reduction of norbornadiene varies with L in the opposite direction; i.e. for the $[Rh(TFB)\{P(C_6H_4\text{-}4\text{-}R)_3\}_2]^+$ systems (R = MeO-, Me-, F-, Cl-), the catalyst precursor containing $P(C_6H_4\text{-}4\text{-}Cl)_3$ gives the most active system (Table *2.4*) [29a]. One might therefore suspect that path A could not operate in this case. Strong support for this hypothesis derives from the fact that the hydrogenation rate of norbornadiene in the presence of acid is identical to that without added acid.

TABLE 2.4 [29a]
Hydrogenation of 1,3-cyclohexadiene and 2,5-norbornadiene catalysed by the complexes $[Rh(TFB)\{P(C_6H_4\text{-}4\text{-}R)_3\}_2]^+$

R	Initial turnover number a (Maximum % monoene)	
	1,3-Cyclohexadiene	2,5-Norbornadiene
MeO	3.5 (100)	8.3 (62)
Me	6.3 (100)	11.5 (65)
F	8.6 (100)	45.4 (94)
Cl	10.5 (100)	45.5 (95)

a) Mol H_2 (mol Rh)$^{-1}$ min^{-1}

It has also been observed [28] that: i) the rate of diene reduction is independent of norbornadiene concentration until near the endpoint; ii) complexes $[Rh(NBD)L_2]^+$ are the main species under catalytic conditions; iii) the rate at which H_2 reacts with $[Rh(NBD)L_2]^+$ is qualitatively parallel to the hydrogenation rate of the diene. These observations indicate that the $[Rh(NBD)L_2]^+$ complexes are the true catalysts and that the reaction of these compounds with H_2 is the rate-determining step of the process. Therefore, the hydrogenation of norbornadiene proceeds mainly along path C shown in Scheme *2.9*. The hydrogenation of butadienes and cyclohexadienes shows similar behaviour. Thus, other dienes with large complexation constants should also coordinate strongly to give predominantly $[Rh(diene)L_2]^+$ under catalytic conditions, and they should also be reduced largely, if not entirely, along path C.

However, the rate of hydrogenation of some dienes falls progressively as the reaction proceeds when the ancillary ligands L are monodentate; in these cases the catalytic solutions have been found to contain species of the type $[Rh(diene)_2L]^+$, which are inactive. A chelating ligand prevents formation of analogous complexes, so the $[Rh(diene)(chelate)]^+$ compounds are preferred for diene reduction. Thus it has been noted that, for the butadiene reduction, the most successful catalyst precursors are those containing bidentate phosphine or arsine ligands.

Similar results have been observed for the iridium complexes $[Ir(COD)L_2]^+$, e.g., for the reduction of 1,5-cyclooctadiene in dichloromethane; hydrogenation rates in the presence of $[Ir(COD)(PR_3)_2]^+$ fall as the reaction proceeds, whereas in the presence of $[Ir(COD)_2]^+$ they are constant until near the endpoint. When 1,5-cyclooctadiene is hydrogenated in the presence of $[Ir(COD)_2]^+$, the intermediate *cis*-$[IrH_2(COD)_2]^+$ is observed at −80°C. For systems containing monodentate ligands besides the *cis*-$[IrH_2(COD)(PR_3)_2]^+$ the *trans-cis*-isomer is also detected [33]. The latter is most probably formed by reaction of $[IrH_2(\eta^2\text{-}H_2)_2(PR_3)_2]^+$ with 1,5-cyclooctadiene, suggesting that path B may also operate to a certain extent when the catalyst is an iridium complex. In this context, it is interesting to note that the rate of hydrogenation of dienes to monoenes by $[Ir(COD)\{P(C_6H_4\text{-}4\text{-}R)_3\}_2]^+$ systems is dependent on the basicity of the phosphine ligands [29b].

2.2.6. HYDROGENATION OF ALKYNES

The $[M(diene)L_n]^+$ systems catalyse not only the reduction of alkenes and dienes but also the selective hydrogenation of alkynes to alkenes. Thus complexes of the type $[Rh(NBD)L_a]^+$ catalyse the reduction of 2-hexyne. This substrate is hydrogenated at a much faster rate than that at which *cis*-2-hexene is subsequently hydrogenated and isomerized [30].

The overall mechanism of alkyne hydrogenation should be essentially the same as that shown in Scheme *2.9*, substituting the alkyne for the alkene. The reason for the selective reduction of 2-hexyne must be that 2-hexyne simply competes strongly with *cis*-2-hexene for coordination sites on $[RhHS_yL_n]$ and/or $[RhH_2S_xL_n]^+$. *cis*-2-Hexene is the only reasonable primary product. Assume, for the sake of argument, that two hydrides transfer sequentially to coordinated 2-hexyne. The transfer of the first hydride would most reasonably yield a *cis*-alkenyl ligand. The second hydride should transfer to the alkenyl α-carbon, thus retaining the *cis* geometry about the double bond.

For these systems, the hydrogenation rate is maximized by using catalysts containing more basic or less sterically demanding phosphine ligands. In the presence of $H[ClO_4]$, the rate of 2-hexyne hydrogenation is approximately half of that in the absence of $H[ClO_4]$, suggesting that: i) both $[RhH_2S_xL_n]^+$ and $[RhHS_yL_n]$ are active catalysts for the selective hydrogenation of 2-hexyne and, ii) the rate of hydrogenation with the monohydrides is at least twice that with the dihydrides.

The iridium systems are also active catalysts in dichloromethane. Under these conditions the species $[IrH_2(\eta^2\text{-}H_2)_xL_n]^+$ should be formed. It was observed very recently that $[IrH_2(\eta^2\text{-}H_2)(PPhMe_2)_3]^+$ catalyses the hydrogenation of 2-butyne. Treatment of this compound in dichloromethane with 5 equiv. of 2-butyne yields $[Ir(MeC_2Me)(PPhMe_2)_3]^+$ together with a mixture of *cis*-2-butene and 1-butene. Under hydrogen atmosphere, the alkyne complex is converted to $[IrH_2(\eta^2\text{-}H_2)(PPhMe_2)_3]^+$ and butane [41], which closes the cycle shown in Scheme *2.9* (path B).

It is not easy to reduce 1-hexyne selectively to 1-hexene with $[Rh(diene)L_n]^+$ systems as catalysts. The reaction rates fall progressively as the orange solutions of $[Rh(diene)L_n]^+$ turn brown. This suggests that a side reaction destroys the active catalyst. Since terminal alkynes are fairly acidic, one possibility is that shown in equation *2.8*.

$$[RhHS_yL_n] + RC{\equiv}CH \longrightarrow H_2 + [Rh(C{\equiv}CR)S_yL_n] \qquad (2.8)$$

In spite of these difficulties, 1-hexyne can be successfully and selectively reduced to 1-hexene using 2-methoxyethanol and $[Rh(NBD)(PPh_2Me)_2]^+$ as solvent and catalyst respectively, or using dichloromethane and $[Rh(TFB)\{P(C_6H_4\text{-}4\text{-}R)_3\}_2]^+$ type complexes. Figure *2.3* summarizes the course of a typical reaction with $[Rh(TFB)\{P(C_6H_4\text{-}4\text{-}MeO)_3\}_2][ClO_4]$ as catalyst [29a].

The selective hydrogenation of phenylethyne to styrene using $[Ir(diene)(\eta^2\text{-}iPr_2PCH_2\text{-}CH_2OMe)]^+$ complexes has been reported very recently [29c]. Interestingly, kinetic and spectroscopic investigations suggest that the reaction proceeds via the dihydride $[IrH_2(COD)(\eta^2\text{-}iPr_2PCH_2\text{-}CH_2OMe)]^+$ and does not involve the hydrogenation of the coordinated diene.

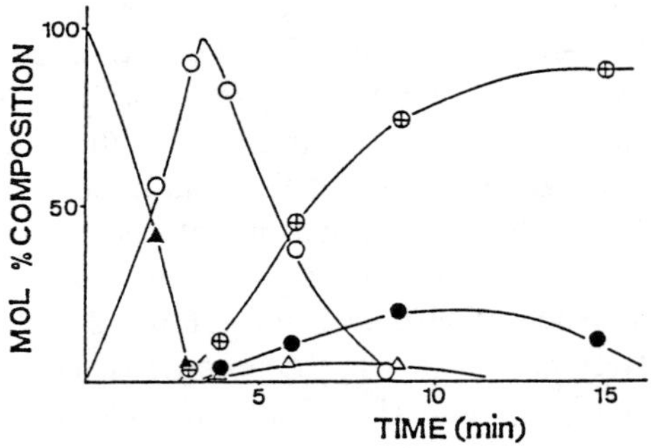

Fig. 2.3. The catalytic hydrogenation of 1-hexyne in dichloromethane with $[Rh(TFB)\{P(C_6H_4\text{-}4\text{-}MeO)_3\}_2][ClO_4]$. 1-hexyne (▲), 1-hexene (○), *trans*-2-hexene (●), *cis*-2-hexene (△), hexane (⊕).

2.2.7. HYDROGENATION OF KETONES [21b]

The previous sections discussed the catalytic activity of the $[M(diene)L_n]^+$ systems in the hydrogenation of alkenes, dienes and alkynes. Interestingly, the rhodium systems containing the most basic phosphines as ligands (e.g. PPh_2Me, $PPhMe_2$, PMe_3) also catalyse the reduction of ketones under mild conditions.

There are some points of interest concerning the catalytic process. It is found that using anhydrous ketones, the initial rate of reduction is extremely slow. Incremental additions of water increase the rate markedly, reaching a maximum after the addition of ca. 1% water by volume. From a mechanistic point of view it is interesting to mention that the reduction of acetone containing 1% water with deuterium yields 2-propanol labelled at the α-carbon, no label being detectable at the β-carbon. Therefore, the enol form of the ketone plays no significant role in the catalytic process. The most commonly accepted mechanism for this catalytic process is shown in Scheme *2.10*.

Replacement of solvent by ketone in *14* yields *15* with coordinated ketone. It is suggested that a step-wise process then occurs, initially involving a 1,3-hydride migration from a *cis*-site on the metal to the carbon of the ketone group. The second proton-transfer step is promoted by small quantities of water. This process is shown below (Figure *2.4*).

Deprotonation may be carried out by hydroxyl ions or by a water molecule, and protonation of the alkoxy-group may occur simultaneously or, more probably, in a subsequent step. The catalytic cycle is

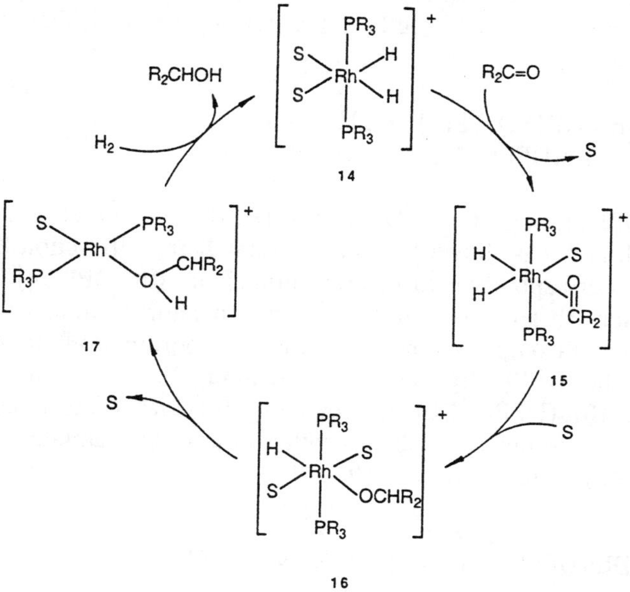

S= solvent

Scheme 2.10.

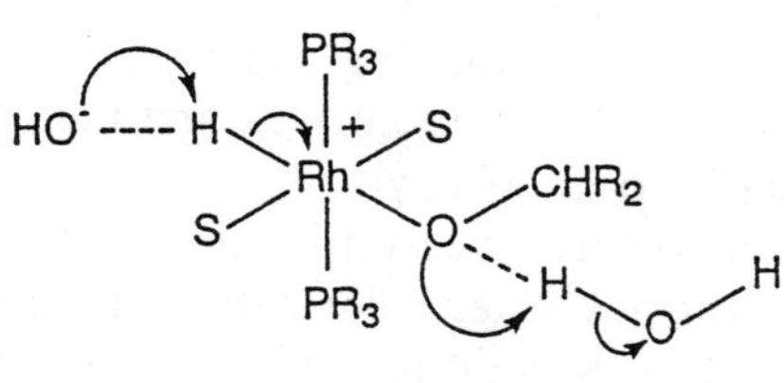

Fig. 2.4.

completed by dissociation of the alcohol product from the catalyst. This may occur either before or after the addition of a mole of hydrogen to the catalyst by a $d^8 \rightarrow d^6$ oxidative addition process to re-form *14*.

2.2.8. HYDROGENATION OF ALKENES CATALYSED BY $[Rh(NBD)(DPPE)]^+$ [14]

The chemistry of the complex $[Rh(NBD)(DPPE)]^+$ differs in some respects from that of complexes containing monodentate phosphine ligands. In methanol solutions, $[Rh(NBD)(DPPE)]^+$ reacts rapidly with a H_2/Rh molar ratio of exactly 2:1 according to the stoichiometry of equation *2.9*.

$$[Rh(NBD)(DPPE)]^+ + 2H_2 \longrightarrow [Rh(DPPE)]^+ + \text{norbornane} \tag{2.9}$$

$[Rh(DPPE)]^+$ can be isolated as a $[BF_4]^-$ salt, which contains no methanol, and was shown by single-crystal X-ray diffraction to have a structure corresponding to discrete binuclear $[Rh_2(DPPE)_2]^{2+}$ ions in which each Rh atom is bonded to two phosphorus atoms of a DPPE ligand and, through symmetrical π-arene coordination, to a phenyl ring of the DPPE ligand of the second Rh atom. In methanol solutions, $[Rh_2(DPPE)_2][BF_4]_2$ apparently dissociates into mononuclear $[Rh(DPPE)]^+$, forming 1:1 adducts with a variety of unsaturated substrates (sub) according to equation *2.10*.

$$[Rh(DPPE)]^+ + \text{sub} \underset{}{\overset{K_{10}}{\rightleftarrows}} [Rh(\text{sub})(DPPE)]^+ \tag{2.10}$$

In the case of benzene, the composition of the adduct was confirmed by isolating the salt $[Rh(\eta^6\text{-}C_6H_6)(DPPE)][BF_4].C_6H_6$. Values of K_{10} in methanol, determined for selected substrates are as follows: benzene ($18M^{-1}$); toluene (97); o-, m-, or p-xylene (500); 1-hexene(2); styrene (20); methyl acrylate (3). The binding constants of arenes are significantly higher than those of simple alkenes and the binding of styrene is clearly due primarily to the phenyl ring.

$[Rh(DPPE)]^+$ is an effective catalyst for the hydrogenation of simple alkenes as well as various alkene derivatives (styrene, acrylic acid, amidoacrylic acids, etc.). Kinetic measurements of the hydrogenation of 1-hexene in conjunction with the equilibrium measurements of the type quoted above lead to the following rate law

$$\frac{-d[h_1]}{dt} = \frac{k_{13}\, K_{12}\, [Rh]_{Tot}\, [h_1]\, [H_2]}{1 + K_{12}\, [h_1]} \tag{2.11}$$

where $[Rh]_{Tot} = [\{Rh(DPPE)\}^+]$ and $[h_1] = [1\text{-hexene}]$

This rate law is consistent with the following set of reactions:

$$[Rh(DPPE)]^+ + h_1 \overset{K_{12}}{\rightleftarrows} [Rh(h_1)(DPPE)]^+ \qquad \text{fast} \tag{2.12}$$

$$[Rh(h_1)(DPPE)]^+ + H_2 \xrightarrow{k_{13}} [Rh(DPPE)]^+ + h_0 \quad \text{slow} \tag{2.13}$$
h_0 = hexane

It is interesting to note that this set of reactions constitutes the general mechanism represented by path C in the Scheme *2.9*, the rate determining step being the reaction shown in equation *2.13*. Thus, the mechanism for the hydrogenation of alkenes catalysed by $[Rh(NBD)(DPPE)]^+$ departs significantly from that invoked for the corresponding reaction catalysed by $[Rh(NBD)(PR_3)_2]^+$ in which the principal paths involve the hydride complexes $[RhH_2S_xL_n]^+$ and $[RhHS_yL_n]$.

2.2.9. SOLVENT EFFECTS

The catalytic activation of the $[MH_2S_x(PR_3)_2]^+$ (M = Rh,Ir) systems requires dissociation of a solvent ligand rather than a phosphine (as in the Wilkinson's catalyst case) before the alkene substrate can gain access to the active site.

The solvent dissociation step is an important factor in the catalytic activity (solvent effect). Table *2.5* illustrates this phenomenon. Cationic rhodium and iridium systems are clearly more active in non-coordinating solvents such as dichloromethane or chloroform than in coordinating solvents such as acetone or alcohols. In order to obtain an idea of the relative stabilities of the different solvent ligands (S) in the $[IrH_2S_2(PR_3)_2]^+$ complexes, Crabtree *et al.* [24] performed displacement studies by adding 2 equivalents of one solvent to a CD_2Cl_2 solution of complexes of type $[IrH_2S_2(PR_3)_2]^+$ containing different solvents as ligands. From these experiments, the following order of stability was obtained: $H_2O \approx$ THF < *t*-BuOH < *i*-PrOH < Me_2CO < EtOH < MeOH < MeCN.

The activity observed in the presence of benzene merits further comment. Benzene, and arenes in general, shows a high tendency to form $[M(\eta^6\text{-arene})L_2]^+$ species [42], which are inactive [43]. The formation of these species also explains why substrates such as styrene or phenylalkynes can not be reduced with these systems as catalysts. The same effect can make $[BPh_4]^-$ a coordinating anion, to the extent that it can even displace a tertiary phosphine [22] (equation *2.14*). For this reason, its use as a counter ion can lead to catalyst deactivation.

$$[IrH_2(Me_2CO)_2(PCy_3)_2]^+ \xrightarrow[Me_2CO]{[BPh_4]^-} [(\eta^6\text{–}C_6H_5BPh_3)IrH_2(PCy_3)] \tag{2.14}$$

TABLE 2.5 [23]
Solvent effect on the hydrogenation of 1-hexene catalysed by $[M(diene)\{P(C_6H_4\text{-}4\text{-}MeO)_3\}_2][ClO_4]$ (M=Rh, diene=NBD, M=Ir, diene=COD)

M	Solvent	Initial turnover number mol H_2 $(molM)^{-1}min^{-1}$
Ir	Dichloromethane	67
	Chloroform	45
	Ethanol	3
	2-Methoxyethanol	1
	Acetone	1.5
	Benzene	3
Rh	Dichloromethane	29
	Chloroform	15
	Ethanol	5.4
	2-Methoxyethanol	4.5
	Acetone	1.1
	Benzene	0.9

2.2.10. CATALYTIC ACTIVITY OF MIXED-LIGAND COMPLEXES OF TYPE $[M(diene)L(PR_3)]^+$ (M = Rh, Ir. L = NITROGEN-DONOR LIGAND)

The $[Ir(COD)(py)_2]^+$ complex is totally inactive as catalyst, apparently because it fails to add hydrogen. However, the mixed-ligand complexes $[Ir(COD)L(PR_3)]^+$ (L = nitrogen-donor ligand) are active catalysts for the reduction of a variety of substrates, particularly tetrasubstituted alkenes [2b, 27, 33, 44, 45]. Thus, in dichloromethane as solvent, $[Ir(COD)(py)(PCy_3)]^+$ catalyses the hydrogenation of 2,3-dimethylbut-2-ene as well as tetrasubstituted prochiral amido-alkenes. Interestingly, even better results are found for the benzonitrile derivatives $[Ir(COD)(bzn)(PR_3)]^+$ (PR_3 = PCy_3, neomenthyl-diphenylphosphine) [45].

The reduction of the more hindered alkenes does not proceed to completion with the $[Ir(COD)(py)(PCy_3)]^+$complex, generally known as Crabtree's catalyst. This is due to an irreversible deactivation process, which occurs even for unhindered alkenes when the substrate has been consumed. The deactivation process produces the cation $[Ir_3H_7(py)_3(PCy_3)_3]^+$ (Figure 2.5) which may be isolated from the catalytic solutions [2b].

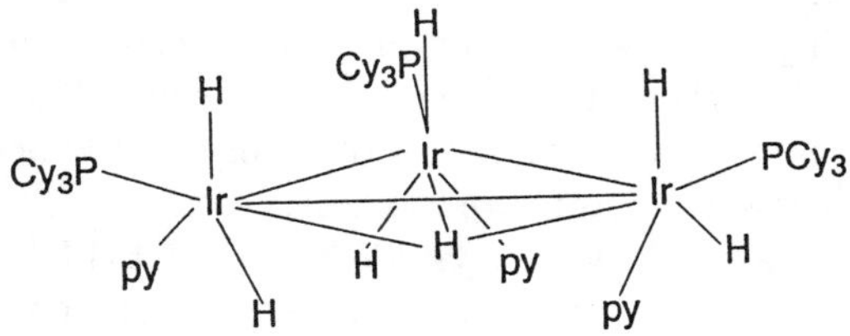

Fig. 2.5.

Rhodium complexes of the type $[Rh(NBD)L_2]^+$ (L = nitrogen-donor ligand) react with hydrogen in acetone to form species which are moderately active in the catalytic hydrogenation of monoenes. During the catalytic process, some decomposition to metallic rhodium takes place. This decomposition decreases as the basicity of the ligand increases. This general tendency of the cationic rhodium complexes containing nitrogen donor ligands partially to decompose to metallic rhodium is markedly reduced for the rhodium mixed-ligand complexes $[Rh(NBD)L(PR_3)]^+$ [27].

2.3. Ruthenium catalysts containing tertiary phosphine ligands

2.3.1. INTRODUCTION

The previous section showed the importance of the steric and electronic properties of the ancillary ligands on catalytic activity of the $[M(diene)L_n]^+$ systems. In general, the ligands can modify the properties of a given metal dramatically, for example, by stabilising different oxidation states or by fine-tuning the electrophilic or nucleophilic properties of the metal. However, it is also true that certain distinctive properties of a given metallic element often persist through quite drastic ligand changes. Probably no element illustrates this better than ruthenium, as reflected in the following distinctive properties that are manifested by a wide range of its complexes containing a variety of ligands: i) propensity for π-back-bonding, reflected in the marked stability of $Ru^{II}(CO)$, $Ru^{II}(\eta^1\text{-}N_2)$ [46] and $Ru(\eta^2\text{-}H_2)$ [47, 48] complexes; ii) tendency to undergo intra and intermolecular metallation [49] and iii) tendency to form polyhydride complexes [50–53].

Another characteristic of ruthenium is the wide range of complexes of this metal which have catalytic properties. Ruthenium systems which catalyse homogeneous hydrogenation have been studied extensively and some interesting reviews of chlororuthenate(II) and tetrachloro (bipyridyl)ruthenate(II) systems have been published [2a].

Very high rates of hydrogenation of terminal alkenes, alkynes or polynuclear heteroaromatic compounds can be achieved when $[RuHCl(PPh_3)_3]$ is used as catalyst. When $[RuCl_2(PPh_3)_3]$ is used, a small induction period is observed, and the solutions change colour to the characteristic deep violet of the hydrido derivative, suggesting that the true catalyst for these reactions is also $[RuHCl(PPh_3)_3]$ [54]. Wilkinson's group [55] reported in 1965 that $[RuCl_2(PPh_3)_3]$ reacted with hydrogen in ethanol-benzene solution to give the hydrido-complex, $[RuHCl(PPh_3)_3]$ according to equation *2.15*.

$$[RuCl_2(PPh_3)_3] + H_2 \longrightarrow [RuHCl(PPh_3)_3] + HCl \qquad (2.15)$$

As expected, this heterolytic activation of molecular hydrogen is accelerated by bases such as NEt_3 and it now seems very likely that the intermediate in this reaction is the dihydrogen complex $[RuCl_2(\eta^2\text{-}H_2)(PPh_3)_3]$ [1g]. The protons of a dihydrogen ligand are known to be more acidic than those of free H_2, and some dihydrogen-complexes can even be deprotonated by diethyl ether [56]. In this way the metal gives the same products that would have been obtained by an oxidative addition/reductive elimination pathway, but by avoiding the oxidative step, the metal avoids becoming Ru(IV), which is not a very stable oxidation state for ruthenium; even $[RuH_4(PPh_3)_3]$, long thought to be Ru(IV), is now known to have the structure $[RuH_2(\eta^2\text{-}H_2)(PPh_3)_3]$ [48].

The complex $[RuHCl(PPh_3)_3]$ also catalyses the reduction of aldehydes and ketones; however, the carbonyl derivative $[RuHCl(CO)(PPh_3)_3]$ is a more efficient catalyst for these processes [57]. From a mechanistic point of view, the catalytic activity of some anionic hydrido- and poly-hydrido-ruthenium complexes in the reduction of arenes, extensively studied by Halpern *et al.* [58, 59], is of special interest.

2.3.2. HYDROGENATION OF ALKENES CATALYSED BY $[RuHCl(PPh_3)_3]$

Chlorohydridotris(triphenylphosphine)ruthenium (II) is a specific hydrogenation catalyst for terminal alkenes under mild conditions. Isotope exchange experiments suggest that 1-alkyl complexes are formed preferentially and that ruthenium secondary alkyl complexes are formed less readily and are much less stable. The small degree of isomerization observed also suggests that 2-alkyl complexes are not usually formed from 1-alkenes [54, 60].

Initial hydrogen uptake rates measured in benzene solutions indicate that the alkene reduction rates are first order in alkene concentration and are inhibited by the addition of PPh_3 [38d]. In the light of these observations, the suggested mechanism is shown in the following set of reactions:

$$[RuHCl(PPh_3)_3] \overset{K_{16}}{\rightleftharpoons} [RuHCl(PPh_3)_2] + PPh_3 \tag{2.16}$$

$$[RuHCl(PPh_3)_2] + \text{alkene} \overset{K_{17}}{\rightleftharpoons} [Ru(alkyl)Cl(PPh_3)_2] \tag{2.17}$$

$$[Ru(alkyl)Cl(PPh_3)_2] + H_2 \xrightarrow{k_{18}} [RuHCl(PPh_3)_2] + \text{alkane} \tag{2.18}$$

The general rate law for this mechanism is of the form shown in equation *2.19*.

$$\frac{-d[H_2]}{dt} = \frac{k_{18}\, K_{17}\, [Ru]_{Tot}\, [\text{alkene}]\, [H_2]}{1 + K_{17}\, [\text{alkene}] + [PPh_3]/K_{16}} \tag{2.19}$$

where $[PPh_3]$ indicates a free concentration; in terms of added PPh_3 this rate law becomes:

$$\frac{-d[H_2]}{dt} = \frac{k_{18}\, K_{17}\, [Ru]_{Tot}\, [\text{alkene}]\, [H_2]}{1 + K_{17}\, [\text{alkene}] + [PPh_3]_{Tot}/K_{16} - 2[Ru]_{Tot}/K_{16}} \tag{2.20}$$

Detailed kinetics of these hydrogenation reactions in benzene solutions have not been studied due to the catalyst's limited solubility. However, detailed kinetic and spectroscopic studies on the hydrogenation of butenedioic acid in N,N-dimethyl-ethanamide solution strongly support the mechanism outlined in equations *2.16–2.18* [2a]. The kinetics were in complete agreement with the expression shown in equation *2.20*. The rate was first order in H_2, between zero and first order in alkene and showed an inverse dependence on PPh_3; the dependence on ruthenium depends on the conditions used, since reactions *2.16* and *2.17* occur to an extent which depends on the ratio of alkene to phosphine concentrations; at high ratios the critical red hydride colour turns yellow, all the ruthenium is present as $[Ru(alkyl)Cl(PPh_3)_2]$ and the rate law reduces to the simple form

$$\frac{-d[H_2]}{dt} = k_{21}[Ru][H_2] \tag{2.21}$$

At lower alkene concentration, reaction *2.17* is not complete and the effect of the dissociation equilibrium shown in equation *2.16* is observed in the ruthenium dependence, which becomes less than one at higher $[Ru]_{Tot}$ in systems where no excess PPh_3 has been added. This is reflected in the $[PPh_3]/K_{16}$ term in equation *2.19* which increases with increasing $[Ru]_{Tot}$.

2.3.3. ALKYNE HYDROGENATION CATALYSED BY $[RuHCl(PPh_3)_3]$

In the presence of the $[RuHCl(PPh_3)_3]$, terminal alkynes such as 1-hexyne are reduced by hydrogen at atmospheric pressures. Benzene/ethanol mixtures are the best solvents for these reactions. Alkynes are reduced more slowly than the corresponding alkenes [54]. However, hydrogenation of a mixture of 1-hexyne and 1-octene can be stopped at a stage showing the conversion of 99% of the alkyne to 1-hexene, with the 1-octene remaining unchanged. Using the reaction Scheme outlined in equations *2.16–2.18*, such selectivity and the relative rates of reduction of the single substrates can be explained qualitatively if K_{17}(alkyne) > K_{17}(alkene) and k_{18}(alkyne) < k_{18}(alkene) [2a].

In the presence of four or more equivalents of 3-hexyne, the compound $[RuHCl(PPh_3)_3]$ is rapidly converted to two new species detectable by ^{31}P NMR spectroscopy, $[\overline{RuCl(H_4C}_6PPh_2)(PPh_3)(alkyne)]$ and $[\overline{RuCl(H_4C}_6PPh_2)(PPh_3)_2]$; furthermore alkene is formed. The latter complex is converted to $[RuHCl(PPh_3)_3]$ by reaction with H_2. The formation of $[\overline{RuCl(H_4C}_6PPh_2)(PPh_3)_2]$ and its reaction with H_2 formally establishes a mechanism for alkyne hydrogenation, pathway B in Scheme *2.11*, which differs from the mechanism proposed for alkene hydrogenation, pathway A [61].

Hydrogenation of alkynes by pathway B requires the incorporation of a hydrogen atom from a phosphine ortho-aryl site in the alkene product. The replacement for this hydrogen atom comes from H_2 gas.

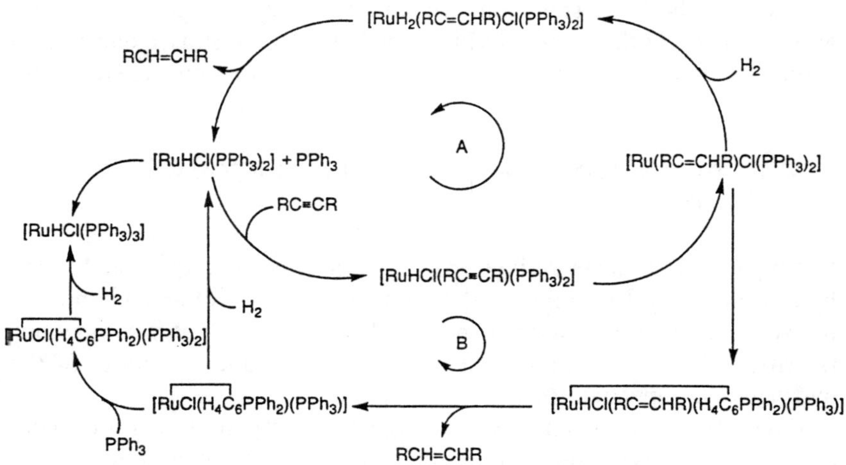

Scheme 2.11.

If pathway B prevails, hydrogenation of alkyne by $[RuHCl(PPh_3)_3]$ requires isotopic exchange between phosphine ortho-aryl sites and H_2 gas to occur at a rate equal to the rate of alkene formation. Examination of the ^1H NMR spectrum of the $[RuHCl(PPh_3)_3]$ complex recovered from alkyne reduction with D_2 reveals that only slightly more deuterium has been incorporated than during exposure to D_2 alone for an identical period of time. Near complete exchange would have been expected if pathway B prevailed, barring substantial kinetic isotope effects or rapid loss of the deuterium label into the large substrate pool. To preclude this possibility, the reincorporation of hydrogen in predeuterated $[RuHCl(PPh_3)_3]$ has been examined after alkyne hydrogenation. Once again, only slightly more label exchange occurs during hydrogenation than during exposure to H_2 alone. Thus, all the evidence is consistent with there being little or no hydrogenation-dependent exchange. So, examination of hydrogen isotope exchange during alkyne hydrogenation establishes that the observed species $[RuCl(H_4C_6PPh_2)(PPh_3)_2]$ are not true intermediates in the hydrogenation reaction.

Stoichiometric hydrogenation of alkenes by $[RuHCl(PPh_3)_3]$ has been observed. For example, the reaction of butenedioic acid with the hydride complex gives the alkane and $[RuCl(H_4C_6PPh_2)(PPh_3)_2]$ [62]. However, this stoichiometric process is again unimportant in catalytic alkene hydrogenation, which is much more rapid. Spectrophotometric data indicate that both the stoichiometric and catalytic hydrogenations probably proceed *via* the same alkyl intermediate but the subsequent hydrogenolysis step is faster than the intramolecular hydrogen transfer step.

2.3.4. HYDROGENATION OF POLYNUCLEAR HETEROAROMATIC COMPOUNDS

Synthetic fuel products derived from coal or oil shale require additional hydroprocessing to minimize their nitrogen and sulphur content. So, the selective hydrogenation of polynuclear heteroaromatic compounds is critical. Thus, it is important to have a basic understanding of the hydrogenation mechanism of these compounds.

Recently, it has been observed that a variety of transition-metal complexes catalyse the regioselective hydrogenation of polynuclear heteroaromatic compounds. Fish *et al.* found that ruthenium [63, 64] and rhodium [65] complexes show the greatest activity in the selective reduction of the heteroaromatic ring in this type of substrates. Interestingly, the hydridochlorotris(triphenylphosphine)ruthenium (II), formed in situ from dichlorotris(triphenylphosphine)ruthenium (II) and hydrogen gas with the heterocycle acting as base, is an exceptionally active catalyst for the hydrogenation of the substrates shown in Scheme *2.12* [64]. Because this compound is three times more active than its rhodium analogue, $[RhH_2Cl(PPh_3)_3]$, and the mechanism proposed for both systems is similar, the mechanistic features will be discussed in this section.

Table *2.6* shows the inital rates for the selective reduction of the

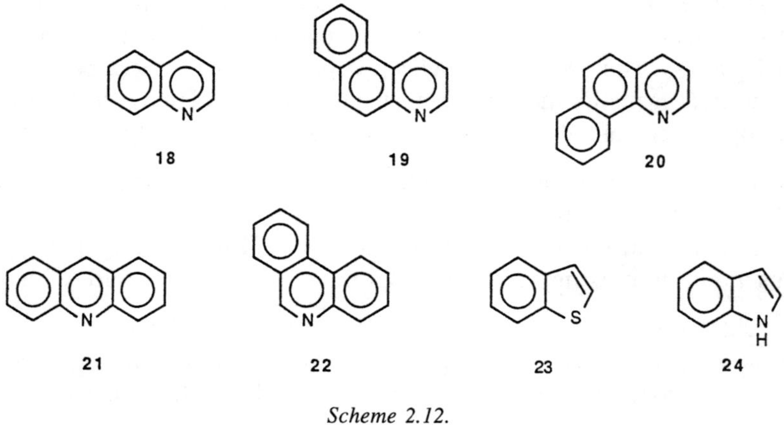

Scheme 2.12.

heterocycle ring in substrates *18–24* catalysed by [RuHCl(PPh_3)$_3$]. The order of individual reduction rates was *22* > *21* >> *18* > *24* > *19* > *23* > *20*, reflecting both steric and electronic effects. In order to obtain mechanistic information about this process, substrate *18* was reduced with D_2, according to equation *2.22*.

TABLE 2.6
Rates of hydrogenation of compounds 18–24 using [RuHCl(PPh_3)$_3$] as catalyst

Comp	Product	rate [a]
18	1,2,3,4-tetrahydroquinoline	1.00
19	1,2,3,4-tetrahydro-5,6-benzoquinoline	0.12
20	1,2,3,4-tetrahydro-7,8-benzoquinoline	0.03
21	9,10-dihydroacridine	9.2
22	9,10-dihydrophenanthridine	>24
23	2,3-dihydrobenzothiophene	0.09
24	2,3-dihydroindole	0.018

[a] Rates are relative to quinoline. Reaction conditions: see reference 64

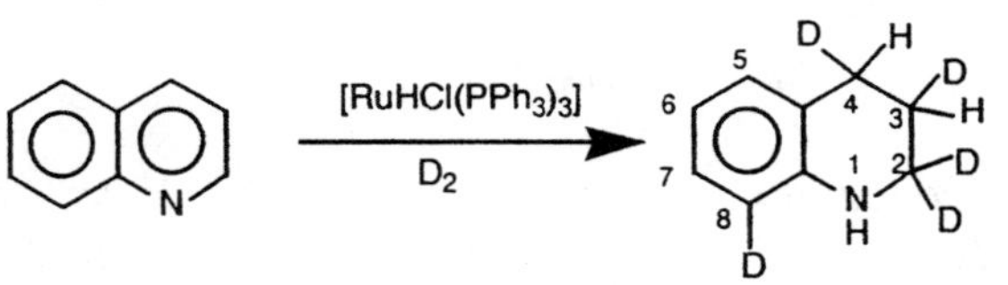

(2.22)

The product of reaction *2.22* contains 1.8 deuteriums at position 2, 1.0 deuterium each at positions 3 and 4, and 0.8 deuteriums at position 8. When this same reduction was carried out to approximately 50% conversion, the deuterium substitution pattern was much the same as in the case of complete reduction and the unreduced quinoline contained 0.5 deuterium substitution at the 2-position. In the same way 1,2,3,4-tetrahydroquinoline was reacted with deuterium and the ruthenium catalyst under the same conditions used to reduce *18*. The results are shown in equation *2.23*.

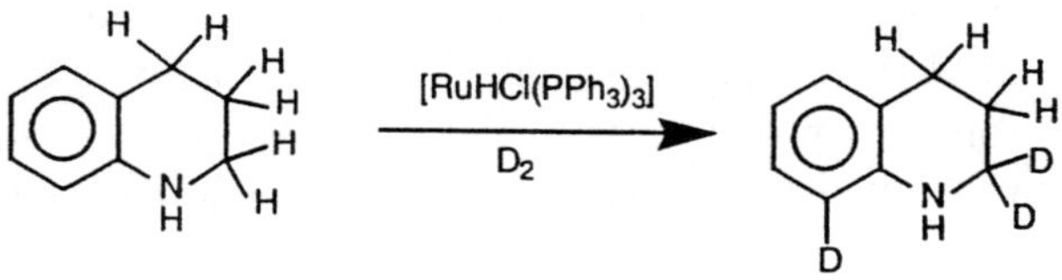

(2.23)

The 2-position was substituted with 1.8 deuteriums, and the 8-position was substituted with 0.1 deuterium.

This deuterium exchange can be accounted for by several plausible intermediates, which are included in the reduction Scheme proposed by Fish *et al.* (Scheme *2.13*).

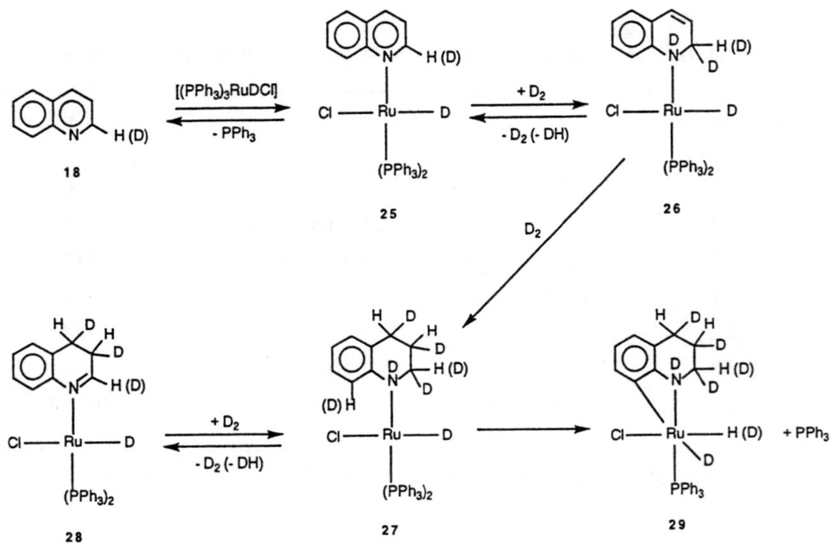

Scheme 2.13.

The overall reduction occurs in the order *18* ⇆ *25* ⇆ *26* ⇆ *27*. The first step, *18* ⇆ *25*, is the necessary step prior to coordination of *18* to the ruthenium metal catalyst. After coordination, the reversible reduction of the C-N double bond occurs, step *25* ⇆ *26*. It is the reversibility of this step which accounts for the incorporation of deuterium into the 2-position of the unreduced quinoline. It also accounts for some of the exchange at the 2-position of the product. However, not all of the exchange at this position in the product can occur by this mechanism. When the reduction was carried out to 50% completion, the reduced *18* was substituted with 1.85 deuteriums. If all the exchange had occurred through this reversible step, no more than 1.5 deuteriums would have been found at the 2-position of the product. The next step is the irreversible reduction of the 3,4 double bond. This step is shown to be irreversible since only 1.0 deuterium was found at the 3- and 4- positions on the product, and also because no *18* was ever observed being formed from 1,2,3,4,-tetrahydroquinoline by dehydrogenation, under reducing conditions. The post-reduction reversible step (*27* ⇆ *28*) is proposed to explain the exchange of deuterium found at the 2-position of both reactant and product, while the cyclometallated intermediate *29* is proposed to explain the deuterium incorporation at position 8.

2.3.5. HYDROGENATION OF ARENES CATALYSED BY ANIONIC- AND NEUTRAL- HYDRIDO (PHOSPHINE) RUTHENIUM COMPLEXES

The starting point for the research in this field was the synthesis of the anionic orthometallated hydridoruthenate complex, $[RuH_2(H_4C_6PPh_2)(PPh_3)_2]^-$, which is a catalyst or catalyst precursor for the selective hydrogenation of certain arenes, e.g. of anthracene to 1,2,3,4-tetrahydroanthracene. This compound was prepared by the slow addition of a solution of potassium naphtalene to a suspension of $[RuHCl(PPh_3)_3]$ in tetrahydroanthracene cooled from –80 to –111°C [49].

Recently, Halpern studied the fundamental coordination chemistry of $[RuH_2(H_4C_6PPh_2)(PPh_3)_2]^-$ and related anionic ruthenium complexes,

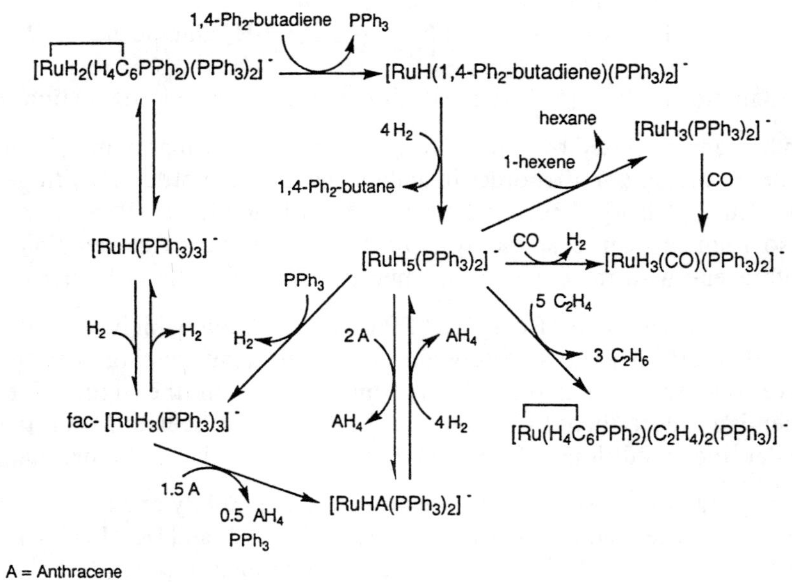

Scheme 2.14.

and the stoichiometric reactions of such complexes with possible relevance to their catalytic chemistry (Scheme *2.14*) [58, 66, 67]. $[RuH_2(H_4C_6PPh_2)(PPh_3)_2]^-$ reacts with hydrogen in THF solution at 25°C, according to equation *2.24*, to form fac-$[RuH_3(PPh_3)_3]^-$ which reacts with anthracene (equation *2.25*) to form a new red complex, $[RuH(anthracene)(PPh_3)_2]^-$, which also reacts rapidly with hydrogen in THF at 25°C (equation *2.26*), to yield $[RuH_5(PPh_3)_2]^-$. This compound reacts with a stoichiometric amount (1:2) of anthracene according to equation *2.27* to yield $[RuH(anthracene)(PPh_3)_2]^-$ quantitatively.

$$[RuH_2(H_4C_6PPh_2)(PPh_3)_2]^- + H_2 \longrightarrow \text{fac-}[RuH_3(PPh_3)_3]^- \quad (2.24)$$

$$\text{fac-}[RuH_3(PPh_3)_3]^- + 1{,}5 \text{ anthracene} \longrightarrow [RuH(\text{anthracene})(PPh_3)_2]^- + 0.5(1{,}2{,}3{,}4\text{-}H_4\text{-anthracene}) + PPh_3 \quad (2.25)$$

$$[RuH(\text{anthracene})(PPh_3)_2]^- + 4H_2 \longrightarrow [RuH_5(PPh_3)_2]^- + 1{,}2{,}3{,}4\text{-}H_4\text{-anthracene} \quad (2.26)$$

$$[RuH_5(PPh_3)_2]^- + 2 \text{ anthracene} \longrightarrow [RuH(\text{anthracene})(PPh_3)_2]^- + 1{,}2{,}3{,}4\text{-}H_4\text{-anthracene} \quad (2.27)$$

Kinetic studies [58] suggest that the rate of hydrogenation of anthracene catalysed by $[RuH_2(H_4C_6PPh_2)(PPh_3)_2]^-$ is approximately first order in ruthenium, first order in anthracene and zero order in hydrogen. fac-$[RuH_3(PPh_3)_3]^-$, $[RuH(anthracene)(PPh_3)_2]^-$ and $[RuH_5(PPh_3)_2]^-$ were also found to serve as catalyst precursors for the hydrogenation of anthracene with rates which, in some cases, were initially higher than those obtained with $[RuH_2(H_4C_6PPh_2)(PPh_3)_2]^-$ but which ultimately levelled off to approximately the same rates, suggesting that they give rise to a common catalytic mechanism. In the light of the chemistry described by equations *2.24–2.27*, it seems likely that, under the conditions of the catalytic reaction, the orthometallated $[RuH_2(H_4C_6PPh_2)(PPh_3)_2]^-$ complex is converted rapidly and irreversibly to other species (notably $[RuH(anthracene)(PPh_3)_2]^-$ and $[RuH_5(PPh_3)_2]^-$) and so is not directly involved in the catalytic mechanism.

The distinctive selectivity of $[RuH_2(H_4C_6PPh_2)(PPh_3)_2]^-$ or its derivatives for the catalytic hydrogenation of anthracene to 1,2,3,4-H_4-anthracene would appear to reflect the ability of these compounds to bind arenes in the "η^4" mode, which does not appear to be related to their anionic character.

The complexes $[RuH_3(PPh_3)_3]^-$ and $[RuH_2(\eta^2\text{-}H_2)(PPh_3)_3]$ were also found to be effective catalysts for the hydrogenation of 9-methyl-anthracene to 1,2,3,4,5,6,7,8-H_8-methyl-anthracene [59]. By using a gas-uptake method, the kinetics of the $[RuH_2(\eta^2\text{-}H_2)(PPh_3)_3]$-catalysed hydrogenation of 9-methyl-anthracene in toluene were found to obey a rate law (equation *2.28*) similar to that described for the $[RuH_3(PPh_3)_3]^-$-catalysed hydrogenation of anthracene.

$$-d[An]/dt = k_{28}[\{RuH_2(\eta^2\text{-}H_2)(PPh_3)_3\}]\,[An] \qquad (2.28)$$
$$An = \text{9-methyl-anthracene}$$

The catalytic activities of $[RuH_2(\eta^2\text{-}H_2)(PPh_3)_3]$ and $[RuH_3(PPh_3)_3]^-$ were compared directly by measuring the rates of hydrogenation of 9-methyl-anthracene at 55°C under 4 atm of H_2 in THF solutions. Although $[RuH_3(PPh_3)_3]^-$ is itself a catalyst for this reaction, its activity is only about half that of $[RuH_2(\eta^2\text{-}H_2)(PPh_3)_3]$. The details of the mechanism of the $[RuH_2(\eta^2\text{-}H_2)(PPh_3)_3]$-catalysed reaction have yet to be elucidated.

2 3.6. HYDROGENATION OF ALDEHYDES AND KETONES

Unlike other homogeneous hydrogenation catalysts, $[RuHCl(PPh_3)_3]$ becomes active towards a wide range of substrates under more severe operating conditions. For example, both aldehydes and ketones can be reduced to alcohols [68–70].

An interesting example of reduction achieved using $[RuHCl(PPh_3)_3]$ as catalyst is the reduction of aldehydic sugars. Glucose is reduced to sorbitol under severe conditions, although much lower hydrogen pressures and reaction temperatures are required if the hydrogenation is run in dimethylethanamide. The activity of $[RuHCl(PPh_3)_3]$ is impaired by carbonyl abstraction from glucose, which forms the complex $[RuHCl(CO)(PPh_3)_3]$ [71].

Sanchez-Delgado *et al.* studied the kinetics of the hydrogenation of acetone and propanal catalysed by $[RuHCl(CO)(PPh_3)_3]$ [72–74]. Results indicate that for both reactions, the rate is first order with respect to the concentration of catalyst and substrate and first order with respect to the hydrogen pressure. On the basis of these data and other considerations, they have proposed a general schematic mechanism for the $[RuHCl(CO)(PPh_3)_3]$-catalysed hydrogenation of the C=O bond, as shown in Scheme *2.15* [57].

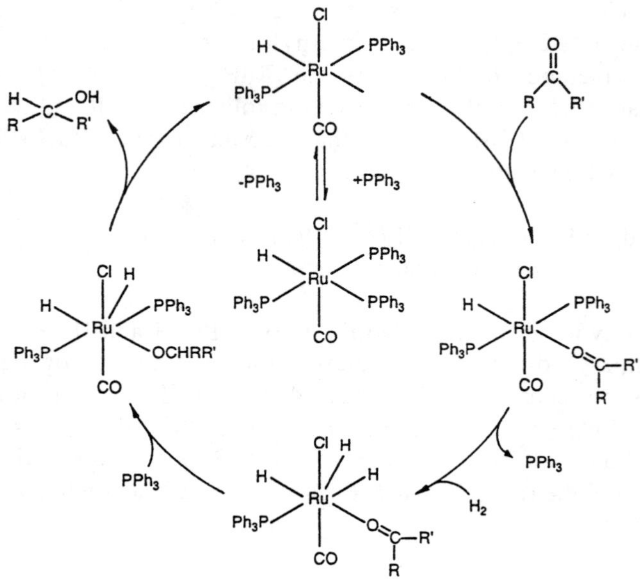

Scheme 2.15.

The dependence of the rate on the concentration of catalyst and substrate, and on the hydrogen pressure is consistent with the mechanism shown in Scheme *2.15*. Water accelerates the reaction; the reaction rate for acetone hydrogenation is first order with respect to the concentration of water. This indicates that water assists either the formation of the catalytically active species or its decomposition into the alcohol product [74].

It has also been observed that the addition of small amounts of ethanoic acid to the reaction mixture caused an increase in the hydrogenation rate [57]. One possible explanation for this effect is the formation of the carboxylate complex $[RuCl(O_2CCH_3)(CO)(PPh_3)_2]$, according to equation *2.29* [75].

$$[RuHCl(CO)(PPh_3)_3] + CH_3COOH \longrightarrow [RuCl(O_2CCH_3)(CO)(PPh_3)_2] + H_2 + PPh_3 \qquad (2.29)$$

However, the [RuCl(O_2CCH_3)(CO)(PPh_3)$_2$] complex is less active than the [RuHCl(CO)(PPh_3)$_3$] hydride. So, the increased activity observed on adding small amounts of ethanoic acid to the hydride precursor is probably best explained in terms of a hydrolytic cleavage of an alkoxymetal intermediate. The lower activity of this carboxylate complex is explained in terms of the mechanism shown in Scheme *2.16*. It is postulated that the coordination site occupied by the aldehyde or ketone molecule in species *31* is made available by opening the carboxylate chelate in *30*. Oxidative addition of H_2 followed by hydride transfer to the carbon atom (*31*→*33*) implies the formation of the 7-coordinate 18-electron Ru(IV)-alkoxy intermediate which, upon reductive elimination of the alcohol product, regenerates the active species *30*. However, the most recent results suggest that Ru (IV)-intermediates are rather improbable, and it is now known that the [$RuH_4(PPh_3)_3$] complex is actually a Ru(II)-dihydride-dihydrogen complex [48].

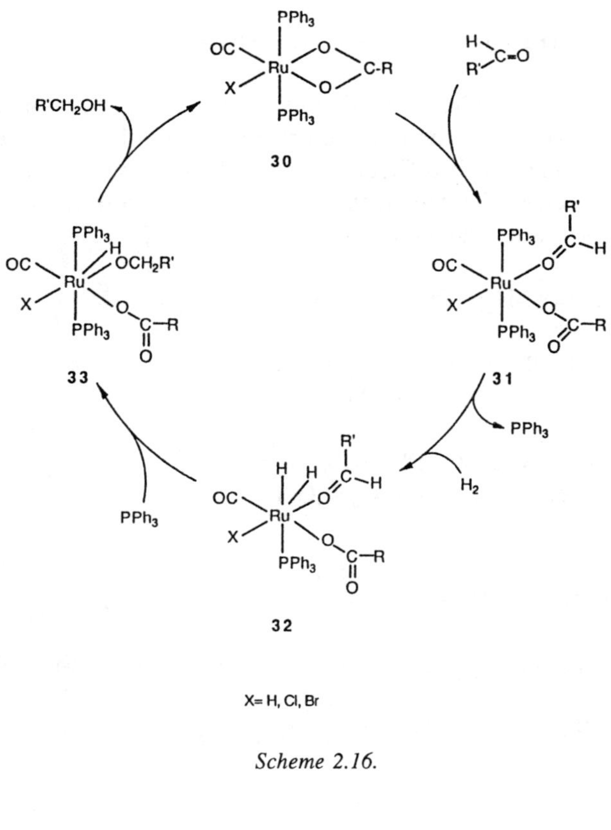

Scheme 2.16.

This dihydride-dihydrogen compound catalyses the hydrogenation of cyclohexanone in toluene as solvent [59]. Kinetic studies on the reaction lead to the rate law shown in equation *2.30*.

$$-d[c\text{-}C_6H_{10}O]/dt = k_{30}\ [\{RuH_2(\eta^2\text{-}H_2)(PPh_3)_3\}]\ [c\text{-}C_6H_{10}O] \quad (2.30)$$

This catalytic rate law is essentially identical to that observed for the stoichiometric reaction between $[RuH_2(\eta^2\text{-}H_2)(PPh_3)_3]$ and cyclohexanone ($k_{30} = k_{31}$), confirming that the mechanism shown in equations *2.31* and *2.32* is the only catalytic mechanism possible for this reaction, in which the reaction of $[RuH_2(\eta^2\text{-}H_2)(PPh_3)_3]$ with cyclohexanone is the rate-determining step.

$$[RuH_2(\eta^2\text{-}H_2)(PPh_3)_3] + c\text{-}C_6H_{10}O \xrightarrow{k_{31}} [RuH_2(PPh_3)_3] + c\text{-}C_6H_{11}OH \quad \text{slow} \quad (2.31)$$

$$[RuH_2(PPh_3)_3] + H_2 \longrightarrow [RuH_2(\eta^2\text{-}H_2)(PPh_3)_3] \quad \text{fast} \quad (2.32)$$

The detailed mechanism of this step is unclear since $[RuH_2(\eta^2\text{-}H_2)(PPh_3)_3]$ is coordinately saturated. A rate-determining or pre-equilibrium loss of H_2 is difficult to reconcile with the observed kinetics unless the inverse $[H_2]$ dependence of the latter step is compensated by a subsequent H_2-dependent step (equations *2.33–2.35*).

$$[RuH_2(\eta^2\text{-}H_2)(PPh_3)_3] \rightleftharpoons [RuH_2(PPh_3)_3] + H_2 \quad (2.33)$$

$$[RuH_2(PPh_3)_3] + c\text{-}C_6H_{10}O \rightleftharpoons [RuH_2(c\text{-}C_6H_{10}O)(PPh_3)_3] \quad (2.34)$$

$$[RuH_2(c\text{-}C_6H_{10}O)(PPh_3)_3] + H_2 \rightleftharpoons [RuH_2(PPh_3)_3] + c\text{-}C_6H_{11}OH \quad (2.35)$$

$$c\text{-}C_6H_{10}O + H_2 \xrightarrow{[RuH_2(\eta^2\text{-}H_2)(PPh_3)_3]} c\text{-}C_6H_{11}OH \quad (2.36)$$

To support this hypothesis, Halpern *et al.* found that $[RuH_2(\eta^2\text{-}H_2)(PPh_3)_3]$ undergoes rapid substitution by N_2 or PPh_3 to form $[RuH_2(N_2)(PPh_3)_3]$ or $[RuH_2(PPh_3)_4]$, respectively.

2.3.7. HYDROGENATION OF NITROCOMPOUNDS

The complex $[RuHCl(PPh_3)_3]$ also catalyses the hydrogenation of nitrocompounds [76]. Basic reaction conditions favour faster reduction rate by deprotonation of the nitrocompound to its anionic form, thereby shifting the equilibrium shown in equation *2.37* further to the right. The formation of this anion has been confirmed spectroscopically.

$$RR'CHNO_2 + B: \rightleftharpoons (RR'CNO_2)^- + BH^+ \quad (2.37)$$

The suggested mechanism for hydrogenation of nitroalkanes to give amines (equations *2.16, 2.38, 2.39*) has several features in common with that proposed above for alkene hydrogenation. Thus, initial dissociation of $[RuHCl(PPh_3)_3]$ to give $[RuHCl(PPh_3)_2]$ is consistent with the observed inhibition by excess triphenylphosphine.

$$[RuHCl(PPh_3)_2] + (RR'CNO_2)^- \longrightarrow [RuCl(PPh_3)_2(RR'CNO)] + OH^- \quad (2.38)$$

$$[RuCl(PPh_3)_2(RR'CNO) + 3H_2 \longrightarrow [RuHCl(PPh_3)_2] + RR'CHNH_2 + H_2O \quad (2.39)$$

A variety of ruthenium complexes with π-bonding ligands capable of forming hydrido species of differing lability have been tested and found active for hydrogenation of nitroalkanes. Bis(triphenylphosphine) iron tricarbonyl and iron pentacarbonyl both yield some amine but are generally less effective and less stable in the basic media.

2.4. Osmium catalysts containing tertiary phosphine-type ligands

2.4.1. INTRODUCTION

The fact that the *5d* metals form stronger bonds than their *3d* and *4d* counterparts with the ligands typically involved in catalytic transformations has led in the past to the general assumption that reactions involving third-row transition metal complexes are too slow for catalytic cycles and are thus of no practical use in catalysis. A number of second-row metal compounds with excellent catalytic properties have been discovered and, in addition, various iridium and osmium complexes have been synthesized which could serve as stable models of reactive intermediates proposed for catalytic transformations involving *4d* metal species. Representative examples of second-row metal catalysts are the neutral chloro and hydrido phosphine rhodium and ruthenium complexes mentioned above. Furthermore, cationic diene rhodium complexes are also active and prove to be particularly selective catalysts.

Besides these rhodium derivatives, it was shown above that some cationic iridium compounds behave similarly provided that appropriate ancillary ligands are linked to the metal. As an example, it was mentioned that cationic cyclo-octadiene iridium complexes containing both P and N donor ligands, e.g. PCy_3 and bzn, are more active than their rhodium counterparts and are able to reduce tetrasubstituted prochiral alkenes. Therefore, the once widely held view that third-row transition metal compounds are unimportant in catalysis has had to be revised, at least for the case of iridium as metal centre.

A similar situation could be anticipated for osmium if the ligands and reaction conditions are selected appropriately. Following earlier reports [77], Sánchez-Delgado *et al.* observed that the complex $[OsHBr(CO)(PPh_3)_3]$ not only catalyses the isomerisation of allylic alcohols, but also the hydrogenation of acyclic and cyclic alkenes, of dienes, alkynes, α,β-unsaturated aldehydes, ketones, etc. [78]. The synthesis, reactivity and catalytic activity of the complexes $[OsHCl(CO)(PR_3)_2]$ and $[OsH_2Cl_2(PR_3)_2]$ (PR_3 = P*i*Pr_3, P*t*Bu_2Me) has been reported in the last few years [79–89]. The complex $[OsHCl(CO)(PiPr_3)_2]$ which, unlike $[OsHBr(CO)(PPh_3)_3]$, is coordinatively unsaturated, behaves similarly, and under hydrogen it catalyses the reduction of cyclohexene, 1,3-and 1,4-cyclohexadiene, styrene, diphenyl or phenylethyne and benzylideneacetone [83, 86]. It has been shown that in the hydrogenation of phenylethyne, the formation of styryl derivatives is the step which determines the selectivity for hydrogenation to the alkene [86]. This is one of the few catalytic cycles in which each of the postulated steps has been checked, so it is possible to follow the mechanism of the sequential hydrogenation of phenylethyne in some detail.

2.4.2. $[OsHCl(CO)(PR_3)_2]$ COMPLEXES (PR_3 = P*i*Pr_3, P*t*Bu_2Me): THE ZARAGOZA-WÜRZBURG CATALYSTS

These complexes were prepared in 96% yield heating under reflux $OsCl_3.xH_2O$ with the phosphine ligand in methanol or 2-methoxy-ethanol.

During the reaction, the alcohol is dehydrogenated by the metal trichloride to give methanal, which is probably the source of the carbonyl ligand [79, 81]. The complex [OsHCl(CO)(P*i*Pr$_3$)$_2$] reacts readily with hydrogen, oxygen, triethylsilane, alkynes, ethene, methylacrylate, acrylonitrile and methyl vinyl ketone to give dihydrogen, dioxygen, alkenyl-metal and hydride-alkene metal compounds (Scheme *2.17*) [79, 80, 83, 87] which can be considered as models for catalytic intermediates.

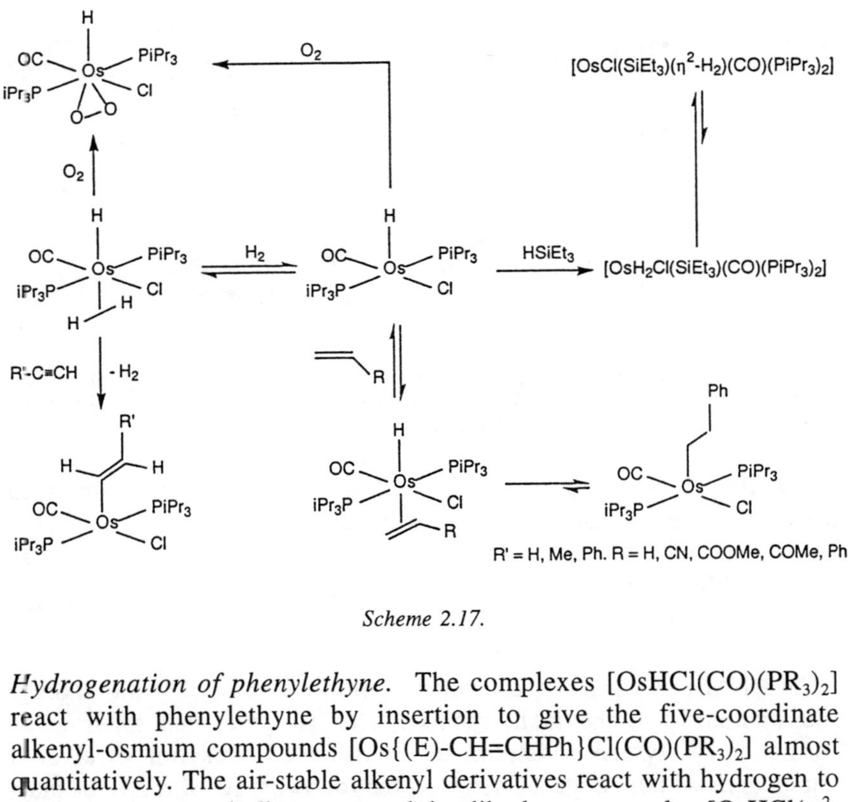

Scheme 2.17.

Hydrogenation of phenylethyne. The complexes [OsHCl(CO)(PR$_3$)$_2$] react with phenylethyne by insertion to give the five-coordinate alkenyl-osmium compounds [Os{(E)-CH=CHPh}Cl(CO)(PR$_3$)$_2$] almost quantitatively. The air-stable alkenyl derivatives react with hydrogen to produce styrene, ethylbenzene, and the dihydrogen complex [OsHCl(η^2-H$_2$)(CO)(PR$_3$)$_2$]. This hydrogenation reaction, together with the formation of the alkenyl complexes [Os{(E)-CH=CHPh}Cl(CO)(PR$_3$)$_2$], constitutes a catalytic cycle for the reduction of phenylethyne to styrene [86]. As expected from the coordination chemistry, the [OsHCl(CO)(PR$_3$)$_2$] complexes are efficient catalysts for the sequential hydrogenation of phenylethyne in 2-propanol solution. At 60°C and atmospheric pressure, selectivities close to 100% are achieved for the hydrogenation of the alkyne to alkene, as illustrated in Figure *2.6* . Reduction of the double bond only begins to take place when most of the alkyne has been consumed.

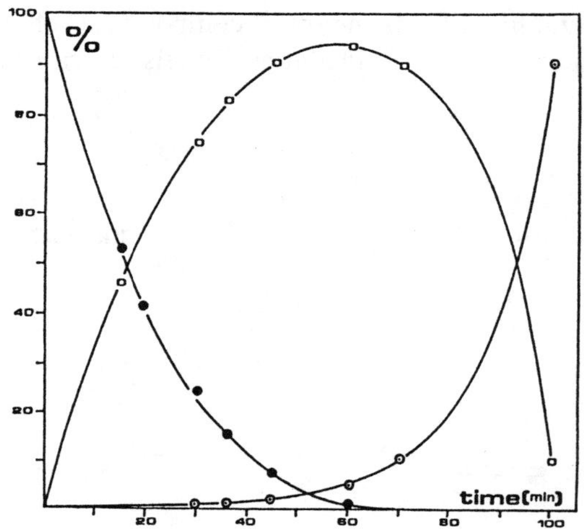

Fig. 2.6. Hydrogenation of phenylethyne catalysed by $[OsHCl(CO)(P\textit{i}Pr_3)_2]$ in 2-propanol at 60°C (1 atm of H_2; 2.5×10^{-3}M $[OsHCl(CO)(P\textit{i}Pr_3)_2]$; 0.25M HC≡CPh). (●) phenylethyne, (○) styrene, (⊙) ethylbenzene.

Detailed kinetic studies of the hydrogenation of phenylethyne to styrene lead to the rate law shown in equation *2.40.*

$$\frac{d[\text{Styrene}]}{dt} = \frac{-d[\text{PhC}_2\text{H}]}{dt} = k_{40}[\text{Cat}][\text{PhC}_2\text{H}]P(\text{H}_2) \qquad (2.40)$$

The NMR spectra of the catalytic solutions show that the alkenyl intermediates are the main species. This suggests that the rate of formation of styrene is determined by the rate of reaction of alkenyl compounds with hydrogen. Therefore, the following set of reactions must be consistent with the catalytic cycle:

$$\text{Os-H} + \text{PhC}_2\text{H} \xrightleftharpoons{K_{41}} \text{Os-alkenyl} \qquad \text{fast} \qquad (2.41)$$

$$\text{Os-alkenyl} + \text{H}_2 \xrightarrow{k_{42}} \text{Os-H} + \text{styrene} \qquad \text{slow} \qquad (2.42)$$

Equations *2.40–2.42* are consistent with the mechanism shown in Scheme *2.18* [86]. The reaction of the monohydride or the dihydrogen with the alkyne is fast and leads to stable 16-electron alkenyl complexes. The elementary steps involved in the formation of styryl derivatives [Os{(E)-CH=CHPh}Cl(CO)$(PR_3)_2$] are too fast to be observed by spectroscopic methods. However, NMR spectroscopy has shown that alkyne dicarboxylic methyl ester coordinates to [OsHCl(CO)$(P\textit{i}Pr_3)_2$] *trans* to the hydride at room temperature; then rearrangement to the *cis*-isomer takes place, followed by insertion to yield the corresponding alkenyl species [86]. It is reasonable to assume that the same sequence of events is operative in the formation of the styryl compounds. The slow step of this catalytic cycle is the reaction of these five-coordinate complexes with hydrogen to yield the alkene and regenerate the monohydrides in equilibrium with the dihydrogen complexes. Although more intimate details of this cycle remain to be elucidated, the reaction of the alkenyl compounds with hydrogen is likely to involve a series of elementary steps. One plausible sequence is the oxidative addition of H_2 – perhaps *via* a dihydrogen-alkenyl-osmium intermediate – to yield the 18-electron Os(IV) species [OsH_2(CH=CHPh)Cl(CO)$(PR_3)_2$] followed by reductive elimination of styrene.

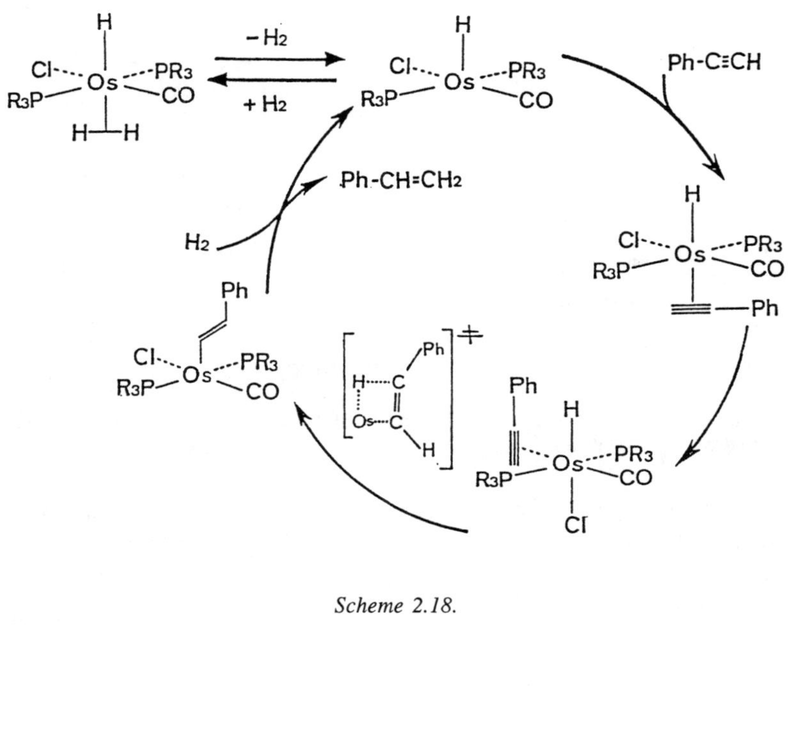

Scheme 2.18.

The hydrogenation of styrene to ethylbenzene is less clear-cut from a mechanistic point of view. In the light of the coordination chemistry presented in Scheme *2.17*, the mechanism shown in Scheme *2.18* may also be operative for this reaction, but another possible route is the initial coordination of hydrogen.

The high selectivity observed for the hydrogenation of phenylethyne to styrene merits further comment. The independent study of the reduction of C≡C and C=C bonds indicates that the latter are kinetically favoured (in the absence of phenylethyne, $[OsHCl(CO)(PR_3)_2]$ catalyse the hydrogenation of styrene to ethylbenzene at rates about 1 order of magnitude faster than those for C≡C bond reduction), and thus the origin of this selectivity cannot be kinetic. Under catalytic conditions, the alkenyl compounds are the main species. These alkenyl complexes represent a thermodynamic sink which causes virtually all the osmium present in solution to be tied up in this form, and therefore the kinetically unfavourable pathway becomes essentially the only one available in the presence of alkyne. This thermodynamic difference, illustrated qualitatively in Scheme *2.19*, may be at the origin of the high selectivity for the hydrogenation of the C≡C bond.

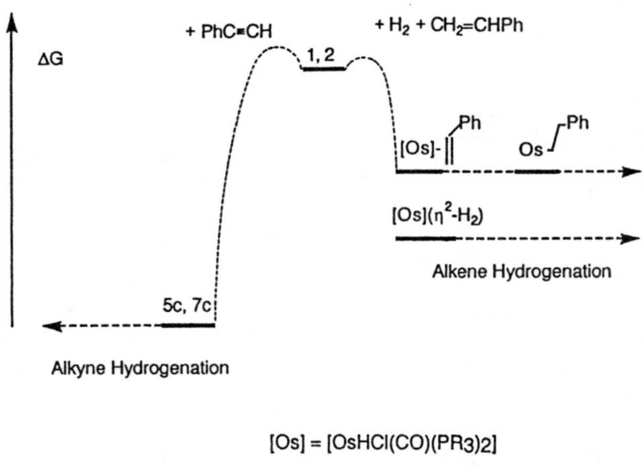

Scheme 2.19. Qualitative diagram of free energy for the hydrogenation of phenylethyne catalysed by the complexes $[OsHCl(CO)(PR_3)_2]$ (PR_3 = $P\textit{i}Pr_3$, $P\textit{t}Bu_2Me$).

Hydrogenation of benzylideneacetone [83b]. In 2-propanol solutions, the complexes $[OsHCl(CO)(PR_3)_2]$ (PR_3 = $P\textit{i}Pr_3$,$P\textit{t}Bu_2Me$) catalyse the hydrogenation of benzylideneacetone to 4-phenylbutan-2-one with selectivities close to 100%.

In the presence of [OsHCl(CO)(P*i*Pr$_3$)$_2$] the reaction is first order with respect to the concentration of catalysts and substrate and independent of the pressure of hydrogen. The mechanism deduced for this reaction based on these kinetic data and on spectroscopic observations is shown in Scheme *2.20*. As the insertion of the substrate in the Os-H bond of [OsHCl(CO)(P*i*Pr$_3$)$_2$] is not favoured due most probably to the trans disposition of the hydride ligand and the coordination vacancy, formation of *trans(hydride,dihydrogen)*-[OsHCl(η^2-H$_2$)(CO)(P*i*Pr$_3$)$_2$] following isomerization to *cis(hydride,dihydrogen)*-[OsHCl(η^2-H$_2$)(CO)(P*i*Pr$_3$)$_2$] and subsequent hydrogen dissociation produce a rearrangement of the catalyst ligands to give a new mono-hydride isomer which contains the hydride ligand and the coordination vacancy in a *cis* disposition. Then, coordination of the substrate and its subsequent insertion in the Os-H bond must lead to an alkyl intermediate which, by reaction with molecular hydrogen gives 4-phenylbutan-2-one and regenerates the catalyst.

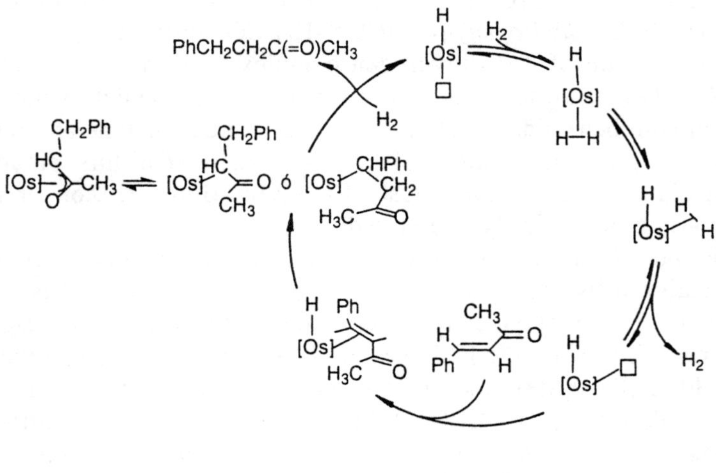

Scheme 2.20.

In the presence of [OsHCl(CO)(P*t*Bu$_2$Me)$_2$] the reaction is second-order with respect to the catalyst and first-order with respect to hydrogen and benzylideneacetone. Scheme *2.21* shows a catalytic cycle that is consistent with these kinetic data and which contains, in equilibrium, the species spectroscopically detected in the reaction of

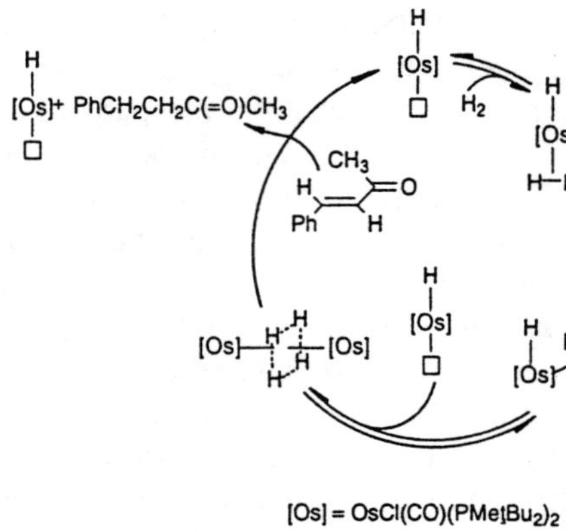

Scheme 2.21.

$[OsHCl(CO)(PtBu_2Me)_2]$ with molecular hydrogen. Under catalytic conditions this mono-hydride is in a dynamic equilibrium with *trans-(hydride-dihydrogen)*-$[OsHCl(\eta^2\text{-}H_2)(CO)(PtBu_2Me)_2]$ which isomerizes to *cis-(hydride-dihydrogen)*-$[OsHCl(\eta^2\text{-}H_2)(CO)(PtBu_2Me)_2]$. The subsequent reaction of this dihydrogen complex with the mono-hydride $[OsHCl(CO)(PtBu_2Me)_2]$ leads to a binuclear intermediate which, by reaction with benzylideneacetone gives the saturated ketone and regenerates the catalyst. Theoretical works suggest that this binuclear intermediate could be *trans*-$[\{OsCl(CO)(PtBu_2Me)_2\}_2H_4]$ containing a planar 4-gon of cyclically bound hydrogen atoms.

It is interesting to mention that the mechanism of this hydrogenation catalyzed by the complex containing $PiPr_3$ as phosphorus-donor ligand involves mononuclear species during the full catalytic cycle (Scheme *2.20*), while the hydrogenation carried out in the presence of $[OsHCl(CO)(PtBu_2Me)_2]$ must probably takes place through the binuclear intermediate *trans*-$[\{OsCl(CO)(PtBu_2Me)_2\}_2H_4]$. It is also important to note that two isoelectronic and isostructural complexes with phosphines of similar basicities and steric requirements catalyze the same reaction via a completely different mechanism, thus illustrating different roles of the dihydrogen complexes in catalytic hydrogenation reactions. In the light of this, it is clear that it is only possible to propose a sensible catalytic cycle on the basis of kinetic and spectroscopic studies of the reactions. Generally, the catalytic mechanism involves multistep reactions where the intermediates are connected by equilibriums that are highly dependent on the electronic properties and on the steric requirements of the catalyst ligands, as well as on the characteristics of the substrates. Thus, slight modifications of these factors can change completely the direction of the equilibriums and therefore, the contribution of a particular species to the overall catalytic process. Scheme *2.22* sumarizes Schemes *2.18* and *2.20* and *2.21* and illustrates this phenomenon for the hydrogenation of phenylethyne to styrene and benzylideneacetone to 4-phenylbutan-2-one, catalysed by the $[OsHCl(CO)(PR_3)_2]$ (PR_3 = P*i*Pr_3, P*t*Bu_2Me).

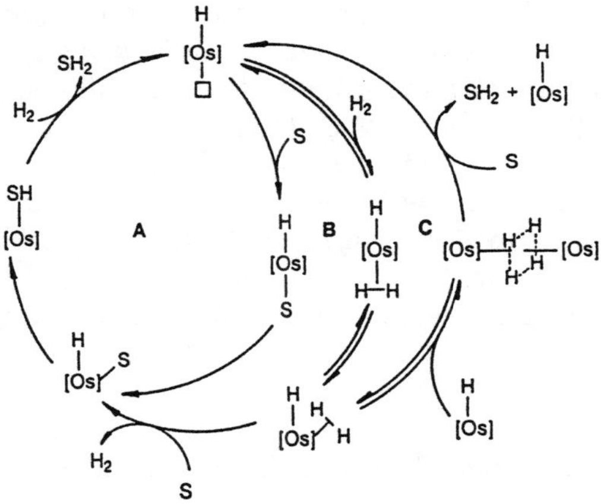

[Os] = $OsCl(CO)(PR_3)_2$

	PR_3	S
A	P*i*Pr_3, PMe*t*Bu_2	HC≡CPh
B	P*i*Pr_3	PhHC=CHC(=O)CH_3
C	PMe*t*Bu_2	PhHC=CHC(=O)CH_3

Scheme 2.22.

2.4.3. $[OsH_2Cl_2(PiPr_3)_2]$ AS CATALYST [89]

Provided the assumption that the *CO* ligand of the hydrido (carbonyl)compound $[OsHCl(CO)(PiPr_3)_2]$ is generated from methanol through methanal is correct, the same preparative reaction in 2-propanol instead of methanol should lead to complexes without a carbonyl ligand. In fact, $OsCl_3.xH_2O$ reacts with P*i*Pr_3 in boiling 2-propanol to give the dihydride dichloro complex $[OsH_2Cl_2(PiPr_3)_2]$ in ca. 80% yield [89].

Interestingly, under hydrogen atmosphere, solutions of this complex in 2-propanol, 1,2-dichloroethane or toluene catalyse the hydrogenation of styrene, methylstyrene, cyclohexene and cyclooctene at considerable initial rates, which depend both on the solvent (for cyclooctene:v_0 [*i*PrOH] > v_0 [C_6H_5Me] > v_0 [1,2-$C_2H_4Cl_2$]) and the alkene substrate (in *i*PrOH: cyclooctene > styrene > cyclohexene > methylstyrene).

This complex also catalyses the hydrogenation of the C=C bond of α,β-unsaturated ketones, as well as dienes. 1,5-Cyclooctadiene is more rapidly reduced than the 1,3 isomer. This finding is also true in a competitive sense: 1,3-cyclooctadiene is not hydrogenated until the concentration of the 1,5-isomer is almost zero. Unfortunately, the selectivity for these reactions to give the cyclooctene is poor.

In short, osmium (II) and osmium (IV) complexes with formulae [$OsHCl(CO)(PR_3)_2$] and [$OsH_2Cl_2(PR_3)_2$] respectively catalyse the hydrogenation of alkenes and dienes as well as the selective reduction of benzylideneacetone, benzylideneacetophenone and phenylethyne. Therefore, there is now increasing evidence that for the metals in the iron triad not only ruthenium but also osmium forms a variety of complexes which are good catalysts for the reduction of unsaturated organic substrates.

2.5. Binuclear complexes as catalysts

2.5.1. CATALYTIC SYNERGISM

Studies of multi-component metal systems under homogeneous conditions have revealed the existence of catalytic activity enhancement when compared to that of the individual components [90]; this enhancement effect is known as synergism [91, 92], and it has played an important role in the development of homogeneous polymetallic catalysis [93]. From a mechanistic point of view, the concept of synergism has been applied to metal atoms operating either individually [92] or, more rigorously, in concert [94–97] in reaction sequences. A genuine bimetallic mechanism in which two metal centres act in concert has been postulated by Kalck's group for alkene hydroformylation catalysed by compounds of the type *cis*-[$Rh_2(\mu\text{-S}t\text{Bu})_2(CO)_2(PR_3)_2$], on the basis of theoretical calculations and spectroscopic studies [94]. Bimetallic catalytic pathways involving the concerted action of two metals have been also proposed for processes such as the hydration of acrylonitrile to acrylamide catalysed by binuclear palladium complexes [95], alkene hydroformylation catalysed by compounds of the type [$Ru_2(\mu\text{-O}_2CR)_2(CO)_4(PR_3)_2$] [96], or hydrogen transfer from alcohols to ketones in the presence of

$[Fe_3H(CO)_{11}]^-$ [97]. These proposals have been postulated on the basis of two ideas: i) the flexibility of M-L-M′ bridges or M-M bonds permits the transfer of an atom or ligand from one metal centre to the other; ii) an electronic cooperative effect can take place between metallic centres *via* orbital interactions with the bridging ligands.

Bimetallic catalytic pathways have been also proposed for homogeneous hydrogenation. For example Muetterties has suggested that the dimeric hydride $[\{RhH(P\{O\textit{i}Pr\}_3)_2\}_2]$ catalyses alkene and alkyne hydrogenation *via* binuclear intermediates [98]. However, no kinetic evidence has been reported to prove the integrity of the catalysts during the reactions. On the other hand, Maitlis has found kinetic evidence in favour of the cleavage of the bridges between the metallic atoms of compounds $[\{(\eta^5\text{-}C_5Me_5)MCl_2\}_2]$ (M = Rh, Ir) or $[\{(\eta^5\text{-}C_5Me_5)IrHCl\}_2]$, which catalyse alkene hydrogenation [99].

Binuclear compounds of formulae $[H(CO)(PPh_3)_2Ru(\mu\text{-bim})M(COD)]$ (M = Rh,Ir; bim = 2,2′-biimidazolate), $[H(CO)(PPh_3)_2Ru(\mu\text{-pz})_2Ir(TFB)]$ (pz = pyrazolate) and $[\{Ir(\mu\text{-pz})(diene)\}_2]$ (diene = TFB,COD) have been recently reported [100, 101]. These compounds are more active catalysts for the hydrogenation of cyclohexene than the mononuclear parent complexes $[RuH(Hbim)(CO)(PPh_3)_2]$, [Ir(Hbim)(COD)], $[RuH(pz)(CO)(Hpz)(PPh_3)_2]$ and $[Ir(TFB)(Hpz)_2][BF_4]$ [102, 103]. Studies of the kinetics of the hydrogenation of cyclohexene catalysed by these binuclear compounds suggest that the full catalytic cycle involves binuclear intermediates [104].

2.5.2. HYDROGENATION OF ALKENES AND ALKYNES CATALYSED BY $[\{RhH(P\{O\textit{i}Pr\}_3)_2\}_2]$

This dimer has the bridged square planar geometry shown in Figure *2.7*. It reacts in a very slow and complex manner with alkenes. However, it reacts instantaneously with one mole of hydrogen to give a new dimeric

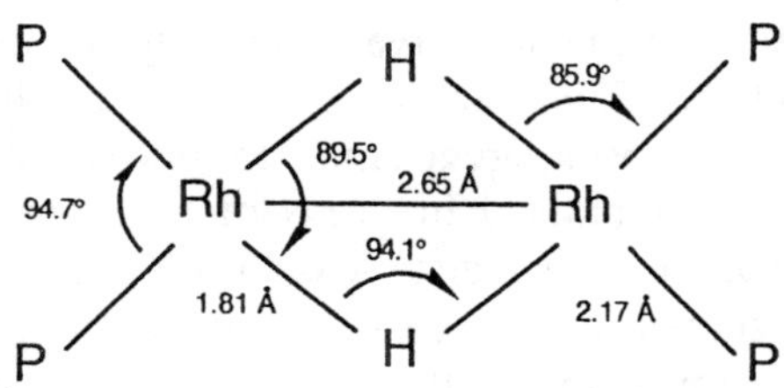

Fig. 2.7.

hydride (equation *2.43*) with three bridging hydrides. This tetrahydride is stereochemically non-rigid and in solution it evolves to a new dimeric tetrahydride containing two bridging hydrides (equation *2.44*). This polyhydride is presumed to be the species that interacts with the alkene to give an alkene complex (equation *2.45*), which rapidly yields alkane and the original binuclear rhodium hydride (equations *2.46* and *2.47*) [98].

(2.43)

(2.44)

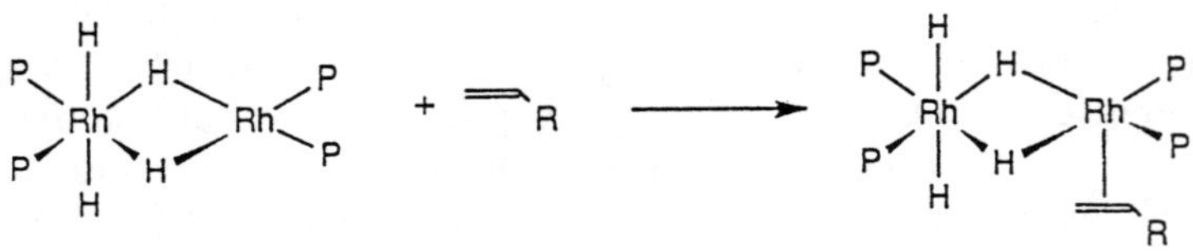

(2.45)

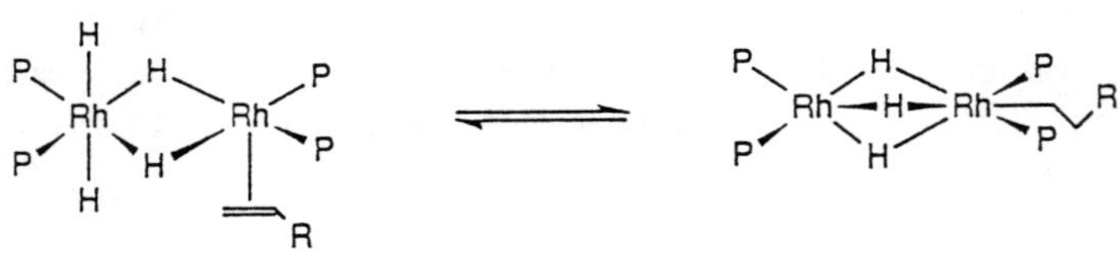

(2.46)

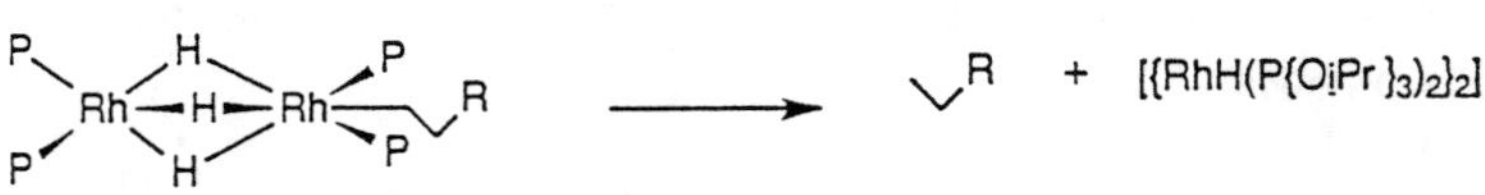

(2.47)

In contrast to alkenes, alkynes react instantaneously with the binuclear rhodium hydride. It has been proposed that the product is the alkenyl bridged dimer shown in Figure *2.8*. Reaction of this alkenyl compound with hydrogen leads to the selective formation of the alkene even in the presence of alkene. This reaction is slower than alkene hydrogenation. The slow but selective hydrogenation of alkynes with this compound as catalyst can be rationalised in terms of a higher thermodynamic stability of the alkenyl intermediate shown in Figure *2.8* compared to the intermediates shown in reaction sequences *2.43–2.47*.

The reductive elimination of the alkene in the intermediate shown in Figure *2.8* is not observed at least at rates which are consistent with

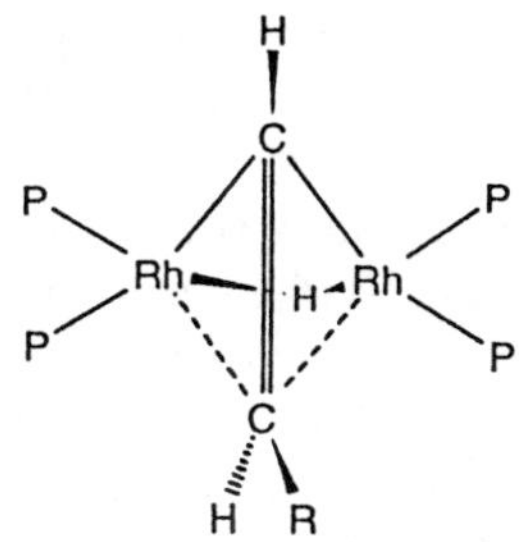

Fig. 2.8.

the observed rate of catalytic alkyne hydrogenation. The rate-determining step in the catalytic sequence is apparently hydrogen addition to this alkenyl species, where the product should be a simple gamma-alkenyl derivative analogous to the ethyl intermediate shown in the alkene hydrogenation sequence (equation *2.46*). Thus, reductive elimination of alkene can directly yield only the *cis*-alkene and $[\{RhH(P\{OiPr_3\}_3)_2\}_2]$. The latter does not interact with an alkene; and accordingly, neither alkene isomerisation nor hydrogenation can occur as long as an alkyne is present to capture the dimeric hydride [98].

In the overall proposed mechanism (equations *2.43–2.47*), it is not possible to exclude mononuclear $[RhH_3\{P(OR)_3\}_2]$ species as viable intermediates in alkene or alkyne hydrogenation. These monomeric species may be formed by reaction of the dimer with hydrogen, and they are also the formal result of adding hydrogen to monohydrides, $[RhHL_n]$, which catalyse the hydrogenation of alkenes and alkynes *via* pathway A of Scheme *2.9*.

2.5.3. HYDROGENATION OF ALKENES CATALYSED BY (PENTAMETHYLCYCLOPENTADIENYL) RHODIUM AND IRIDIUM COMPLEXES

The complexes $[\{(\eta^5\text{-}C_5Me_5)MX_2\}_2]$, $[\{(\eta^5\text{-}C_5Me_5)M\}_2HX_3]$ (M = Rh,Ir; X = Cl,Br or I) and $[\{(\eta^5\text{-}C_5Me_5)IrHCl\}_2]$ have a high activity as alkene hydrogenation catalysts at 20°C and 1 atm. hydrogen pressure. The activity is highest in 2-propanol and with NEt_3 as co-catalyst. The base is required mainly to suppress reactions such as

$$[M]HCl + HCl \longrightarrow [M]Cl_2 + H_2 \qquad (2.48)$$

$$[M] \equiv M(C_5Me_5)$$

since acid is generated in the heterolytic activation step, and needs to be removed before maximum catalyst efficiency can be attained [99].

Kinetic studies of these reactions suggest that the bridges are absent in the most active catalysts. Evidence in favour of this comes from the observation that the most active systems are those based on the dichloride dimers $[\{(\eta^5\text{-}C_5Me_5)MCl_2\}_2]$ which showed activity proportional to $[\text{catalyst}]^{1/2}$, implying that the actual catalysts involved are monomer complexes of the type $[M(\eta^5\text{-}C_5Me_5)HXS]$ (S = solvent). Measurements during the first 10–20% of reaction indicated that the reaction is second order with respect to hydrogen pressure and that the alkene dependence is first order at low, and zero order at high alkene concentration. The same dependence on catalyst and alkene concentration is observed for the complex $[\{(\eta^5\text{-}C_5Me_5)IrHCl\}_2]$, but here the dependence on hydrogen pressure is first order. These data suggest that for the compounds $[\{(\eta^5\text{-}C_5Me_5)MCl_2\}_2]$ and $[\{(\eta^5\text{-}C_5Me_5)IrHCl\}_2]$ the reaction paths shown in equations *2.49–2.55* are probably the most significant:

$$[\{(\eta^5\text{-}C_5Me_5)MCl_2\}_2] \rightleftharpoons 2\ [M(\eta^5\text{-}C_5Me_5)Cl_2S] \qquad (2.49)$$

$$[\{(\eta^5\text{-}C_5Me_5)IrHCl\}_2] \rightleftharpoons 2\ [Ir(\eta^5\text{-}C_5Me_5)HClS] \qquad (2.50)$$

$$[M(\eta^5\text{-}C_5Me_5)Cl_2S] + H_2 \rightleftharpoons [M(\eta^5\text{-}C_5Me_5)HClS] + HCl \qquad (2.51)$$

$$[M(\eta^5\text{-}C_5Me_5)HClS] + H_2 \rightleftharpoons 2\ [M(\eta^5\text{-}C_5Me_5)H_2S] + HCl \qquad (2.52)$$

$$\boxed{[M(\eta^5\text{-}C_5Me_5)H_2S]} + \text{alkene} \xrightleftharpoons{\text{slow}} [M(\eta^5\text{-}C_5Me_5)H_2(\text{alkene})] \qquad (2.53)$$

$$[M(\eta^5\text{-}C_5Me_5)H_2(\text{alkene})] \xrightarrow{\text{fast}} [M(\eta^5\text{-}C_5Me_5)S_3] + \text{alkane} \qquad (2.54)$$

$$[M(\eta^5\text{-}C_5Me_5)S_3] + H_2 \longrightarrow \boxed{[M(\eta^5\text{-}C_5Me_5)H_2S]} \qquad (2.55)$$

Solvent-assisted cleavage into monomers occurs in the first steps (equations *2.49* and *2.50*). In equations *2.51* and *2.52* , the heterolytic hydrogen activation gives what is probably the true catalyst, which then reacts with alkene and hydrogen in a cycle (equations *2.53–2.55*).

The monohydride $[\{Ir(\eta^5\text{-}C_5Me_5)\}_2HCl_3]$ shows a first order dependence on catalyst concentration, and it appears to react by a path where binuclear intermediates play a key role, and where dissociation to monomer does not occur to a significant extent in the initial stages.

2.5.4. HYDROGENATION OF ALKENES CATALYSED BY BINUCLEAR COMPLEXES CONTAINING AZOLATES AS BRIDGING LIGANDS [103]

Clear evidence has been found for electronic communication between the iridium centres in iridium pyrazolate complexes *via* orbital interaction with the bridging ligands [105]. Interestingly, a clear enhancement of catalysis in binuclear complexes of formula $[H(CO)(PPh_3)_2Ru(\mu\text{-bim})M(diene)]$ and $[H(CO)(PPh_3)_2Ru(\mu\text{-pz})_2M(TFB)]$ (M = Rh,Ir) has been observed [102].

These binuclear compounds are more active than the parent mononuclear complexes in catalysing the hydrogenation of cyclohexene. In the presence of $[H(CO)(PPh_3)_2Ru(\mu\text{-bim})Ir(COD)]$ the reaction rate is first order with respect to the catalyst and cyclohexene concentrations, second order with respect to hydrogen pressure, and inversely proportional to the concentration of added phosphine [104].

The ^1H NMR spectrum of the catalytic solution shows that the complex $[H(CO)(PPh_3)_2Ru(\mu\text{-bim})Ir(COD)]$ is the main species; this, together with the first order dependence of the rate on the concentration of catalysts, eliminates the possibility of $[H(CO)(PPh_3)_2Ru(\mu\text{-bim})Ir(COD)]$ breaking down to mononuclear catalytically active species. Furthermore, the complex $[H(CO)(PPh_3)_2Ru(\mu\text{-bim})Ir(COD)]$ is recovered unchanged after the catalytic reactions, indicating that the 1,5-cyclooctadiene ligand coordinated to the iridium atom is not hydrogenated. This suggests therefore, that the ruthenium atom is the only active centre. Because the ruthenium atom is the only active centre and it is coordinatively saturated, coordination of cyclohexene implies displacement of PPh_3. In support of this, the rate of formation of cyclohexane is found to be reduced by the addition of PPh_3.

The second order dependence of the rate on the hydrogen pressure can be rationalised in terms of the equilibrium shown in equation *2.56*, which is strongly supported by ^1H NMR spectroscopic observations [104].

$$[H(CO)(PPh_3)_2Ru(\mu\text{-bim})Ir(COD)] + H_2 \overset{K_{56}}{\rightleftharpoons} [H(CO)(PPh_3)_2Ru(\mu\text{-bim})IrH_2(COD)] \quad (2.56)$$

In the light of the above, Scheme *2.23* shows the most reasonable proposal for the catalytic cycle.

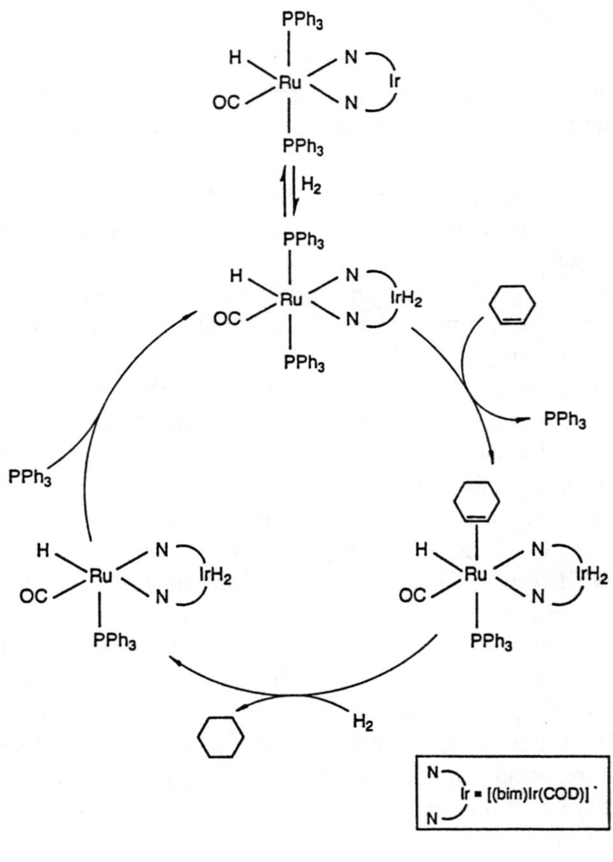

Scheme 2.23.

The rate determining step of this process is the reaction between $[H(CO)(\eta^2\text{-}C_6H_{10})(PPh_3)Ru(\mu\text{-bim})IrH_2(COD)]$ and hydrogen. Some more details of this cycle remain to be elucidated; the reaction of $[H(CO)(\eta^2\text{-}C_6H_{10})(PPh_3)Ru(\mu\text{-bim})IrH_2(COD)]$ with hydrogen is likely to involve a series of several elementary steps. One plausible sequence would be the insertion of cyclohexene into the hydride-ruthenium bond with formation of a 16e$^-$ cyclohexyl-Ru species, which would allow H_2 to coordinate with and protonate the cyclohexyl group to give [H(CO)(PPh_3)Ru(μ-bim)IrH_2(COD)]. The reaction of [H(CO)(PPh_3)$_2$Ru(μ-bim)IrH_2(COD)] with cyclohexene to form [H(CO)(η^2-C_6H_{10})(PPh_3)Ru(μ-bim)IrH_2(COD)] is less clear-cut from an intimate mechanistic point of view, even though the relatively long distance observed for the Ru-N bond *trans* to the hydride ligand in the related compound [H(CO)(PPh_3)$_2$Ru(μ-bim)Rh(COD)] (2.283Å [100]) suggests that this bond might be broken during the reaction.

The enhancement of catalytic activity of the mononuclear fragment [RuH(Hbim)(CO)(PPh_3)$_2$] by bonding to "Ir(COD)" must be due to electronic factors. Replacement of the acidic proton of the [Hbim]$^-$ ligand in [RuH(Hbim)(CO)(PPh_3)$_2$] by "Ir(COD)" produces a significant decrease in the electron density on the ruthenium atom as evidenced by the displacement of the ν_{CO} absorption towards higher frequencies [100]. The addition of molecular hydrogen to the iridium atom must lead to a further decrease in the electron density on the ruthenium atom as previously has been observed for similar cases [106], which agrees well with theoretical studies. Interestingly, the most active catalysts for the reduction of unsaturated organic compounds are mononuclear second- and third-row metal complexes, such as [M(COD)(PR_3)$_2$]$^+$ (M = Rh,Ir), [Ir(COD)L(PR_3)]$^+$ (L = nitrogen donor ligand) or [OsHCl(CO)(P*i*Pr_3)$_2$], all of which show Lewis acid behaviour.

The hydrogenation of cyclohexene catalysed by [H(CO)(PPh_3)$_2$ Ru(μ-pz)$_2$Ir(TFB)] has an induction period that disappears on treating the solution of [H(CO)(PPh_3)$_2$Ru(μ-pz)$_2$Ir(TFB)] with hydrogen for 1h at 60°C. This induction period is most probably related to the reduction of the diene coordinated to the iridium atom. Thus, under catalytic conditions, the species [H(CO)(PPh_3)$_2$Ru(μ-pz)$_2$IrS_x] is formed [104]. These observations suggest that the iridium centre plays a fundamental role in this binuclear catalyst.

For this reaction the experimental data [104] fit a rate expression of the form:

$$\frac{-d[\text{Cyclohexene}]}{dt} = k_{57}[\text{Catalyst}][\text{Cyclohexene}]\ P(H_2) \qquad (2.57)$$

Equation *2.57* is consistent with the mechanism shown in Scheme *2.24*. This mechanism is a standard unsaturated mechanism and it was

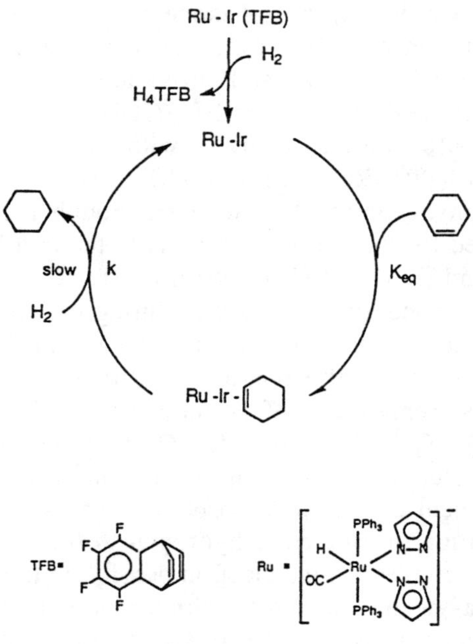

Scheme 2.24.

previously proposed, for example, for the hydrogenation of alkenes catalysed by the mononuclear complex $[Rh(DPPE)]^+$ (Section 2.2).

The significant modification of catalytic activity of the mononuclear iridium fragment by bonding to the "$RuH(CO)(PPh_3)_2$" unit merits further comment. The replacement of the acid protons of the pyrazole ligands in the mononuclear iridium complex, $[Ir(TFB)(Hpz)_2]^+$, by the ruthenium unit produces a decrease in the capacity of the pyrazole to donate electrons to the iridium atom, and causes a flow of electron density from the iridium atom to the ruthenium atom, as evidenced by the reduction in ν_{CO} of 13 cm^{-1} in $[H(CO)(PPh_3)_2Ru(\mu\text{-}pz)_2Ir(TFB)]$ as compared with $[RuH(pz)(CO)(Hpz)(PPh_3)_2]$ [100]. This reduction of electron density on the iridium atom is most probably responsible for its activation for catalysis, particularly for the formal oxidative addition of hydrogen. Thus, it was mentioned that the addition of hydrogen to mononuclear iridium -diene compounds is inhibited by electron-donor ligands but occurs with relatively electron-accepting ligands.

To summarise, the dinuclear compounds $[H(CO)(PPh_3)_2Ru(\mu\text{-}bim)_2Ir(COD)]$ and $[H(CO)(PPh_3)_2Ru(\mu\text{-}pz)_2Ir(TFB)]$ are more active catalysts for the hydrogenation of cyclohexene than the mononuclear parent compounds [RuH(Hbim)(CO)(PPh_3)$_2$], [Ir(Hbim)(COD)], [RuH(pz)(CO)(Hpz)(PPh_3)$_2$] and [Ir(TFB)(Hpz)$_2$][BF_4]. This catalytic synergism is produced by electronic communication between the metal centres through the bridging ligands. The hydrogenation proceeds mainly *via* one metal, and the other metal acts as the core of a metal-ligand complex of variable electron density.

Studies on the hydrogenation of cyclohexene catalysed by [{Ir(μ-pz)(diene)}$_2$] (diene = COD, TFB) also prove that the full catalytic cycle involves binuclear species. However, there is currently no evidence as to whether only one or both of the metal centres is directly involved in the catalytic reaction [104].

2.5.5. COMMENTS ON THE MECHANISTIC PROPOSALS

Three distinct mechanistic proposals for binuclear catalysts have been discussed above. The first one involves the participation of binuclear species derived from [{RhH(P{O*i*Pr}$_3$)$_2$}$_2$] during the full catalytic cycle; furthermore, the transfer of hydride ligands from one metal to the other (equations *2.44* and *2.46*) is suggested. However, there has been insufficient detailed research into the kinetics to prove the integrity of dimer as the catalyst during the reaction. A similar mechanism to that described in equations *2.43–2.47* has been proposed by Bianchini *et al.* for the hydrogenation of alkenes catalysed by the tetrahydride [(triphos)RhH(μ-H)$_2$HRh(triphos)]$^{2+}$ [107].

The second proposal involves mononuclear intermediates formed by cleavage of the bridges in the binuclear precursor. Under catalytic conditions, [{(η^5-C_5Me_5)MCl_2}$_2$] dimers are converted to the active species, the monomers [M(η^5-C_5Me_5)H_2S], which catalyse alkene hydrogenation by a process (equations *2.53–2.55*) which is broadly similar to the mechanism deduced for alkene hydrogenation catalysed by the complex [RhCl(PPh_3)$_3$].

The third proposal suggests that the full catalytic cycle involves binuclear intermediates. However, the catalysis proceeds *via* one metal, and the other metal acts as the core of a metallo-ligand complex supplying variable electron density. Because the catalysis proceeds only *via* one metal, the corresponding mechanisms are similar to those reviewed for mononuclear compounds. Thus, for the complex [H(CO)(PPh_3)$_2$Ru(μ-bim)IrH_2(COD)], the active centre is the ruthenium atom, and consequently, the mechanism for the hydrogenation of cyclohexene catalysed by this binuclear complex is similar to the mechanism deduced for alkene hydrogenation catalysed by the compound $[RuHCl(PPh_3)_3]$. The same relationship is found between $[H(CO)(PPh_3)_2Ru(\mu\text{-pz})_2IrS_x]$ and $[Rh(DPPE)S_x]^+$.

2.6. Homogeneous hydrogenation catalysed by clusters

2.6.1. INTRODUCTION

The complexity of the surface chemistry and physics of heterogeneous catalysts makes it difficult to understand the detailed mechanisms of catalytic action. One result of this complexity is that the development of new, improved heterogeneous catalysts depends too often on empirical studies rather than on chemical insight derived from systematic investigation of these systems. Therefore, methods for circumventing the difficulties inherent in the direct investigation of heterogeneous catalysts could be very useful in catalyst development.

It has been proposed that discrete molecular metal clusters may be reasonable models of metal surfaces in the processes of chemisorption and catalysis [108–117]. This proposal is based on the assumption that many heterogeneous catalytic reactions require multiple-site catalysis. A metal cluster can be considered to be a polynuclear compound which contains at least one metal-metal bond [115], so homogeneous cluster catalysts should be capable of multiple-site catalysis. A detailed knowledge of homogeneous cluster catalysis should help in the advancement of surface science because the many details of structure, kinetic and mechanistic features of mobility for bound molecules or atoms, and the mechanistic features of stoichiometric or catalytic reactions are more readily established for cluster chemistry than for surface chemistry.

In addition to serving as potential models of heterogeneous catalysts, homogeneous cluster catalysts also hold considerable potential as catalysts in their *own* right. They are capable of providing novel catalyst chemistry which is unavailable with mononuclear homogeneous catalysts. In a metal cluster the sites of catalyst activity are affected both electronically and sterically by the surrounding metals and the ligands on those metals. The surrounding metals and the ligands attached to them can be considered as metallo-ligand-clusters. These metallo-ligand-clusters of variable electron density will affect the chemistry of the metals that serve as the active sites for catalysis in ways which are not available with the simple ligands used to modify the catalyst activity of mononuclear catalysts. Furthermore, the availability of two or more active sites in the clusters can also provide novel chemistry *via* multiple-site reactivity, since multiple sites will offer unusual substrate-catalyst interaction geometries as well as electronic effects which are not possible with mononuclear catalysts.

2.6.2. CLUSTER CATALYSIS

A strict definition of cluster catalysis can be formulated as a reaction in which two or more metal atoms in the cluster are mechanistically necessary for catalysis to occur [115]. However, this definition is probably too strict when one considers the number of catalytic reactions that are possible using only one metal site in the cluster as the active site.

The second, more general, definition of cluster catalysis is: a reaction in which at least one site of the cluster molecule is mechanistically necessary for catalysis [117]. One of the benefits of this definition is that it is possible to evaluate the effects of the metallo-ligand-clusters on the active site in the clusters and to compare their effects to those of simple ligands in reactions catalysed by mononuclear catalysts.

The third and lower-limit definition of cluster catalysis arises from the fact that transition metal cluster chemistry frequently involves cluster fragmentation to other clusters of lower nuclearity and/or to mononuclear species. Thus, there is a type cluster catalysis when the precursor cluster or some cluster of lower nuclearity is needed for at least one step in the catalytic cycle.

These definitions include, as particular cases, some of the binuclear complexes considered in the previous section.

According to Laine, the definitions provide the basis for five separate criteria which are useful in identifying cluster catalysis [115]:

i) Catalyst concentration studies in which the turnover frequency increases with increasing catalyst concentration are indicative of cluster catalysis.
ii) If the product selectivities obtained in a given catalytic reaction using cluster catalyst precursors are different from those obtained using mononuclear catalytic precursors, or the products themselves can not be reconciled with mechanisms involving only mononuclear species, then cluster catalysis is indicated.
iii) If a specific combination of two or more different transition metals can be used to significantly enhance the catalysis rates or to change the product selectivity of a given reaction which is normally catalysed by either of the metals, or if the combination enables the

catalysis of a reaction which is not catalysed independently by either of the metals, then mixed-metal cluster catalysis is suggested.

iv) If it is possible to modify the catalyst or the reaction conditions for a given reaction to favour metal-metal bond formation and the modification results in increased catalytic activity, then metal cluster catalysis should be suspected.

v) Asymmetric catalytic induction with chiral metal clusters in which the asymmetry resides in the metal framework or which is a basic skeletal property of the cluster is indicative of cluster catalysis.

Each of these criteria can be applied to different homogeneous catalytic reactions [116] and in particular to homogeneous hydrogenation. For these reactions, three mechanisms have been proposed, which are related to the three limit definitions mentioned above. For example, Shapley *et al.* [119] suggested that the hydrogenation of alkenes catalysed by $[Os_3H_2(CO)_{10}]$ proceeds *via* a multiple-site mechanism. A similar proposal was made by Basset *et al.* [120] for the hydrogenation of ethene catalysed by the silica-supported cluster $[Os_3(CO)_{10}(\mu\text{-}H)(\mu\text{-}OSi\equiv)]$. However, for the hydrogenation of alkenes catalysed by $[(\mu\text{-}H)_2Ru_3(\mu_3\text{-}O)(CO)_5(DPPM)_2]$ [121] or $[Ru_4H_4(CO)_{12}]$ [122] clusters, only one site is mechanistically necessary. Finally, cluster-fragmentation has been proposed by Sanchez-Delgado *et al.* [123] in the hydrogenation of styrene catalysed by $[Os_4H_3I(CO)_{12}]$, $[Os_4H_4(CO)_{12}]$, $[Os_4H_2I(CO)_{12}]^-$ or $[Os_4H_3(CO)_{12}]^-$. The rate of styrene hydrogenation depends on the structure of the cluster and is first order with respect to the concentrations of styrene and hydrogen; the turnover frequency, however, increases with decreasing cluster concentration.

2.6.3. HYDROGENATION OF ALKENES CATALYSED BY $[Os_3H_2(CO)_{10}]$

The unsaturated metal cluster, $[Os_3H_2(CO)_{10}]$, reacts with two equivalents of ethene to give ethane and the $[Os_3H(CH{=}CH_2)(CO)_{10}]$ hydridoalkenyl cluster [124].

The reactivity of an alkene towards $[Os_3H_2(CO)_{10}]$ is quite sensitive to steric and electronic effects. For example, ethyl acrylate and vinyl acetate react faster than ethene, whereas propene and isobutylene react more slowly, and no reaction is observed for cyclooctene or norbornene. The interaction of an excess of 1-alkene with $[Os_3H_2(CO)_{10}]$ normally provides the corresponding alkane and the cluster $[Os_3H(alkenyl)(CO)_{10}]$ [119a].

Treatment of $[Os_3H_2(CO)_{10}]$ with 1 equivalent of diethyl fumarate or diethyl maleate readily forms the complex, $[Os_3H\{CH(CH_2CO_2Et)CO_2Et\}(CO)_{10}]$ hydridoalkyl. A related compound, $[Os_3H\{CH(CH_2\overline{CO)COO}\}(CO)_{10}]$, is obtained from maleic anhydride, and ethyl acrylate provides $[Os_3H\{(CH(CH_3)CO_2Et\}(CO)_{10}]$. In solution these compounds decompose to give the alkane. Decomposition of $[Os_3H\{CH(CH_2CO_2Et)CO_2Et\}(CO)_{10}]$ at 25–50°C in the presence of an excess of 1-alkene, such as ethene, propene, isobutene, styrene or 1-hexene, provides the corresponding $[Os_3H(alkenyl)(CO)_{10}]$ cluster in high yield. When this decomposition is conducted under hydrogen, $[Os_3H_2(CO)_{10}]$ and diethyl succinate are the only products detected and both are formed in high yield. This activation of molecular hydrogen closes the cycle of hydrogen transfer from $[Os_3H_2(CO)_{10}]$ to diethyl fumarate. Catalytic hydrogenation of diethyl fumarate or ethyl acrylate by $[Os_3H_2(CO)_{10}]$ is observed under similar conditions, but the number of cycles is limited by destructive side reactions. A much cleaner catalytic system is established with 1-hexene. Interaction of $[Os_3H_2(CO)_{10}]$ with 100 equivalents of 1-hexene under hydrogen results, after 3h, in 31 equivalents of hexane and 69 equivalents of internal hexenes, and essentially all the catalyst is recovered at the end of the reaction.

These observations on the interaction of $[Os_3H_2(CO)_{10}]$ with alkenes are elegantly accommodated by the cycle shown in Scheme *2.25*. The high degree of isomerization indicates the insertion and coordination steps are reversible. The key intermediate in this Scheme is presumed to be the highly unsaturated 44-electron species, $[Os_3(CO)_{10}]$, which reacts with molecular hydrogen or vinylic C-H bonds to give $[Os_3H_2(CO)_{10}]$ or $[Os_3H(alkenyl)(CO)_{10}]$ respectively [119a].

2.6.4. HYDROGENATION OF ALKENES BY CLUSTERS WITH ONLY ONE ACTIVE SITE

The trinuclear cluster $[(\mu\text{-}H)_2Ru_3(\mu_3\text{-}O)(CO)_5(DPPM)_2]$ is found to be an efficient catalyst for alkene hydrogenation reactions. A detailed kinetic study reveals that the reaction is first order in catalyst and alkene. This, together with spectroscopic data, leads to the proposal of the catalytic cycle shown in Scheme *2.26* [121].

The isomerization of 1-hexene, which is found to occur with and without hydrogen, suggests that the first step may be alkene activation by the hydride cluster. The second step is the formation of an alkyl cluster

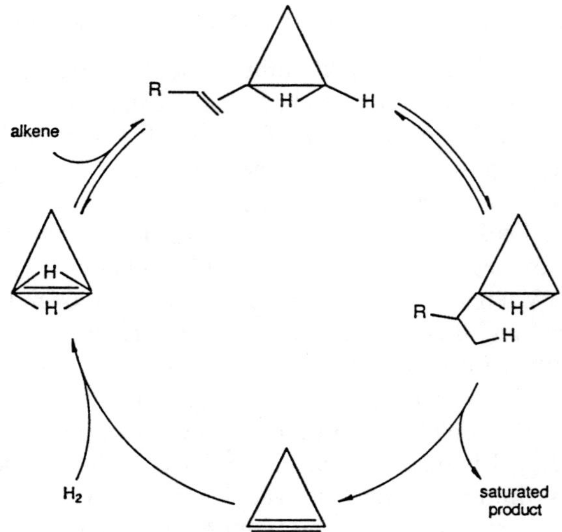

Scheme 2.25.

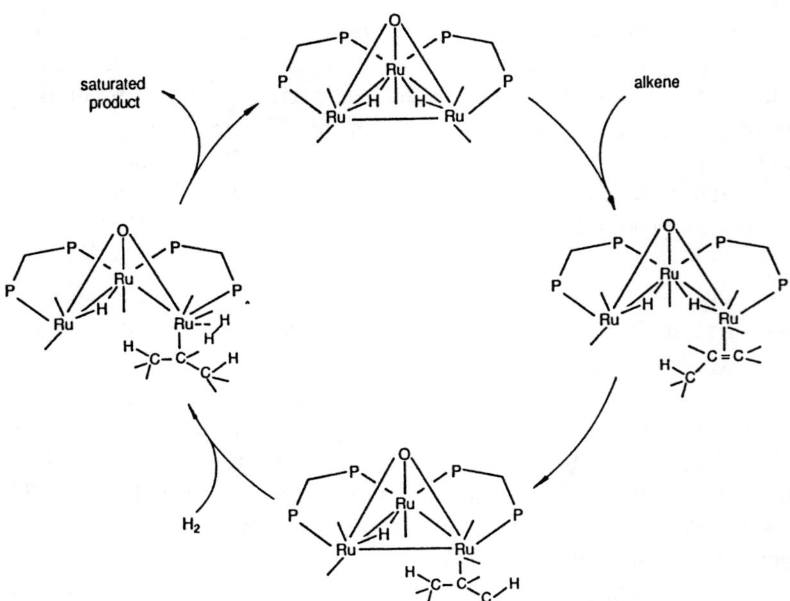

Scheme 2.26.

intermediate. Such a species has been observed by ^{1}H NMR spectroscopy in the stoichiometric reaction of the dihydride cluster with acrylonitrile. The last step then uses hydrogen to eliminate alkane and regenerate the catalyst. The interaction between the molecular hydrogen and the alkyl group involves the same ruthenium atom as coordinated the free alkene above. Therefore, only one ruthenium atom is involved throughout the catalytic cycle.

Tetraruthenium hydride cluster, $[Ru_4H_4(CO)_{12}]$, reacts with ethene at 27°C to give two molecules of ethane per cluster molecule. In the presence of an excess of hydrogen, this cluster complex acts as a catalyst for the hydrogenation of ethene in heptane solution. The rate of this catalytic hydrogenation is proportional to the concentration of $[Ru_4H_4(CO)_{12}]$ [122]. Therefore, it is reasonable to assume that the tetranuclear cluster frameworks themselves provide the catalytic sites. Because the catalyst precursor, $[Ru_4H_4(CO)_{12}]$, is coordinatively saturated, the first step in the catalytic cycle must involve the creation of vacant coordination sites without breaking up the cluster into fragments of lower nuclearity. The formation of such vacant coordination sites may be achieved in two ways: i) cleavage of one ruthenium-ruthenium bond and ii) dissociation of a ligand, specifically carbon monoxide (equation *2.58*).

$$[Ru_4H_4(CO)_{12}] \overset{K_{58}}{\rightleftharpoons} [Ru_4H_4(CO)_{11}] + CO \qquad (2.58)$$

Catalyst activation *via* the first path should lead to a new cluster containing two coordinatively unsaturated ruthenium atoms which could catalyse the reaction *via* a multi-site mechanism, whereas the catalyst activation *via* the second path must lead to a new cluster in which only one coordinatively unsaturated ruthenium atom would be involved in the catalysis.

Activation involving CO dissociation is fully consistent with the kinetic data. Thus, plots of $[Ru_4]_{Tot}$ $(d[C_2H_6]/dt)^{-1}$ vs. P_{CO} are linear.

The mechanism proposed by Doi *et al.* for the hydrogenation of ethene in the presence of $[Ru_4H_4(CO)_{11}]$ is outlined in equations *2.59–2.61.*

$$[Ru_4H_4(CO)_{11}] + H_2 \overset{K_{59}}{\rightleftharpoons} [H_6Ru_4(CO)_{11}] \qquad (2.59)$$

$$[Ru_4H_4(CO)_{11}] + C_2H_4 \overset{K_{60}}{\rightleftharpoons} [Ru_4H_3(C_2H_5)(CO)_{11}] \qquad (2.60)$$

$$[Ru_4H_3(C_2H_5)(CO)_{11}] + H_2 \xrightarrow{k_{61}} [Ru_4H_4(CO)_{11}] + C_2H_6 \quad (2.61)$$

This mechanism is supported by the kinetic study of the reaction which is consistent with the rate law of the equation *2.62* and by the following experimental results: i) the reversible I.R. spectral change of the heptane solution of $[Ru_4H_4(CO)_{12}]$ under argon and hydrogen appears to support the reversible reaction *2.59*, giving $[Ru_4H_6(CO)_{11}]$ under hydrogen. In addition, a rapid H_2/D_2 exchange reaction to give HD in the presence of $[Ru_4D_4(CO)_{12}]$ can be explained in terms of hydride exchange on the ruthenium cluster framework through reaction *2.59*; ii) the reaction of ethene with molecular deuterium reveals that the exchange reaction of hydrogen-deuterium atoms between reactants takes place to give C_2H_3D and HD at a more rapid rate than that of ethane formation. Therefore, it is reasonable to assume that an ethyltetraruthenium intermediate is present in the catalytic cycle and that the reverse reaction of equation *2.60* to give a free ethene molecule is substantially faster than the formation of ethane *via* reaction *2.61*; and iii) the observed acceleration of the hydrogenation rate by the presence of hydrogen suggests that a hydrogenolysis reaction *2.61* is the predominant final step in the catalytic cycle of ethene hydrogenation [122].

$$\frac{d[C_2H_6]}{dt} = \frac{k_{61}K_{60}P_{C2H4}P_{H2}[Ru_4]_{Tot}}{1 + K_{59}P_{H2} + K_{60}P_{C2H4} + P_{CO}/K_{59}} \quad (2.62)$$

2.7. Complexes of non-platinum group metals as hydrogenation catalysts

2.7.1. INTRODUCTION

The most important catalytic cycles for homogeneous hydrogenation catalysed by platinum group metal complexes were reviewed above. Hydrogenation catalysts from non-platinum group metals are also known [2c]. Thus, the organolanthanides $[\{(\eta^5\text{-}C_5Me_5)_2MH\}_2]$ and $[\{(\eta^5\text{-}C_5Me_4SiMe_3)_2MH\}_2]$ (M = La, Nd, Sm and Lu) are active catalysts for alkene hydrogenation. A comparison of $[\{(\eta^5\text{-}C_5Me_5)_2LuH\}_2]$ with some representative *d*-block phosphine complexes under similar conditions shows that the turnover numbers per hour at 25°C and 1 atm. H_2 for 1-hexene are $[\{(\eta^5\text{-}C_5Me_5)_2LuH\}_2]$ (120000) $[Ir(COD)(py)(PCy_3)][PF_6]$ (6400) $[Rh(COD)(PPh_3)_2][PF_6]$ (4000), $[RuHCl(PPh_3)_3]$ (3000) and $[RhCl(PPh_3)_3]$ (650) [125a]. The complex

[Y(η^5-C_5Me_5)$_2$Me(THF)] has been developed as an efficient catalyst for the selective reduction of substituted dienes [125b]. The niobium tris(4-methylbenzyl) compound [Nb(OC_6H_3Ph-2,6)$_2$($CH_2C_6H_4$-4Me] acts as a catalyst precursor for the hydrogenation of benzene and a variety of polynuclear aromatic hydrocarbons [125c].

In the presence of alkyl derivatives of aluminium, alkaline earths and alkali metals, various transition metal salts constitute the Ziegler catalyst systems. Most known Ziegler catalysts for the dimerization and polymerization of alkenes are heterogeneous [126], but homogeneous analogues have been developed and many are effective, especially for reducing monoenes, dienes and aromatic substrates [127].

The chromium complexes [Cr(CO)$_6$], [{(η^5-C_5H_5)Cr(CO)$_3$}$_2$] and [Cr(CO)$_3$(arene)] behave similarly in this respect. When the arene is phenanthrene, naphthalene or anthracene, the [Cr(CO)$_3$(arene)] complexes are more active than when the arene is a substituted benzene, and this is attributed to easier displacement of the arene by the diene substrate [128]. These complexes are useful catalysts for conversions of dienes to *cis*-alkenes.

The complexes [Fe(CO)$_5$], [Fe(CO)$_3$(diene)] and [Fe(CO)$_3$(triene)] behave similarly as catalysts in this respect. For these systems, the active transient species is thought to be the metal tricarbonyl. The dihydride [FeH_2(CO)$_3$] has been detected, but combined unsaturated and dihydride routes (Scheme *2.1*) have been proposed. The reductions are not very selective due to accompanying isomerization *via* a π-allyl hydride intermediate. A final step involving molecular hydrogen which results in the formation of the monoene and regeneration of a dihydride complex has sometimes been suggested instead of transfer of the second hydrogen [2c]. More recently, it has been observed that the complexes [FeH(η^2-H_2)(PP_3)]$^+$ and [FeH(η^1-N_2)(PP_3)]$^+$ (PP_3 = P($CH_2CH_2PPh_2$)$_3$) catalyse the selective hydrogenation of terminal alkynes [129].

Aromatic substrates such as thiophene are reduced completely to the corresponding thiolane with [Co_2(CO)$_8$] as catalyst [130]. Some allylcobalt (I) complexes [Co(η^3-C_3H_5)L_3] have been found to catalyse the hydrogenation of benzene to cyclohexane under ambient conditions. With L = P(OMe)$_3$, C_6D_6 yielded *cis*-$C_6D_6H_6$ only, with no competing hydrogen exchange. The catalysts show a slight selectivity for hydrogenation of arene when using benzene/cyclohexene or benzene/hexene mixtures as substrates, and because neither cyclohexadiene nor cyclohexene were detected during hydrogenation of benzene, a mechanism was suggested in which the C_6 moiety remains attached to the cobalt until a cyclohexyl species is formed [131].

For hydrogenation in water, the $[CoH(CN)_5]^{3-}$ catalyst is still extremely useful and will be considered in some detail in Chapter 5. The system is selective for hydrogenation of a carbon-carbon double bond which is conjugated with another C=C or with C=O, C≡N or phenyl groups, although reduction of NO_2, -NO and C=NOH groups is quite common at high pressures, and hydrogenolysis of carbon-halogen bonds is very easy. Selective reductions of carvone, other terpenes and a Δ^1-3-ketosteroid to the 3-one have been reported. Carbon-carbon triple bonds are reduced only when they form part of a conjugated system. $[CoH(CN)_5]^-$ is relatively unreactive towards unconjugated dienes such as 1,5-cyclooctadiene [2c].

$[Ni_2(CN)_6]^{4-}$ is reported to be effective for the hydrogenation of allenes to monoenes. Borohydride reduction of $NiCl_2$ and $CoCl_2$ in N,N-dimethylmethanamide or N,N-dimethyl-ethanamide solution provides a simple way of generating very active catalysts under ambient conditions. They show remarkable versatility in hydrogenation of monoalkenes, of cyclic dienes to monoenes, and of unsaturated fats, alkynes, and saturated aldehydes and ketones [2c].

2.7.2. MECHANISM FOR CYCLOHEXENE HYDROGENATION CATALYSED BY $[\{(\eta^5\text{-}C_5Me_5)_2MH\}_2]$ (M = La, Nd)

The rate law for cyclohexene hydrogenation catalysed by the title complex (equation *2.63*) is first order in alkene and zero-order in $P(H_2)$. However, the rate is one-half-order in the lanthanide complex.

$$-d[\text{Cyclohexene}]/dt = k_{63}\,[\text{Lanthanide}]^{1/2}\,[\text{Cyclohexene}] \qquad (2.63)$$

The most reasonable interpretation of the kinetic order of the reaction involving the lanthanide complex is the dissociation of the dimer (equation *2.64*).

$$[\{(\eta^5\text{-}C_5Me_5)_2MH\}_2] \underset{}{\overset{K_{64}}{\rightleftharpoons}} 2\,[\{(\eta^5\text{-}C_5Me_5)_2MH] \qquad (2.64)$$

M = La, Nd.

Thus, the kinetic behaviour can be most simply interpreted if the addition of the lanthanide hydride to the cyclohexene double bond is the rate-limiting step of the cycle shown in equations *2.65* and *2.66*.

$$[(\eta^5\text{-}C_5Me_5)_2MH] + C_{10}H_{10} \overset{k_{65}}{\rightleftarrows} [\{(\eta^5\text{-}C_5Me_5)_2M(C_{10}H_{11})] \quad \text{slow} \qquad (2.65)$$

$$[(\eta^5\text{-}C_5Me_5)_2M(C_{10}H_{11})] + H_2 \longrightarrow [(\eta^5\text{-}C_5Me_5)_2MH] + C_{10}H_{12} \quad \text{fast} \qquad (2.66)$$

When the $[\{(\eta^5\text{-}C_5Me_5)_2LaH\}_2]$-catalysed hydrogenation is carried out with molecular deuterium, the cyclohexane produced under non-mass-transport-limited conditions is ≥ 98% cyclohexane-1,2-d_2. If the reaction is stopped at either 15% or 80% conversion, the unreacted cyclohexene contains traces (ca. 1%) of cyclohexene-d_1. This isotopic cyclohexene is suggestive of a minor β-hydride elimination pathway [125].

This mechanistic proposal involves mononuclear intermediates formed by cleavage of the bridges in the binuclear precursor. Under catalytic conditions, $[\{(\eta^5\text{-}C_5Me_5)_2MH\}_2]$ dimers dissociate to give the active species (equations *2.64*) which catalyses the alkene hydrogenation by a process that is broadly similar to the mechanism deduced for alkene hydrogenation catalysed by $[RuHCl(PPh_3)_3]$ or for alkyne hydrogenation catalysed by the Zaragoza-Würzburg catalysts, $[OsHCl(CO)(PR_3)_2]$, although in the case of the lanthanide catalyst the rate-determining step is the interaction of the monohydride with the cyclohexene.

2.7.3. MECHANISM FOR THE SELECTIVE HYDROGENATION OF TERMINAL ALKYNES CATALYSED BY $[FeH(\eta^2\text{-}H_2)(PP_3)]^+$

The $[FeH(\eta^1\text{-}N_2)(PP_3)]^+$ and $[FeH(\eta^2\text{-}H_2)(PP_3)]^+$ react with terminal alkynes in the absence of hydrogen to give alkenyl complexes which may or may not be isolated depending on the alkyne substituent. In particular, reactions with $HC{\equiv}CSiMe_3$ give the alkenyl derivative $[Fe\{(E)\text{-}CH{=}CHSiMe_3\}(PP_3)]^+$ as the predominant product together with some σ-alkynyl complex, $[Fe(C{\equiv}C\text{-}SiMe)(PP_3)]^+$ and free vinyltrimethylsilane. Treatment of $[Fe\{(E)\text{-}CH{=}CHSiMe_3\}(PP_3)]^+$ with more $HC{\equiv}CSiMe_3$ yields $[Fe(C{\equiv}C\text{-}SiMe_3)(PP_3)]^+$ and $H_2C{=}CHSiMe_3$ quantitatively. At reflux temperature, a small amount of free 1,4-bis(trimethylsilyl)butadiene is produced. Other terminal alkynes such as $HC{\equiv}CPh$ or $HC{\equiv}C\text{-}C_3H_7$ give the σ-alkynyl compounds $[Fe(C{\equiv}C\text{-}R)(PP_3)]^+$ directly and the corresponding alkenes with no formation of any stable alkenyl complex or free butadiene. Alkene formation is observed also when the σ-alkynyl compounds are treated with molecular hydrogen [129].

Under 1 atmosphere of hydrogen the reactions between $[FeH(\eta^2\text{-}H_2)(PP_3)]^+$ (or $[FeH(\eta^1\text{-}N_2)(PP_3)]^+$) and alkynes are catalytic and produce only alkenes [129, 132] except for HC≡CSiMe$_3$. In this case, the reductive dimerization of the alkyne to 1,4-bis(trimethylsilyl)butadiene predominates over hydrogenation at high temperature [129]. In no case is any appreciable formation of alkanes observed, even with very long reaction times [129, 132].

The rate-law for the hydrogenation of phenylethyne in 1,2-dichloroethane catalysed by $[FeH(\eta^2\text{-}H_2)(PP_3)]^+$ (or $[FeH(\eta^1\text{-}N_2)(PP_3)]^+$) (equation *2.67*) is first order in alkyne and catalyst and zero-order in hydrogen pressure.

$$-d[PhC_2H]/dt = k_{67}[Fe][PhC_2H] \qquad (2.67)$$

From the temperature dependence of k_{67} the activation parameters listed in Table *2.7* are deduced, suggesting a common mechanism for both catalysts [132]. Equation *2.67* and the above-mentioned coordination chemistry are consistent with the mechanism shown in Scheme *2.27* [133a].

TABLE 2.7
Activation parameters for the hydrogenation of phenylethyne catalysed by $[FeH(\eta^1\text{-}N_2)(PP_3)]^+$ or $[FeH(\eta^2\text{-}H_2)(PP_3)]^+$

	$[FeH(\eta^1\text{-}N_2)(PP_3)]^+$	$[FeH(\eta^2\text{-}H_2)(PP_3)]^+$
Ea (Kcal mol^{-1})	11.8 ± 0.81	12.71 ± 1.99
$\Delta H^{\neq}$ (Kcal mol^{-1})	11.2 ± 0.81	12.12 ± 1.99
$\Delta S^{\neq}$ (u.e.)	-26.7 ± 2.59	-24.45 ± 6.25
$\Delta G^{\neq}$ (Kcal mol^{-1})	19.16 ± 1.58	19.41 ± 3.85

The related ruthenium derivative $[RuH(\eta^2\text{-}H_2)(PP_3)]^+$ also catalyses the selective hydrogenation of phenylethyne to styrene. The kinetic study of the reaction shows that the rate is proportional to the initial concentration of the catalyst precursor, is second-order with respect to hydrogen

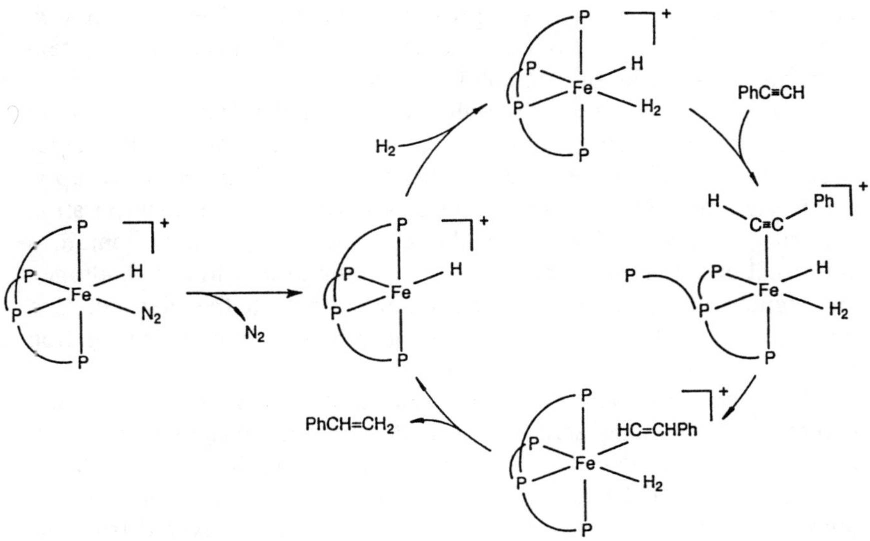

Scheme 2.27.

pressure and independent of the concentration of substrate. At very low concentrations of phenylethyne (<0.12M) the order of the catalytic rate with respect to the substrate concentration tends to 1. In the light of these kinetic data, a reaction mechanism is proposed which essentially involves the usual cycle adopted by mono(hydride) metal catalysts (Scheme 2.2.). Thus, a free site for the incoming alkyne molecule is provided by dihydrogen decoordination rather than by unfastening of one arm of the PP_3 ligand as it was proposed to occur for the iron analog, where the dihydrogen ligand remains strongly bound to the metal centre in the course of the catalytic cycle [133b].

Reactivity studies have clearly shown that the solution chemistry of these compounds is dominated by the different metal-dihydrogen bond strengths, which increase in the order Ru<Fe. Experimental evidence for a larger $d\pi(M)–\sigma^*(H_2)$ back-donation for Fe has recently been obtained by inelastic neutron scattering experiments [133c].

2.7.4. MECHANISM FOR THE HYDROGENATION OF AROMATIC HYDROCARBONS CATALYSED BY $[Co(\eta^3\text{-}C_3H_5)\{P(OMe)_3\}_3]$

In 1973, $[Co(\eta^3\text{-}C_3H_5)\{P(OMe)_3\}_3]$ was shown to be the precursor of a catalyst for the homogeneous hydrogenation of arenes under mild conditions of temperature and pressure [134–135]. The absolute *cis*-stereoselectivity in the cyclohexane products distinguishes this system from all known solution and solid-state catalysts.

In the catalytic reaction of benzene with molecular deuterium, the only hydrogenation product is all *cis*-$C_6D_6H_6$. Reactions of molecular deuterium with alkyl-substituted benzenes in which the alkyl group is linear yield alkylcyclohexanes with extensive deuterium incorporation into the alkyl side chain. The alkylcyclohexane products contain a statistical distribution of species with differing proportions of hydrogen and deuterium. The five ring hydrogen atoms on the alkylbenzene molecule do not undergo H-D exchange during the hydrogenation process.

The reaction of toluene with molecular deuterium shows that the ring carbons possess either zero or one deuterium but never two deuterium atoms, confirming that the ring hydrogen atoms on the starting toluene do not undergo H-D exchange. No H-D exchange is observed at either aliphatic or aromatic sites in the unreduced arenes recovered from the catalytic reaction systems.

Catalytic reactions of molecular deuterium with alkyl-substituted benzenes in which the alkyl substituent is branched provide obvious mechanistic information about the exchange process. With tert-butylbenzene, there is no deuterium incorporation into the tert-butyl substituent. With neopentylbenzene, a maximum of two deuterium atoms are introduced into the neopentyl group. With isopropylbenzene, there is extensive deuterium incorporation throughout the side chain. Quantitatively, there is more deuterium exchange than in the isomeric *n*-propylbenzene. The deuterium distribution is bimodal, the most abundant species being $C_6H_5D_6(i\text{-}C_3DH_6)$, $C_6H_5D_6(i\text{-}C_3D_5H_2)$, and $C_6H_5D_6(i\text{-}C_3D_6H)$.

The allyl group of the catalyst precursor is hydrogenated to propane containing a statistical distribution of species ranging from C_3H_8 to C_3D_8. However, $[Co(\eta^3\text{-}C_3H_5)\{P(OMe)_3\}_3]$ recovered after a catalytic run contains no deuterium in the allyl ligand.

The proposed mechanism for the hydrogenation of arenes and the exchange of hydrogen for deuterium atoms in the alkyl side chains is shown in Scheme *2.28*. The relative stabilities of the proposed intermediates are unknown, and the reversibility of some steps has not been demonstrated.

Initial complexation of the arene probably occurs away from the alkyl side chain for steric reasons. As shown in Scheme *2.28*, the ring is probably hydrogenated before there is any reaction involving the side

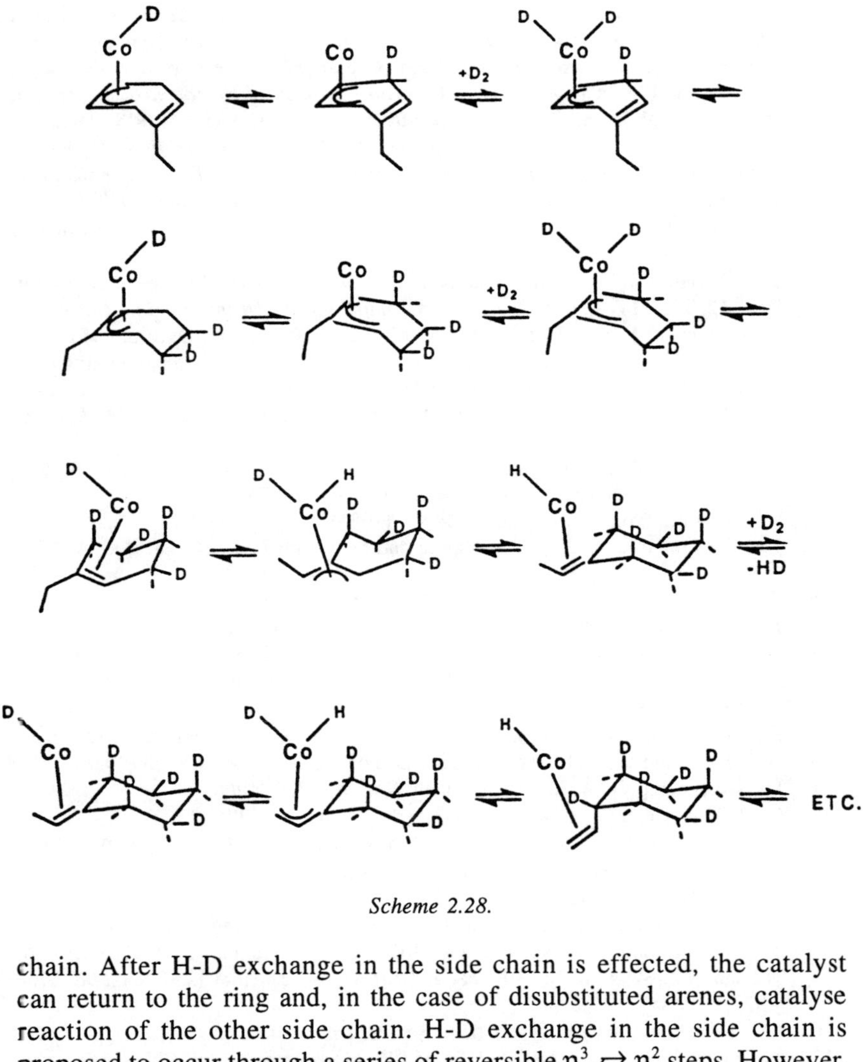

Scheme 2.28.

chain. After H-D exchange in the side chain is effected, the catalyst can return to the ring and, in the case of disubstituted arenes, catalyse reaction of the other side chain. H-D exchange in the side chain is proposed to occur through a series of reversible $\eta^3 \rightleftarrows \eta^2$ steps. However, $\eta^1 \leftrightarrows \eta^2$ interconversions may also play a role in this reaction.

References

1. See for example: a) B.R. James, *Homogeneous Hydrogenation*, John Wiley and Sons, New York, 1973. b) M. Freifelder, *Catalytic Hydrogenation in Organic Synthesis: Procedures and Commentary*, John Wiley and Sons, New York, 1978.

c) L.H. Pignolet, *Homogeneous Catalysis with Metal Phosphine Complexes*, Plenum Press, New York, 1983. d) J.P. Collman and L.S. Hegedus, *Principles and Applications of Organotransition Metal Chemistry*, University Science Books, Mill Valley, 1987, Chapter 6. e) R. Ugo (Ed.), *Aspects of Homogeneous Catalysis*, D. Reidel Publishing Company, Dordrecht, Vol. 4 (1981). f) R.S. Dickson, *Homogeneous Catalysis with Compounds of Rhodium and Iridium*, D. Reidel Publishing Company, Dordrecht, 1985. g) R.H. Crabtree, *The Organometallic Chemistry of the Transition Metals*, J. Wiley and Sons, New York, 1988, Chapter 9. h) G. Henrici-Olive and S. Olive, *Coordination and Catalysis*, Verlag Chemie, Weinheim, 1977.

2. a) B.R. James, Inorg. Chim. Acta Rew., 73(1970). b) R.H. Crabtree, Acc. Chem. Res., **12**, 331, (1979). c) B.R. James, in *Comprehensive Organometallic Chemistry*, G. Wilkinson, F.G.A. Stone and E.A. Abel Eds., Pergamon, Oxford, 1982, Vol. 8, Chapter 51. d) L. Marko, J. Organomet. Chem., **432**, 1 (1992) and previous annual surveys. e) A. Spencer, *in Comprehensive Coordination Chemistry*, G. Wilkinson, R.D. Gillard and J.A. McCleverty Ed., Pergamon, Oxford, 1987, Vol.6, Chapter 61.2.
3. a) B.R. James, Avd. Organometal. Chem., **17**, 319 (1979). b) S.I. Hommeltoft, D.H. Berry and R. Eisenberg, J. Am. Chem. Soc., **108**, 5345 (1986).
4. F.H. Jardine, Prog. Inorg. Chem., **28**, 64 (1981).
5. J.A. Osborn, F.H. Jardine, J.F. Young and G. Wilkinson, J. Chem Soc. **A**, 1711 (1966).
6. R.S. Coffey, British Patent No.1121642 (18 February 1965)
7. J.A. Osborn and G. Wilkinson, Inorg. Synth., **10**, 67 (1967)
8. R.S. Dickson, *Homogeneous Catalysis with Compounds of Rhodium and Iridium*, D. Reidel Publishing Company, Dordrecht, 1985. Chapter 3.1 p 46 and the references therein.
9. R.L. Augustine and J.F. Van Peppen, J. Chem. Soc., Chem. Commun., 495 (1970)
10. G.C. Bond and R.A. Hillyard, Faraday Discuss. Chem. Soc., **46**, 20 (1968)
11. J.T. Mague and G. Wilkinson, J. Chem. Soc. **A**, 1736 (1966)
12. C. O'Connor and G. Wilkinson, Tetrahedron Lett., **18**, 1375 (1969)
13. a) M.A. Bennett and D.L. Milner, J. Am. Chem. Soc., **91**, 6983(1969) b) M.A. Esteruelas, J. Herrero, A.M. L6pez, L.A. Oro, M. Schulz and H. Werner, Inorg. Chem., **31**, 4013 (1992)
14. J. Halpern, Inorg. Chim. Acta, **50**, 11 (1981) and the references therein.
15. a) J. Halpern and S. Wong, J. Chem. Soc., Chem. Commun., 629(1973). b) C.A. Tolman, P.Z. Meakin, D.L. Lindner and J.P. Jesson, J. Am. Chem. Soc., **96**, 2762 (1974).
16. J.M. Brown and A.R. Lucy, J.Chem. Soc. Chem. Commun., 914 (1981)
17. C. Daniel, N. Koga, J. Han, X.F. Fu and K. Morokuma, J. Am. Chem. Soc., **110**, 3773 (1988)
18. G.J. Kubas, R.R. Ryan, B.I. Swanson, P.J. Vergamini and H.J. Wasserman, J. Am. Chem. Soc., **106**, 451 (1984)
19. P.J. Hay, J. Am. Chem. Soc., **109**, 705 (1987)
20. Theoretical studies on the insertion step had been previously reported. See: A. Dedieu, Inorg. Chem., **20**, 2803 (1981)
21. a) J.R. Shapley, R.R. Schrock and J.A. Osborn, J. Am. Chem. Soc., **91**, 2816 (1969). b) R.R. Schrock and J.A. Osborn, J. Chem. Soc., Chem. Commun., 567 (1970). c) R.R. Schrock and J.A. Osborn, J. Am. Chem Soc., **93**, 2397 (1971). d) R.R.

Schrock and J.A. Osborn, J. Am. Chem Soc., **93**, 3089 (1971). e) R.R. Schrock and J.A. Osborn, J. Am. Chem Soc., **98**, 2134 (1976).
22. R.H. Crabtree in *Homogeneous Catalysis with Metal Phosphine Complexes*, L.H. Pignolet Ed., Plenum Press, New York, 1983.
23. R. Uson, L.A. Oro, M.J. Fernández and R. Sariego, Rev. Acad. Cienc. Zaragoza, **35**, 87 (1980).
24. R.H. Crabtree, P.C. Demon, D. Eden, J.M. Mihelcic, C.A. Parnell, J.M. Quirk and G.E. Morris, J. Am. Chem. Soc., **104**, 6994 (1982).
25. R. Uson, L.A. Oro, J. Artigas and R. Sariego, J. Organomet. Chem., **179**, 65, (1979)
26. R. Uson, L.A. Oro, M.A. Garralda, M.C. Claver and P. Lahuerta, Transition Met. Chem., **4**, 55 (1979).
27. R. Uson, L.A. Oro and M.A. Esteruelas, Transition Met. Chem., **7**, 242 (1982).
28. R.R. Schrock and J.A. Osborn, J. Am. Chem. Soc., **98**, 4450 (1976)
29. a) R. Uson, L.A. Oro, R. Sariego, M. Valderrama and C. Rebullida, J. Organomet. Chem., **197**, 87 (1980). b) R. Usón, L.A. Oro and M.J. Fernández, J. Organomet. Chem., **193**, 127 (1980). c) M.A. Esteruelas, A.M. López, L.A. Oro, A. Pérez, M. Schulz and H. Werner, Organometallics, **12**, 1823 (1993).
30. R.R. Schrock and J.A. Osborn, J. Am. Chem. Soc., **98**, 2143 (1976)
31. M.A. Garralda and L.A. Oro, Transition Met. Chem., **5**, 65 (1980), and the references therein.
32. M.A. Esteruelas Ph. D. Thesis, University of Zaragoza, 1983 and the references therein.
33. R.H. Crabtree, H. Felkin, T. Fillebeen-Khan and G.E. Morris, J. Organomet. Chem., **168**, 183 (1982)
34. a) R.H. Crabtree, G.C. Hlatky, C.P. Parnell, B.E. Segmüller and R.J. Uriarte, Inorg. Chem., **23**, 354 (1984). b) R.H. Crabtree and M. Lavin, J. Chem. Soc., Chem., Commun., 1661 (1985)
35. R.H. Crabtree and J.M. Quirk, J. Organomet. Chem., **199**, 99 (1980)
36. a) R.H. Crabtree, J.M. Quirk, T. Fillebeen-Khan and G.E. Morris, J. Organomet. Chem., **157**, C13 (1978). b) R.H. Crabtree, J.M. Quirk, T. Fillebeen-Khan and G.E. Morris, J. Organomet. Chem., **181**, 203 (1979).
37. R.H. Crabtree, M.F. Mellea, J.M. Mihelcic and J.M. Quirk, J. Am. Chem. Soc., **104**, 107 (1982).
38. a) D. Evans, J.A. Osborn and G. Wilkinson, J. Chem. Soc., A, 3133 (1968). b) C. O'Connor and G. Wilkinson, J. Chem. Soc., A, 2665(1968). c) G. Yagupsky and G. Wilkinson, J. Chem. Soc., A, 725 (1969). d) P.S. Hallman, B.R. McGarvey and G. Wilkinson, J. Chem. Soc., A, 3143 (1968).
39. B.E.R. Schilling, R. Hoffman and J.W. Faller, J. Am. Chem. Soc., **101**, 592 (1979)
40. E.G. Lundquist, J.C. Huffman, K. Folting and K.G. Caulton, Angew. Chem. Int. Ed. Engl. **27**, 1165 (1988)
41. G. Marinelli, I.E. Rachidi, W.E. Streib, O. Eisenstein and K.G. Caulton, J. Am. Chem. Soc., **111**, 2346 (1989)
42. a) E.L. Muetterties, J.R. Blecke, E.J. Wucherer and T.A. Albright, Chem. Rev., **82**, 499 (1982). b) R.G. Gastinger abd K.J. Klabunde, Transition Met. Chem., **4**, 1(1979). c) R. Uson, L.A. Oro, C. Foces-Foces, F.H. Cano, S. García-Blanco and M. Valderrama, J. Organomet. Chem., **229**, 293 (1982). d) E.L. Muetterties, J.R. Bleeke and A.C. Sievert, J. Organomet. Chem., **178**, 197 (1979). e) R. Uson, L.A. Oro, C. Foces-Foces, F.H. Cano, A. Vegas and M. Valderrama, J. Organomet. Chem, **251**, 241 (1981). f) L.A. Oro, C. Foces-Foces, F.H. Cano and S. García-Blanco,

J. Organomet. Chem., **236**, 385 (1982). g) R. Uson, L.A. Oro, D. Carmona, M.A. Esteruelas, C. Foces-Foces, F.H. Cano, S. García-Blanco and A. Vázquez de Miguel, J. Organomet. Chem., **273**, 111 (1984)
43. R. Uson, L.A. Oro, D. Carmona, M.A. Esteruelas, C. Foces-Foces, F.H. Cano and S. García-Blanco, J. Organomet. Chem., **254**, 249 (1983)
44. R.H. Crabtree, H. Felkin and G.E. Morris, J. Organomet. Chem., **141**, 205 (1977)
45. L.A. Oro, J.A. Cabeza, C. Cativiela, M.D. Diaz de Villegas and E. Melendez, J. Chem. Soc., Chem. Commun., 1383 (1983).
46. H. Taube, Survey Prog. Chem., **6**, 1 (1973) and the references therein
47. G.J. Kubas, Comments Inorg. Chem., **7**, 17 (1988)
48. R.H. Crabtree and D.G. Hamilton, J. Am. Chem. Soc., **108**, 3124 (1986)
49. G.P. Pez, R.A. Grey and J. Corsi, J. Am. Chem. Soc., **103**, 7528 (1981)
50. G.L. Geoffroy and J.R. Lehman, Advan. Inorg. Chem. Radiochem., **120**, 290 (1977)
51. B. Chaudret and R. Poilblanc, Organometallics, **4**, 1722 (1985)
52. L.S. Van Der Sluys, G.J. Kubas and K.G. Caulton, Organometallics, **10**, 1033 (1991).
53. G.G. Hlatky and R.H. Crabtree, Coord. Chem. Rev., **65**, 1 (1985)
54. F.H. Jardine, Prog. Inorg. Chem., **32**, 265 (1984) and the references therein
55. D. Evans, J.A. Osborn, F.H. Jardine and G. Wilkinson, Nature, **208**, 1203 (1965)
56. M.S. Chinn, D.M. Heinekey, N.G. Payne and C.D. Sofield, Organometallics, **8**, 1824 (1989)
57. R.A. Sanchez-Delgado, N. Valencia, R.L. Marquez-Silva, A. Andriollo and M. Medina, Inorg. Chem., **25**, 1106 (1986)
58. J. Halpern, Pure Appl. Chem., **59**, 173 (1987)
59. D.E. Linn and J. Halpern, J. Am. Chem. Soc., **109**, 2969 (1987) and the references therein
60. P.S. Hallman, D. Evans, J.A. Osborn and G. Wilkinson, J. Chem. Soc., Chem. Commun., 305 (1967)
61. A.M. Stolzenberg and E.L. Muetterties, Organometallics, **4**, 1739 (1985)
62. B.R. James, L.D. Markham and D.K.W. Wang, J. Chem. Soc., Chem. Commun., 439 (1974)
63. R.H. Fish, A.D. Thormodsen and G.A. Cremer, J. Am. Chem. Soc., **104**, 5234 (1982)
64. R.H. Fish, J.L. Tan and A.D. Thormodsen, Organometallics, **4**, 1743, (1985)
65. R.H. Fish, J.L. Tan and A.D. Thormodsen,J. Org. Chem., **49**, 4500 (1984)
66. R. Wilizynski, W.A. Fordyce and J. Halpern, J. Am. Chem. Soc., **105**, 2066 (1983)
67. W.A. Fordyce, R. Wilizynski and J. Halpern, J. Organomet. Chem., **296**, 115 (1985)
68. Y. Sasson, P. Albin and J. Blum, Tetrahedron Lett., 833 (1974)
69. W. Strohmeier and K. Holke, J. Organomet. Chem., **193**, C63 (1980)
70. J. Tsuji and H. Suzuki, Chem. Lett., 1085 (1977)
71. W.M. Kruse and L.W. Wright, Carbohydr. Res., **64**, 293 (1978)
72. R.A. Sanchez-Delgado and O.L. de Ochoa, J. Mol. Catal., **6**, 303 (1979)
73. R.A. Sanchez-Delgado and O.L. de Ochoa, J. Organomet. Chem., **202**, 427 (1980)
74. R.A. Sanchez-Delgado, A. Andriollo, O.L. de Ochoa, T. Suárez and N. Valencia, J. Organomet. Chem., **209**, 77 (1981)
75. S.D. Robinson and M.F. Uttley, J. Chem. Soc., Dalton Trans., 1912 (1973)
76. J.F. Knifton, J. Org. Chem., **40**, 519 (1975)
77. a) L. Vaska, Inorg. Nucl. Chem. Lett., **1**, 89 (1965). b) T.R.B. Mitchell, J. Chem. Soc. B, 823 (1970). c) B. Bell, J. Chatt and G.J. Leigh, J. Chem. Soc., Dalton Trans., 977 (1973)

78. R.A. Sánchez-Delgado, A Andriollo, E. González, N. Valencia, V. Leon and J. Espidel, J. Chem. Soc., Dalton Trans., 1859 (1985)
79. M.A. Esteruelas and H. Werner, J. Organomet. Chem., **303**, 221 (1986)
80. H. Werner, M.A. Esteruelas and H. Otto, Organometallics, **5**, 2295 (1986)
81. H. Werner, M.A. Esteruelas, U Meyer and B. Wrackmeyer, Chem. Ber., **120**, 11 (1987)
82. M.A. Esteruelas, E. Sola, L.A. Oro, H. Werner and U. Meyer, J. Mol. Catal., **45**, 1 (1988)
83. a) M.A. Esteruelas, E. Sola, L.A. Oro, U. Meyer and H. Werner, Angew. Chem., Int. Ed. Engl., **27**, 1563 (1988). b) M.A. Esteruelas, L.A. Oro and C. Valero, Organometallics, **11**, 3362 (1992)
84. H. Werner, U. Meyer, M.A. Esteruelas, E. Sola and L.A. Oro, J. Organomet. Chem., **366**, 187 (1989)
85. M.A. Esteruelas, E. Sola, L.A. Oro, H. Werner and U. Meyer, J.Mol. Catal., **53**, 43 (1989)
86. A. Andriollo, M.A. Esteruelas, U. Meyer, L.A. Oro, R. Sánchez-Delgado, E. Sola, C. Valero and. H. Werner, J. Am. Chem. Soc., **111**, 7431 (1989)
87. M.A. Esteruelas, L.A. Oro, and C. Valero, Organometallics, **10**, 462 (1991)
88. M.A. Esteruelas, C. Valero, L.A. Oro, U. Meyer and H. Werner, Inorg. Chem., **30** 1159 (1991)
89. M. Aracama, M.A. Esteruelas, F.J. Lahoz, J.A. López, U. Meyer, L.A. Oro and H. Werner, Inorg. Chem., **30**, 288 (1991)
90. See for example: a) R. Choukroun, D. Gervais, P. Kalck and F. Senocq, J. Organomet. Chem., **335**, C9 (1987). b) R Choukroun, A. Iraqui, D. Gervais, J. Daran and Y. Jeannin, Organometallics, **6**, 1197 (1987). c) F. Dany, R. Mutin, C. Lucas, V. J. Thivolle-Cazat and J.M. Basset, J. Mol. Catal., **51**, L 15 (1989). d) R. Mutin, C. Lucas, J. Thivolle-Cazat, V. Dufand, F. Dany and J.M. Basset, J. Chem. Soc., Chem. Commun., 896 (1988). e) F. Senocq, C. Randrianalimanana, A. Thorez, P. Kalck, R. Choukroun and D. Gervais, J. Chem. Soc., Chem. Commun., 1376 (1988). f) F. Senocq, C. Randrianalimanana, A. Thorez, P. Kalck, R. Choukroun and D. Gervais, J. Mol. Catal., **35**, 213 (1986). g) R. Choukroun, D. Gervais, J. Jand, P. Kalck and. F. Senocq, Organometallics, **5**, 67 (1986). h) J. Evans and J. Jinxing, J. Chem. Soc., Chem. Commun., 39 (1985). i) R. Choukroun, D. Gervais, and C. Rifai, Polyhedron, **8**, 1760 (1989). j) M. Hidai and H. Matsuzuka, Polyhedron, **7**, 2369 (1988). k) I. Ojima, M. Okabe, K. Kato, H.B. Kwon and I.T. Horvath, J. Am. Chem. Soc., **110**, 150 (1988)
91. R.D. Adams, Polyhedron, **7**, 2251 (1988)
92. B.D. Dombeck, Organometallics, **4**, 1707 (1985)
93. a) S. Bhaduri and K. Sharma, J. Chem. Soc., Chem. Commun., 173 (1988). b) M. Okoroafor, L. Shen, R.V. Honeychuck and C.H. Jr. Brubaker, Organometallics, **7**, 849 (1988)
94. a) A. Dedieu, P. Escaffre, J.M. Frances, P. Kalck and A. Thorez, Nouv. J. Chem., **10**, 631, (1986). b) P. Kalck, Polyhedron, **7**, 2441 (1988)
95. C. Mc Kenzie and R. Robson, J. Chem. Soc., Chem. Commun., 112 (1988)
96. J. Jenck, P. Kalck, E. Pinelli, M. Siani, and A. Thorez, J. Chem. Soc., Chem. Commun., 1428 (1988)
97. K. Jothimony and K. Vancheesan, J. Mol. Catal., **52**, 301 (1989)
98. a) A.J. Sivak and E.L. Muetterties, J. Am. Chem. Soc., **101**, 4878 (1979). b) E.L. Muetterties, Inorg. Chim. Acta, **50**, 9 (1981)

99. P.M. Maitlis, Acc. Chem. Res., **11**, 301 (1978)
100. M.P. García, A.M. López, M.A. Esteruelas, F.J. Lahoz and L.A. Oro, J. Chem. Soc., Dalton Trans., 3465 (1990)
101. M.P. García, A.M. López, M.A. Esteruelas, F.J. Lahoz and L.A. Oro, J. Organomet. Chem., **388**, 365 (1990)
102. M.P. García, A.M. López, M.A. Esteruelas, F.J. Lahoz and L.A. Oro, J. Chem. Soc., Chem. Commun., 793 (1988).
103. A.M. López, Ph. D. Thesis, University of Zaragoza, 1991
104. M.A. Esteruelas, M.P. García, A.M. López and. L.A. Oro, Organometallics, **10**, 129 (1991)
105. a) K.A. Beveridge, G.W. Bushnell, K.R. Dixon, D.T. Eadie and S.R. Stobart, J. Am. Chem. Soc., **104**, 920 (1982). b) A.W. Coleman, D.T. Eadie and S.R. Stobart, J. Am. Chem. Soc., **104**, 922 (1982).c) G.W. Bushnell, D.O.K. Fjeldsted, S.R. Stobart and M.J. Zaworotko, Organometallics, **4**, 1107 (1985). d) D.L. Lichtenberger, A.S. Copenhaver, H.B. Gray, J.L. Marshall and M.D. Hopkins, Inorg. Chem., **27**, 4488(1988). e) R.D. Brost, D.O.K. Fjeldsted and R.S. Stobart, J. Chem. Soc., Chem. Commun., 488 (1989)
106. T.C. Schenck, C.R.C. Milne, J.F. Sawyer and B. Bosnich, Inorg. Chem., **24**, 2338 (1985)
107. C. Bianchini, A. Meli, F. Laschi, J.A. Ramírez, P. Zanello and A. Vacca, Inorg. Chem., **27**, 4429 (1988)
108. R. Ugo, Catal. Rev., **11**, 225 (1975)
109. E.L. Muetterties, Bull. Soc. Chim. Belg., **84**, 959 (1975); **85**, 451 (1976)
110. J. Lewis and B.F.G. Johnson, Pure Appl. Chem., **44**, 43 (1975)
111. E.L. Muetterties, Science, **196**, 839 (1977)
112. E.L. Muetterties, Angew. Chem., Int. Ed. Engl., **17**, 545 (1978)
113. E.L. Muetterties, Pure Appl. Chem., **50**, 941 (1978)
114. E.L. Muetterties, T.N. Rhodin, E. Band, C.F. Brucker and W.R. Pretzer, Chem. Rev., **79**, 91 (1979)
115. R.M. Laine, J. Mol. Catal., **14**, 137 (1982)
116. a) J.L. Zuffa and W.L. Gladfelter, J. Am. Chem. Soc., **108**, 4669 (1986). b) G. Süss-Fink and G.F. Schmidt, J. Mol. Catal., **42**, 361 (1986)
117. J.M. Basset, in *Industrial Applications of Homogeneous Catalysis*, A. Mortreux and F. Petit Eds., D. Reidel Publishing Company; Dordrecht, 1988. p. 293–332.
118. A.K. Smith, A. Theolier, J.M. Basset, R. Ugo, D. Commerenc and Y. Chauvin, J. Am. Chem. Soc., **100**, 2590 (1978)
119. a) J.B. Keister and J.R. Shapley, J. Am. Chem. Soc., **98**, 1056 (1976). b) M. Cree-Uchigama, J.R. Shapley and G.M. St. George, J. Am. Chem. Soc., **108**, 1316 (1986)
120. A. Choplin, B. Besson, L. D'Ornelas, R. Sánchez-Delgado and J.M. Basset, J. Am. Chem. Soc., **100**, 2783 (1988)
121. G. Bergounhou, P. Fompegrine, G. Commenges and J.J. Bonnett, J. Mol. Catal., **48**, 285 (1988)
122. Y. Doi, K. Koshizuka and T. Keii, Inorg. Chem., **21**, 2732 (1982)
123. R. Sánchez-Delgado, A. Andriollo, J. Puga and G. Martín, Inorg. Chem., **26**, 1867 (1987)
124. J.B. Keister and J.R. Shapley, J. Organometal. Chem., **85**, C29 (1975)
125. G. Jeske, H, Lauke, H, Mauermann, H. Schumann and T.J. Marks, J. Am. Chem. Soc., **107**, 8111 (1985)
126. a) J. Boor, *Ziegler-Natta Catalysts and Polymerizations*, Academic Press, New York

1978. b) G.A. Molander and J.O. Hoberg, J. Org. Chem., **57**, 3266 (1992). c) J.S. Yu, B.C. Ankianiec, M.T. Nguyen and I.P. Rothwell, J. Am. Chem. Soc., **114**, 1927 (1992)

127. D.V. Sokolskii and N.F. Noskova, *Katalizatory Tipa Tsiglera-Natta v Reaktsii Gidriroyaniha*. Nauka Alma-Ata, Kaz.SSR, 1977 Chem. Abs. B 166927d
128. a) P. Le Maux, J. Y. Saillard, D. Grandjean and G. Jaouen, J. Org. Chem., **45**, 4524 (1980). b) E.N. Frankel, Catal. Org. Synth., 185 (1978). c) N. Sodeoka and M. Shipasaki, J. Org. Chem., **50**, 1147 (1985)
129. C. Bianchini, A. Meli, M. Peruzzini, F. Vizza and F. Zanobini, Organometallics, **8**, 2080 (1989)
130. S. Friedman, H.F. Kauffman and I. Wender, J. Org. Chem., **24**, 1287 (1959)
131. J.R. Bleeke and E.L. Muetterties, J. Am. Chem. Soc., **103**, 566 (1981)
132. C. Bohanna, Dissertation, University of Zaragoza 1991
133. a) C. Bianchini, A. Meli, M. Peruzzini, P. Frediani, C. Bohanna, M.A. Esteruelas and L.A. Oro, Organometallics, **11**, 138 (1992). b) C. Bianchini, C. Bohanna, M.A. Esteruelas, P. Frediani, A. Meli, L.A. Oro and Peruzzini, Organometallics, **11**, 3837 (1992). c) J. Eckert, A. Albinati, R.P. White, C. Bianchini and M. Peruzzini, Inorg. Chem., **31**, 4241 (1992).
134. L.S. Stuhl, M. Dubois, F.J. Hirsekorn, J.R. Bleeke, A.E. Sterens and E.L. Muetterties, J. Am. Chem. Soc., **100**, 2405 (1978)
135. E.L. Muetterties and F.J. Hirsekorn, J. Am. Chem. Soc., **96**, 4063 (1974)

CHAPTER 3

HOMOGENEOUS TRANSFER HYDROGENATION CATALYSED BY METAL COMPLEXES

The preceeding chapter described many homogeneous reductions involving molecular hydrogen (hydrogenation reactions). However, there is an alternative type of reduction reaction where the hydrogen is supplied by a donor molecule, DH_2, which itself undergoes dehydrogenation during the course of the reaction (equation *3.1*).

$$A + DH_2 \rightleftharpoons AH_2 + D \qquad (3.1)$$

The equilibrium between a donor (DH_2 or AH_2) and an acceptor (A or D) compound shown in equation *3.1* is known as a hydrogen transfer reaction. The essential difference between hydrogenation and hydrogen transfer is that in the latter there is no change in the overall unsaturation of the system, in its simplest form. In practice, the donor molecules, DH_2, are usually organic compounds which are present as solvents; this is not, however, a prerequisite of the reaction. Their essential feature is that they are compounds which are able to give up hydrogen, and suitable donors include alcohols, glycols, aldehydes, amides, ethers, amines and even aromatic hydrocarbons. Acceptor molecules, A, are unsaturated organic compounds such as alkenes, alkynes, carbonyl compounds, nitriles, imines, hydroxylamines, hydrazones, azo compounds or nitro compounds. Hydrogen transfer reactions involving water soluble donors as methanoate, ascorbate, etc. are discussed in Chapter 5.

The conditions which must be fulfilled in all hydrogen transfer reactions are quite restrictive. The hydrogen donor must be able to complex to a metal and then transfer hydrogen to it. The hydrogen oxidised donor must then be released from the metal coordination sphere before the back transfer can take place. This dissociation of the oxidised donor is also essential if the hydrogen acceptor is to be able to coordinate to the metal next. The hydrogen acceptor must not itself undergo hydrogen abstraction under the reaction conditions which are sometimes quite severe.

P.A. Chaloner, M.A. Esteruelas, F. Joó and L.A. Oro, Homogeneous Hydrogenation 87–118.

Conventional hydrogenation catalysts derived from the platinum group metals are also frequently active as hydrogen transfer catalysts. This is not surprising because one would expect the necessary mechanistic similarities. On the other hand, the solubility of the hydrogen donors may be much higher than that of hydrogen in the same solvent, thus facilitating the reaction.

3.1. Nature of the donor

3.1.1. INTRODUCTION

Donor compounds, DH_2, can, in principle, be any organic compound whose oxidation potential is sufficiently low that hydrogen transfer can occur under mild conditions [1]. The choice of donor is generally determined by its ease of reaction and its availability; the specific choice also depends on the nature of the functional group to be reduced. For example, the reduction of carbonyl groups generally requires the use of alcohols as donors.

The most commonly used donor molecules mentioned above can be separated for our purposes into two groups, hydrocarbons and other donors which contain heteroatoms. The hydrocarbon group includes alkanes, alkenes and aromatic hydrocarbons which interact with the catalyst *via* C-H activation, while donors of the second type comprise amines, cyclic ethers (usually dioxane) and alcohols, where the coordination of the heteroatom of these molecules to the metal is the initial step before C-H activation. C-H activation is an important field of transition metal catalysis in its *own* right and we would therefore refer the reader to Shilov's book in this series [2].

3.1.2. HYDROCARBONS AS DONORS

Crabtree *et al.* [3–5] have reported the dehydrogenation of cyclopentane and cyclooctane by $[IrH_2(Me_2CO)_2(PPh_3)_2]^+$ and 3,3-dimethyl-1-butene in chlorinated solvents (equations *3.2* and *3.3*).

$$\begin{array}{l}[IrH_2(Me_2CO)_2(PPh_3)_2]^+ + c\text{-}C_5H_{10} + 3Bu_tCH{=}CH_2 \longrightarrow \\ \quad [(\eta^5\text{-}C_5H_5)IrH(PPh_3)_2]^+ + 3Bu_tCH_2\text{-}CH_3 + 2Me_2CO\end{array} \qquad (3.2)$$

$$[IrH_2(Me_2CO)_2(PPh_3)_2]^+ + c\text{-}C_8H_{16} + 3Bu_tCH{=}CH_2 \longrightarrow [Ir(COD)(PPh_3)_2]^+ + 3Bu_tCH_2\text{-}CH_3 + 2Me_2CO \quad (3.3)$$

They propose that the initial step in the reaction is oxidative addition of an alkane C-H bond to the metal. The rationale for this approach is the belief that any intermediate alkyl hydride formed by oxidative addition would be unstable and would be forced to undergo rapid β-elimination to give the corresponding alkene. The resulting metal dihydride would then be dehydrogenated by the 3,3-dimethyl-1-butene and the cycle could continue until a stable product is formed (Scheme *3.1*) [6]. Arylalkenes are ineffective hydrogen acceptors in these instances because they undergo an unusual rearrangement; thus, for example, styrene gives $[Ir(\eta^6\text{-}C_6H_5Et)(PPh_3)_2]^+$ [7]. The above mentioned systems are not catalytic; however, $[IrH_2(O_2CCF_3)(PCy_3)_2]$ is a moderately active system [3]. Up to 28 turnovers of cyclooctene can be obtained with this complex at 20°C under photolytic conditions.

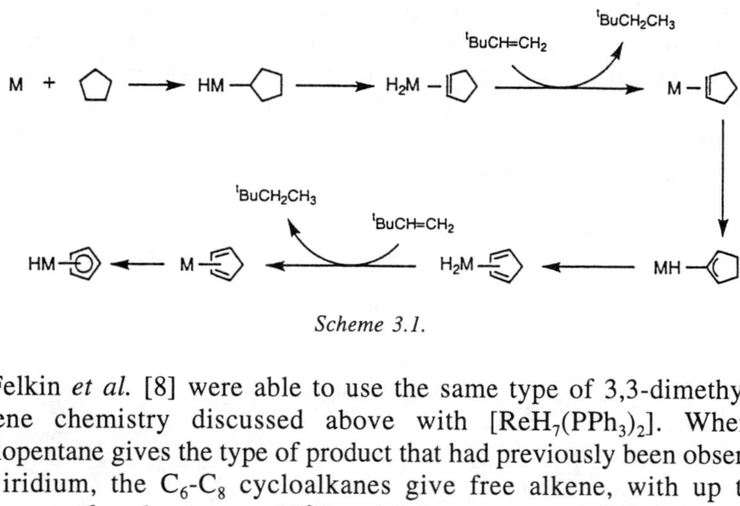

Scheme 3.1.

Felkin *et al.* [8] were able to use the same type of 3,3-dimethyl-1-butene chemistry discussed above with $[ReH_7(PPh_3)_2]$. Whereas cyclopentane gives the type of product that had previously been observed for iridium, the C_6-C_8 cycloalkanes give free alkene, with up to 9 turnovers of cyclooctene at 80°C and 1.6 turnovers at 30°C. Other polyhydrides have also been investigated; $[IrH_5(PR_3)_2]$ and $[RuH_4(PR_3)_2]$ (PR_3 = $P(C_6H_4\text{-}4\text{-}F)_3$) proved to be active for hydrogen transfer from cyclooctane to 3,3-dimethyl-1-butene at 150°C; 40–70 catalytic turnovers are observed over several days [9].

It is not yet known which step in the catalytic systems discussed above is turnover limiting. The β-elimination reaction of the initial alkyl hydride is a possible candidate [3].

Iridium pentahydride complexes [IrH$_5$(PR$_3$)$_2$] (PR$_3$ = PPh$_3$, P*i*Pr$_3$) catalyse the reactions of carbon-hydrogen bonds in linear alkenes under mild conditions. With alkenes having C $\geq$ 4, such as 1-pentene, 1-hexene and 2-hexene, a catalytic intramolecular hydrogen transfer reaction is observed. This reaction gives rise to equimolar amounts of diene and saturated hydrocarbon. The dienes formed are linear, conjugated and predominantly in the *trans,trans* configuration. In these reactions the alkene itself acts as both hydrogen acceptor and hydrogen donor, but ethene can also be used as an acceptor. Thus, the hydrogen transfer reaction does not necessarily take place between identical alkenes [10].

The mechanism proposed for these reactions is shown in Scheme *3.2*. The numerical values of m and n have not been determined, but they are unlikely to be greater than two because of steric hindrance. The only species of the type [Ir(alkene)$_n$(L-C)] isolated to date are pentacoordinated ethene complexes [Ir(C$_2$H$_4$)$_2$(L-C)L], but higher alkenes could form tetracoordinated complexes. Species of the type [IrH(alkene)$_m$L$_2$] have not been isolated, but they could be formed under catalytic conditions by reaction of [IrH$_5$L$_2$] with alkenes. Their lability has been taken as a clear indication of their tendency to undergo alkene insertion into the M-H bond, with formation of a metal-σ-alkyl complex. Thus, complexes of the type of [IrH(alkene)$_m$L$_2$] are expected to be

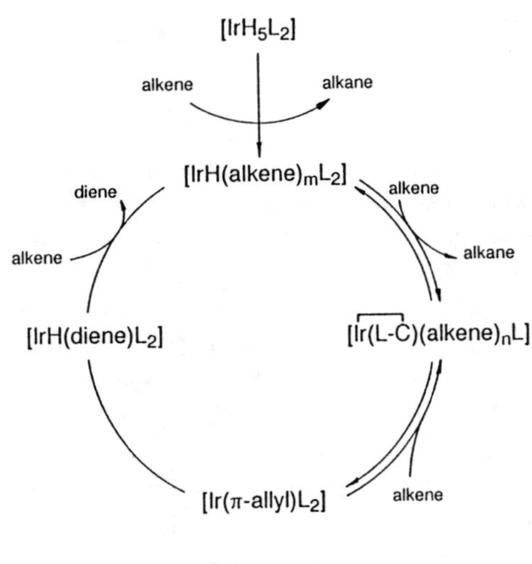

Scheme 3.2.

very reactive, and to give the metallated species $[\overline{Ir(alkene)_n(L\text{-}C})L]$ *via* a number of steps involving iridium alkyls.

The mechanism proposed in Scheme *3.2* is also supported by the following observations; ethene reacts with $[IrH_5L_2]$ to produce ethane and $[\overline{Ir(C_2H_4)_2(L\text{-}C})L]$ (L = $PPh_2(C_6H_4\text{-})$, or $P\mathit{i}Pr_2(C_3H_6\text{-})$). The reaction of these pentahydrides with propene leads to $[Ir(\eta^3\text{-}C_3H_5)L_2]$, whereas the reaction with *cis*-2-butene gives $[IrH(\eta^4\text{-}C_4H_6)L_2]$. The complexes $[Ir(\eta^3\text{-}C_3H_5)(P\mathit{i}Pr_3)_2]$ and $[IrH(\eta^4\text{-}C_4H_6)(P\mathit{i}Pr_3)_2]$ can also be obtained by reaction of the metallated compounds, $[\overline{Ir(C_2H_4)(P\mathit{i}Pr_2C}_3H_6)(P\mathit{i}Pr_3)]$, with propene or *cis*-2-butene, respectively.

Although the dehydrogenation of appropriately substituted aromatic hydrocarbons has been carried out under drastic conditions in the presence of heterogeneous catalysts, Imai *et al.* [11] have reported that the dehydrogenation can occur under mild conditions with homogeneous catalysts. Table *3.1* shows the hydrogen-donating ability of some aromatic hydrocarbons in the reaction of hydrogen transfer to hexanal catalysed by $[RuH_2(PPh_3)_4]$. In substituted ethylbenzenes, the hydrogen-

TABLE 3.1
Hydrogen transfer from aromatic hydrocarbons to hexanal catalysed by $[RuH_2(PPh_3)_4]$

DH_2	Yield of 1-hexanol (%) I[a]	II[b]	D
Diethylbenzene	12	24	c
Indan	13	22	Indene
Triethylbenzene	11	21	c
n-Butylbenzene	10	20	c
Isobutylbenzene	10	20	1-Phenyl-2-methyl-1-propene
4-Ethylanisol	10	20	c
Ethylbenzene	9	18	styrene
n-Propylbenzene	10	18	1-Phenyl-1-propene
Isopropylbenzene	9	16	α−Methylstyrene
3,4-Dichloroethylbenzene	7	9	c

a) In bromobenzene as solvent at 36ºC. b) In the designated hydrogen donor as solvent at 120ºC. c) Dehydrogenation product was not identified.

donating ability decreases in the order 4-ethylanisole > ethylbenzene > 3,4-dichloroethylbenzene. Hydrogen-donating ability is reduced by electron-withdrawing substituents and this suggests that a cationic species is being formed in the transition state of the rate-determining step [11].

Table *3.2* shows the reduction of some carbonyl compounds in indan at 120°C. Aliphatic aldehydes are found to be reduced most easily. The yield of alcohols does not vary greatly with the structure of the aliphatic aldehydes, but aldehydes which are moderately sterically hindered tend to give a higher yield of the corresponding alcohols. It is more difficult to reduce aliphatic ketones, perhaps because the secondary alcohols formed are more effective as hydrogen donors than the primary alcohols. Benzaldehyde and acetophenone are barely reduced and this is interpreted in terms of the stabilisation of the carbonyl group by conjugation with the aromatic ring and the fact that the hydrogen donating ability of benzylic alcohols is stronger than that of the other aliphatic alcohols. *N,N*-dimethylethanamide and ethyl-ethanoate are not reduced.

Kinetic studies of the hydrogen transfer reaction from indan to hexanal

TABLE 3.2
Hydrogen transfer from indan to carbonyl compounds (A) catalysed by $[RuH_2(PPh_3)_4]$

A	Yield of AH_2 %	AH_2
Hexanal	20	1-Hexanol
Pentanal	20	1-Pentanol
Butanal	20	1-Butanol
Heptanal	16	1-Heptanol
Propanal	16	1-Propanol
Octanal	15	1-Octanol
2-Ethyl-1-hexaldehyde	14	2-Ethyl-1-hexylalcohol
2-Butanone	7	2-Butylalcohol
3-Pentanone	7	3-Pentylalcohol
Acetone	6	2-Propanol
Benzaldehyde	Trace	Benzylalcohol

In indan as solvent at 120ºC for 4 h.

catalysed by $[RuH_2(PPh_3)_4]$ lead to the rate law shown in equation *3.4* where $[DH_2]$ and $[Cat]_{Tot}$ are the concentrations of hydrogen donor and added catalyst, respectively. As the coordination power of the aldehyde is strong, it is assumed that in practice, all the ruthenium is present as $[Ru(RCHO)(PPh_3)_3]$. Thus, this rate law is consistent with the catalytic cycle shown in Scheme *3.3*. The first step is presumed to be the formation of $[Ru(PPh_3)_3]$ or a related solvated species; according to the rate law, the rate-determining step of this reduction is hydrogen transfer from the donor to the catalytic species [11].

$$\frac{-d[RCHO]}{dt} = \frac{a[DH_2][Cat]_{Tot}}{1 + b[DH_2]} \tag{3.4}$$

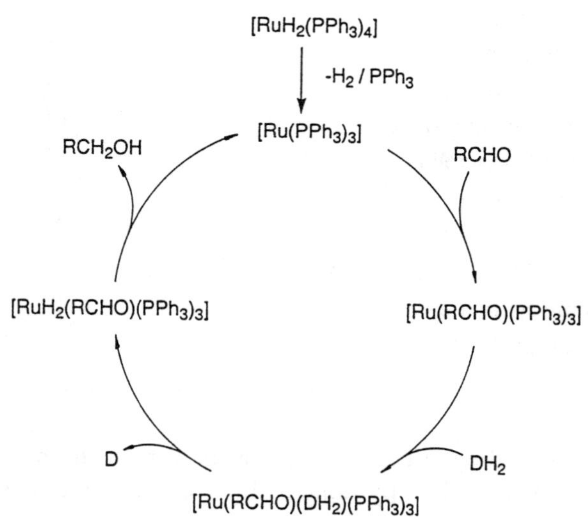

Scheme 3.3.

3.1.3. DONORS CONTAINING HETEROATOMS

Certain amines are good donors in hydrogen transfer reactions. Table *3.3* shows the hydrogen-donating ability of some amines in the hydrogen transfer reaction to cycloheptene catalysed by $[RhCl(PPh_3)_3]$. Unsubstituted cyclic amines such as pyrrolidine, piperidine, indoline and tetrahydroquinoline have a high hydrogen-donating ability; acyclic amines are less effective. The driving force for piperidine and pyrroli-

TABLE 3.3
Hydrogen transfer from amines to cycloheptene catalysed by $[RhCl(PPh_3)_3]$ [12]

Amine	% Conversion
Indoline	100
Pyrrolidine	100
Tetrahydroquinoline	98
Piperidine	94
Morpholine	44
N-Methylmorpholine	40
Benzylamine	26
Piperazine	23
Isopropylamine	18
N-Methylpiperazine	17
N-Methylpyrrolidine	12
Propylamine	8
N-Methylpiperidine	7
N,N′-Dimethylpiperazine	6
Dipropylamine	3
Tripropylamine	2

dine to donate hydrogen can not be due to aromatization, because the products of aromatization, pyridine and pyrrole, are not detected in the reaction mixtures.

The initial rates of the reaction of cyclopentene with indoline have been measured in several solvents by Nishiguchi *et al.* [12]. The results are summarized in Table *3.4*. Solvents having high polarity and strong coordinating power, such as dimethyl sulphoxide, pyridine and propionitrile, dissolved the catalyst easily, but reduction in these solvents was slow. Strongly coordinating solvents may block coordination of the reactants. In most aromatic solvents, such as halogenated benzenes and toluene, the catalyst dissolves well at typical reaction temperatures and the reaction proceeds rapidly.

The mechanism of hydrogen transfer from indoline to cycloheptene in toluene catalysed by $[RhCl(PPh_3)_3]$ has been studied [13]. The rate

TABLE 3.4
Solvent effect in hydrogen transfer from indoline to cyclopentene catalysed by $[RhCl(PPh_3)_3]$

Solvent	Initial rate $M\ min^{-1}$
Chlorobenzene	6.3×10^{-3}
o-Dichlorobenzene	5.1×10^{-3}
Aniline	4.0×10^{-3}
Anisole	3.4×10^{-3}
Toluene	2.9×10^{-3}
Benzene	2.7×10^{-3}
2-Butanone	2.3×10^{-3}
Dimethyl sulfoxide	1.1×10^{-3}
Pyridine	1.6×10^{-3}
Propionitrile	1.1×10^{-3}

[Cyclopentene]=0.50 M, [indoline]=0.50 M and $[\{RhCl(PPh_3)_3\}] = 6.0 \times 10^{-3}$M.

data fit the rate expression shown in equation *3.5* where $[Cat]_{Tot}$, $[DH_2]$ and $[PPh_3]$ are the concentration of the catalyst, indoline and added triphenylphosphine, respectively.

$$-\frac{d[\text{Cycloheptene}]}{dt} = \frac{a[DH_2][Cat]_{Tot}}{b + [PPh_3]} \tag{3.5}$$

Based on equation *3.5*, Scheme *3.4* is proposed for the reaction mechanism, where the rate-determining step is inferred to be the dehydrogenation of indoline. As is the case in the catalytic hydrogenation cycles, the availability of a vacant site for indoline coordination is essential. Intermediates in the reaction could not be isolated even in the absence of cycloalkene. Thus, these amines do react with $[RhCl(PPh_3)_3]$ in sealed-tube reactions at temperatures exceeding 100°C, in the absence of alkene. In these cases, hydrogenolysis of the triphenylphosphine ligands occurs. Pyrrolidine is much more effective in the hydrogenolysis reaction than piperidine, piperazine, indoline, tetrahydroquinoline, propylamine or bis(2-propyl)amine. Gas liquid

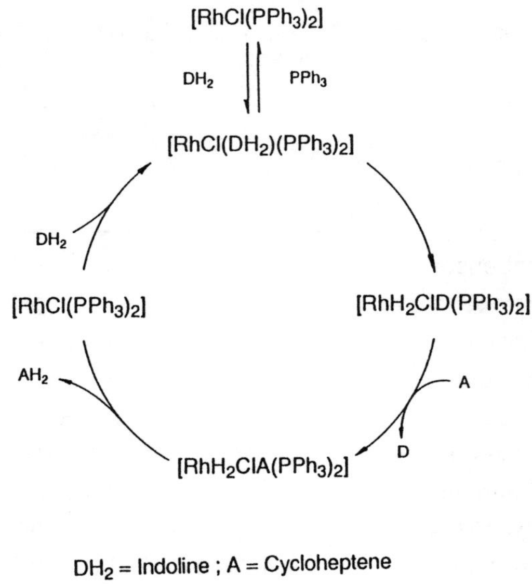

Scheme 3.4.

chromatography shows the amine to be the source of hydrogen. The reaction is therefore as follows:

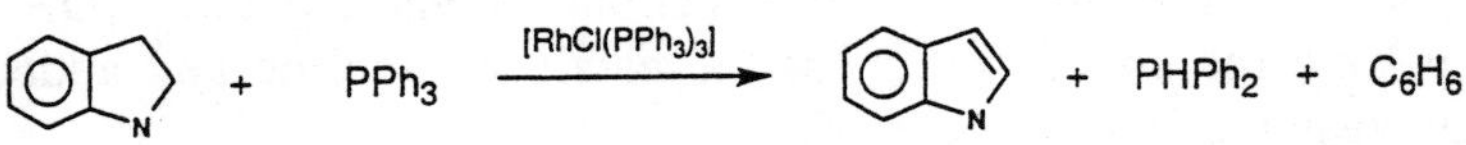

(3.6)

The large yield of benzene implies that further hydrogenolysis to PH_2Ph has occurred, but the primary phosphine was not detected among the reaction products [14, 15].

Dioxane can also be used as a donor in hydrogen transfer reactions [11, 16, 17]. The reduction of cyclopentene can be brought about by $[RhCl(PPh_3)_3]$ catalysed hydrogen transfer from dioxane in sealed-tube reactions at 170°C [16, 17]. Cyclopentane and dioxene are formed in equal amounts. The large isotope effect, $R^H/R^D = 3.1$, with octadeuteriodioxane, together with kinetic results, show that the rate-determining step of the reaction is the dehydrogenation. At lower temperatures the catalyst is known to form a dioxane complex; the following set of reactions have been proposed for the catalytic cycle:

$$[RhCl(PPh_3)_3] + C_4H_8O_2 \rightleftarrows [RhCl(PPh_3)_2(C_4H_8O_2)] + PPh_3 \qquad (3.7)$$

$$[RhCl(PPh_3)_2(C_4H_8O_2)] \xrightarrow{\text{slow}} [RhH_2Cl(PPh_3)_2] + C_4H_6O_2 \quad (3.8)$$

$$[RhH_2Cl(PPh_3)_2] + C_5H_8 \longrightarrow [RhCl(PPh_3)_2] + C_5H_{10} \quad (3.9)$$

$$[RhCl(PPh_3)_2] + PPh_3 \longrightarrow [RhCl(PPh_3)_3] \quad (3.10)$$

Dioxane has also been claimed to cleave the catalytically inactive dinuclear complex $[\{RhCl(PPh_3)_2\}_2]$ to give the dioxane adduct, thereby further improving the effectiveness of the hydrogen transfer [16, 17]. Later kinetic studies disagree with this view, claiming that no dioxane complex is formed and that the supposed complex was merely a mixture of dioxane and the dinuclear complex. It was also found that the reaction is not strictly first order in alkene. This could be due either to formation of the dinuclear complex or to competition between dioxane and cyclopentene for a coordination site. The role of the tertiary phosphine in the reaction has also been investigated. It was found that electron-donating substituents on the aryl groups slightly increase the reaction rate, whereas electron-withdrawing substituents cause a large decrease in the rate [18].

In the presence of $[RuH_2(PPh_3)_4]$, dioxane also reduces aldehydes. The overall rate law can be described by equation *3.4*. The values of the corresponding parameters are nearly equal for the reduction of hexanal by indane and dioxane, suggesting that the two reactions have similar mechanisms (Scheme *3.3*).

Several primary and secondary alcohols have been investigated as hydrogen donors [19, 20]. Table *3.5* shows that 2-propanol is the best donor for most reductions. It is also noteworthy that $[RhH(PPh_3)_4]$ does not form a carbonyl complex on heating in primary alcohols even at high temperatures, in contrast to $[RhCl(PPh_3)_3]$, which abstracts the carbonyl group from primary alcohols [21] and aldehydes [22] to give $[RhCl(CO)(PPh_3)_2]$.

The most widely used hydrogen donors in hydrogen transfer reactions are secondary alcohols, usually 2-propanol [23]. However, some studies using methanol as a source of hydrogen have recently been reported [24, 25]. In fact, the thermodynamics for the reaction shown in equation *3.1* is most favourable using 2-propanol ($\Delta H^\circ(298K) = -25.9$ kJmol^{-1}), the process becomes increasingly less favourable for ethanol, 1-propanol, and methanol ($\Delta H^\circ(298K) = -3.8$, $+ 1.3$ and $+ 21.0$ KJmol^{-1}, respectively).

The next section discusses the widespread use of secondary alcohols, generally 2-propanol, as donors.

TABLE 3.5
Hydrogen transfer from alcohols to cyclopentene catalysed by $[RhH(PPh_3)_4]$

Alcohol	Rate (mol m^{-1})	D
Methanol	6.0×10^{-5}	a
Ethanol	4.6×10^{-4}	Acetaldehyde
1-Propanol	6.9×10^{-5}	Propanal
2-Propanol	1.1×10^{-3}	Acetone
2-Butanol	6.2×10^{-4}	2-Butanone
2-Pentanol	5.6×10^{-4}	3-Pentanone
Benzylalcohol	9.1×10^{-5}	Benzaldehyde

a) Dehydrogenation product was not identified.

3.2. Secondary alcohols as donors: The catalysts

3.2.1. INTRODUCTION

A wide variety of hydrogen donors and acceptors were reviewed above. In general, primary alcohols are not very suitable as sources of hydrogen in hydrogen transfer reactions, probably because in many cases, the metal catalysts abstract carbon monoxide from the resulting aldehydes at the high reaction temperatures. The metal carbonyl complexes are not effective catalysts for further hydrogen transfer. The most common and useful hydrogen donors are secondary alcohols, mainly 2-propanol.

Catalysts are generally cobalt, rhodium, iridium, iron, ruthenium and osmium complexes with tertiary phosphine ligands or nitrogen-containing chelating ligands such as 2,2′-bipyridine, 1,10-phenanthroline and their methyl or methoxy-derivatives. All these ligands are characterized by the presence of unfilled orbitals with π-symmetry and can stabilise low oxidation states of the complex, preventing its decomposition to metal under reducing conditions. The nature of the supporting ligands greatly influences the active centre's variable electronic and steric characteristics of the active metal centre, so that the interactions between a given substrate and the catalysts can lead to high selectivities.

3.2.2. MONONUCLEAR CATALYSTS

Most hydrogen transfer catalysts are mononuclear. Thus, tris(triphenylphosphine) cobalt trihydride is a useful reducing agent for carbonyl groups in organic compounds. It is also a direct precursor of a catalyst (most probably a polymeric cobalt alkoxide) for hydrogen transfer from 2-propanol to cyclohexanone [26]. $[RhCl(PPh_3)_3]$ also catalyses the reduction of cyclic ketones with 2-propanol, and the rates of reaction decrease in the order cyclohexanone > cyclopentanone > cycloheptanone > cyclooctanone. Between 12 and 60 moles of potassium hydroxide are usually added to the system per mole of rhodium. The hydrogen transfer reactions are carried out in refluxing 2-propanol at 83°C. In more concentrated solutions but under much milder conditions, it has been demonstrated that $[RhH(PPh_3)_3]$ can be formed in this way [23b]. This hydride complex is also active in the hydrogen transfer from 2-propanol to cyclopentene [19] and α,β-unsaturated ketones [27]. Another interesting family of catalysts are the cationic rhodium complexes $[Rh(diene)L_2]^+$. These systems catalyse hydrogen transfer from 2-propanol to ketones [23c, 28–30], nitrobenzene [31], alkenes, dienes and alkynes [29]. The catalytic activity depends upon the nature of the mono- or bidentate nitrogen, oxygen or phosphorous-donor ligands; the most active for the reduction of carbonyl groups are those containing chelating-nitrogen-donor ligands such as Me_x-2,2′-bipyridine and Me_x-phenanthroline [23c, 29]. Similar iridium systems also catalyse the reduction of α,β-unsaturated carbonyl compounds. α,β-Unsaturated ketones can be selectively reduced to the saturated forms, while α,β-unsaturated aldehydes can be reduced to the unsaturated alcohols [24]. The species formed *in situ* from $[\{IrCl(COD)\}_2]$ and $P(C_6H_4\text{-}2\text{-}OMe)_3$ catalyse the selective reduction of the carbonyl group in 5-hexen-2-one, the yield in unsaturated alcohol being maximised at high P/Ir ratios [32]. A similar effect has been observed for the reduction of cinammic aldehyde [33]. This peculiar behaviour is probably related to an intramolecular interaction between the substrate and the methoxy groups of the phosphine. The same system has also been tested with benzylideneacetone as substrate, yielding a certain amount of unsaturated alcohol together with saturated ketone and alcohol [32]. The presence of potassium hydroxide is necessary for these rhodium and iridium systems. This is related to the formation of a $M\text{-}OCH(CH_3)_2$-intermediate which undergoes a β-elimination process to give the species $[MH_xL_2]$ [29a, 34]. In a similar vein it has been observed that the $[IrH_x(PR_3)_2]$ (x = 3 or 5) catalyses hydrogen transfer from 2-propanol to 3,3-dimethyl-

1-butene [35], alkynes [36], ketones [37] and α,β-unsaturated ketones [37–39].

Hydrogen transfer to alkenes is not easily accomplished. A low yield of cyclopentane is obtained when cyclopentene and 2-propanol are allowed to react in the presence of $[RuCl_2(PPh_3)_3]$ [19]. It has been found that hydrogen transfer to cyclohexene, styrene or phenylethyne is very slow. It is possible that these substrates do not undergo rapid reduction because they coordinate too strongly to the catalyst and poison their *own* reduction by preventing the coordination of the other reagents in the catalytic cycle. Indeed, these hydrocarbons poison hydrogen transfer to cyclohexanone [40]. $[RuCl_2(PPh_3)_3]$ also catalyses the reduction of α,β-unsaturated ketones to ketones [41, 42]. Unsaturated secondary alcohols undergo internal hydrogen transfer. Both $[RuCl_2(PPh_3)_3]$ and $[RuHCl(PPh_3)_3]$ act as catalysts in this reaction [43, 44]. Both pent-1-en-4-ol and pent-2-en-4-ol yield pentan-2-one. Unsaturated glycols react similarly [44].

Hydrogen transfer is also a method for preparing ketones from secondary alcohols. Cyclododecan-1,2-diol, norbornan-2,3-diol, *trans*-cyclohexan-1,2-diol, butan-1,2-diol, 1,2-diphenylethan-1,2-diol and 9,10-dihydroxystearic acid can all be dehydrogenated when allowed to react with benzylideneacetone in the presence of $[RuCl_2(PPh_3)_3]$ [45].

The rate of hydrogen transfer from secondary alcohols to cyclohexanone in the presence of $[RuCl_2(PPh_3)_3]$ depends markedly upon the nature of the secondary alcohol [46]. If the hydroxyl group is shielded by alkyl groups, the rate of the hydrogen transfer is low [47]. The rate declines in the order MeCHOHMe > MeCHOHEt > $MeCHOHBu^i$ > $MeCHOHC_6H_{11}$ > $MeCHOHBu^t$. 3-Pentanol transfers its hydrogen atoms as slowly as $MeCHOHBu^t$. However, electronic effects are also important, since both 1-phenylethanol and 1-phenylpropanol react more rapidly than 2-propanol.

Cyclohexanone is reduced more rapidly than other cycloalkanones, and the reduction of other ketones using 2-propanol is also slow. Substituted cyclohexanones are also reduced more slowly than the parent ketone [48]. Similar behaviour is observed in the reduction of cycloheptanone and its derivatives, and *N*-methylpiperidone and its derivatives [23e]. In most instances, the fall in the rate of hydrogen transfer can be correlated with the shielding of the keto group by alkyl substituents. It is interesting to note that the hydrogen transfer to 4-alkylcyclohexanones yields both *cis*- and *trans*-4 alkylcyclohexanols.

The rate of hydrogen transfer from secondary alcohols to ketones in the presence of $[RuCl_2(PPh_3)_3]$ is enhanced if a small quantity of potassium hydroxide is added to the system [49, 50]. It now seems very likely that the species $[RuH_2(\eta^2\text{-}H_2)(PPh_3)_3]$ is formed under these conditions, since it has been reported that hydrogen transfer reactions from 2-propanol produce metal polyhydrides *via* β-abstraction [51–54]. The activity of $[MHCl(CO)(P\textit{i}Pr_3)_2]$ (M = Ru, Os) in the hydrogen transfer from 2-propanol to cyclohexanone and acetophenone is rather poor in the absence of a cocatalyst. However, the addition of $Na[BH_4]$ increases catalytic activity considerably [54]. The complexes $[MHCl(CO)(P\textit{i}Pr_3)_2]$ react with $Na[BH_4]$ to give initially $[MH(\eta^2\text{-}H_2BH_2)(CO)(P\textit{i}Pr_3)_2]$ (M = Ru, Os), which decompose under catalytic conditions to give $[MH_4(CO)(P\textit{i}Pr_3)_2]$ (M = Ru, Os) [54, 55]. These tetrahydride compounds also catalyse the reduction of α,β-unsaturated ketones [56] and phenylethyne [57]. Initially, the solutions containing $[MH_4(CO)(P\textit{i}Pr_3)_2]$ rapidly reduce phenylethyne. The reaction rates fall progressively as the colourless solutions of $[MH_4(CO)(P\textit{i}Pr_3)_2]$ turn dark red. This points to a modification in the active species, and is consistent with the observation that the colourless tetrahydrides $[MH_4(CO)(P\textit{i}Pr_3)_2]$ react with phenylethyne to give the dark red complexes $[M(C_2Ph)_2(CO)(P\textit{i}Pr_3)_2]$ [57]. Hexahydrides $[OsH_6(PR_3)_2]$ ($PR_3 = P\textit{i}Pr_3$ and $P\textit{t}Bu_2Me$) are also active in hydrogen transfer from 2-propanol to ketones [58]. Alkynes, alkenes and ketones can also be reduced in the presence of other hydrido-ruthenium complexes containing bulky phosphines [59–62].

Interestingly, hydrogen transfer carried out in the presence of a chiral catalyst offers the possibility of obtaining optically active products. If the hydrogen acceptor is a prochiral compound and the hydrogen donor a racemic species, it is in principle possible to carry out an enantioface-discriminating reduction together with an enantiomer-discriminating dehydrogenation (equations *3.11* and *3.12*).

$$\begin{aligned} &n\ RCH(OH)R' + R^1CH{=}C(R^3)R^2 \longrightarrow \\ &\quad (n-1)RC^*H(OH)R' + RCOR' + R^1CH_2C^*H(R^3)R^2 \end{aligned} \qquad (3.11)$$

$$\begin{aligned} &n\ RCH(OH)R' + R^1COR^2 \longrightarrow \\ &\quad (n-1)RC^*H(OH)R' + RCOR' + R^1C^*H(OH)R^2 \end{aligned} \qquad (3.12)$$

Chiral alcohols can be synthesised by asymmetric hydrogen transfer to ketones but only a few catalysts have been described so far. These are rhodium, iridium and ruthenium complexes containing chiral phosphine [63–69], Schiff base [71–74], carboxylate [75] or amine ligands [76, 77].

The systems formed *in situ* from $RhCl_3 \cdot xH_2O$ and chiral amines such as R,S,-(-)-ephedrine catalyse the enantioselective reduction of ketones but with rather low optical yields [76]. Optical yields of up to 15% can be obtained with rhodium complexes containing optically active 2-(2′-pyridyl)pyridines [77]. Enantioselectivities of up to 23% are obtained in the presence of systems generated *in situ* from $[\{Rh(\mu\text{-}Cl)(HD)\}_2]$ (HD = 1,5-hexadiene) and (-)-2-pyridinalphenyl- ethylimine [72]. Similar iridium catalysts lead to enantioselective reduction of prochiral ketones with optical yields of up to 33% [71].

Very recently, some diastereoisomeric pentacoordinate complexes of the type of [IrI(COD)(NNR*)] (NNR* = 2-pyridinal-1-phenylethylimine, 2-acetylpyridine-1-phenylethylimine) have been synthesized. These complexes are active and selective catalysts for asymmetric hydrogen transfer from 2-propanol to prochiral ketones. Optical yields of up to 84% have been obtained in the reduction of t-butyl phenyl ketone [73].

In general, the enantioselectivity obtained with catalysts containing a chiral phosphine ligand is poor or moderate, and obviously depends substantially on the reactants used. For this type of systems the best results have been obtained with the complexes formed *in situ* from [Ir(acac)(COD)] and chiral menthylphenylphosphines. These systems catalyse hydrogen transfer from 2-propanol to acetophenone with optical yields of up to 42% [67].

3.2.3. POLYNUCLEAR CATALYSTS

Chapter 2 describes the mechanisms of some hydrogenation processes catalysed by binuclear compounds or clusters. Systems of this type also catalyse hydrogen transfer reactions. For example, the heterobinuclear complex $[H(CO)(PPh_3)_2Ru(\mu\text{-bim})Ir(COD)]$ catalyses hydrogen transfer from 2-propanol to cyclohexanone, benzylideneacetophenone or a cyclohexanone:styrene (1:1) mixture. Under the reaction conditions the carbon-carbon double bond of the α,β-unsaturated ketone is reduced with close to 95% selectivity. However, in the competitive reduction of the cyclohexanone:styrene, the carbon-oxygen double bond is reduced preferentially. The binuclear complex is recovered unchanged after the catalytic reactions. This, together with a kinetic study of the selective reduction of the α,β-unsaturated ketone, which leads to the rate law shown in equation *3.13*, suggests that the full catalytic cycle involves binuclear species. Similarly to cyclohexene hydrogenation (Section 2.5.4), the hydrogen transfer proceeds via one metal and the other metal acts as the core of a metal-ligand complex of variable elctron density [78].

$$\frac{-d[\text{Benzylideneacetone}]}{dt} = k[\text{Ru-Ir}][\text{Benzylideneacetone}] \qquad (3.13)$$

The heterobinuclear complexes $[H(CO)(PPh_3)_2Ru(\mu\text{-pz})(\mu\text{-Cl})M(diene)]$ (M = Rh,Ir; diene = COD,TFB) catalyse hydrogen transfer from 2-propanol to cyclohexanone [79]. For hydrogenation reactions of cyclohexene catalysed by $[H(CO)(PPh_3)_2Ru(\mu\text{-pz})_2Ir(TFB)]$, mentioned in chapter *2*, kinetic studies suggest that the nuclearity of the catalyst precursor remains unchanged during the catalysed reaction. However, in heterobridged compounds, bridge cleavage or redistribution reactions may occur because the stability of the heterobridged "$M(\mu\text{-pz})(\mu\text{-Cl})M$" framework is lower than that of the homobridged "$M(\mu\text{-pz})_2M$" one [80]. Thus, under hydrogen transfer conditions, the following equilibrium may exist:

$$2[H(CO)(PPh_3)_2Ru(\mu\text{-pz})(\mu\text{-Cl})M(diene)] \rightleftharpoons 2[RuHCl(CO)(PPh_3)_2] + [M(\mu\text{-pz})(diene)]_2 \qquad (3.14)$$

A study of the catalytic activity of the $[H(CO)(PPh_3)_2Ru(\mu\text{-pz})(\mu\text{-Cl})M(COD)]$ (M = Rh,Ir) complexes in the presence of the non-active homobinuclear $[\{M(\mu\text{-pz})(COD)\}_2]$ (M = Rh,Ir) complexes suggests that the equilibrium shown in equation *3.14* does exist for these heterobinuclear compounds, under catalytic conditions.

A different situation arises for the related tetrafluorobenzobarrelene complexes. Thus, the results of adding $[\{M(\mu\text{-pz})_2(TFB)\}_2]$ (M = Rh,Ir) complexes to catalytic solutions of $[H(CO)(PPh_3)_2Ru(\mu\text{-pz})(\mu\text{-Cl})M(TFB)]$ (M = Rh,Ir) do not provide clear evidence of the involvement of equilibrium *3.14*. Furthermore, it has been observed that the initial rate of reduction of cyclohexanone by hydrogen transfer from 2-propanol catalysed by $[H(CO)(PPh_3)_2Ru(\mu\text{-pz})(\mu\text{-Cl})Ir(TFB)]$ is first order in catalyst concentration [79]. Thus, this diene probably stabilizes the heterobridged "$Ru(\mu\text{-pz})(\mu\text{-Cl})M$" by inhibiting its complete cleavage. In this context, it is interesting to note that the diene tetrafluorobenzobarrelene has a good π-acceptor ability, as shown by its general tendency to form pentacoordinated species [81]. Furthermore, heterobinuclear complexes containing the unit "M(TFB)" bonded to another metal atom through chloride and pyrazolate bridges are more stable than those with the "M(COD)" unit [82].

Dinuclear and trinuclear cationic rhodium complexes such as $[\{Rh(NBD)\}_2(bipym)]^{2+}$ (bipym = 2,2′-bipyrimidine) and $[\{Rh(NBD)\}_3(tpt)]^{3+}$ (tpt = 2,4,6-tris(2-pyridyl)-s-triazine) catalyse hydrogen transfer from 2-propanol to acetophenone. The trinuclear complex $[\{Rh(NBD)\}_3(tpt)]^{3+}$ is 1.6 times more active than the dinuclear species $[\{Rh(NBD)\}_2(bipym)]^{2+}$ [83].

Various homogeneous hydrogen transfer reactions are known to be catalysed by tri- and tetranuclear iron or ruthenium clusters [84–89]. Bhaduri and Sharma [88] found that with 2-propanol as donor, $[Ru_4H_4(CO)_8(PnBu_3)_4]$ catalyses the reduction of α,β-unsaturated aldehydes to the corresponding unsaturated alcohols with high selectivities. Preliminary kinetic and deuterium labelling studies indicate involvement of cluster intermediates and a passive role for the cluster hydrides. The rate of formation of crotyl alcohol from crotonaldehyde is found to be inhibited by the addition of phosphine or application of

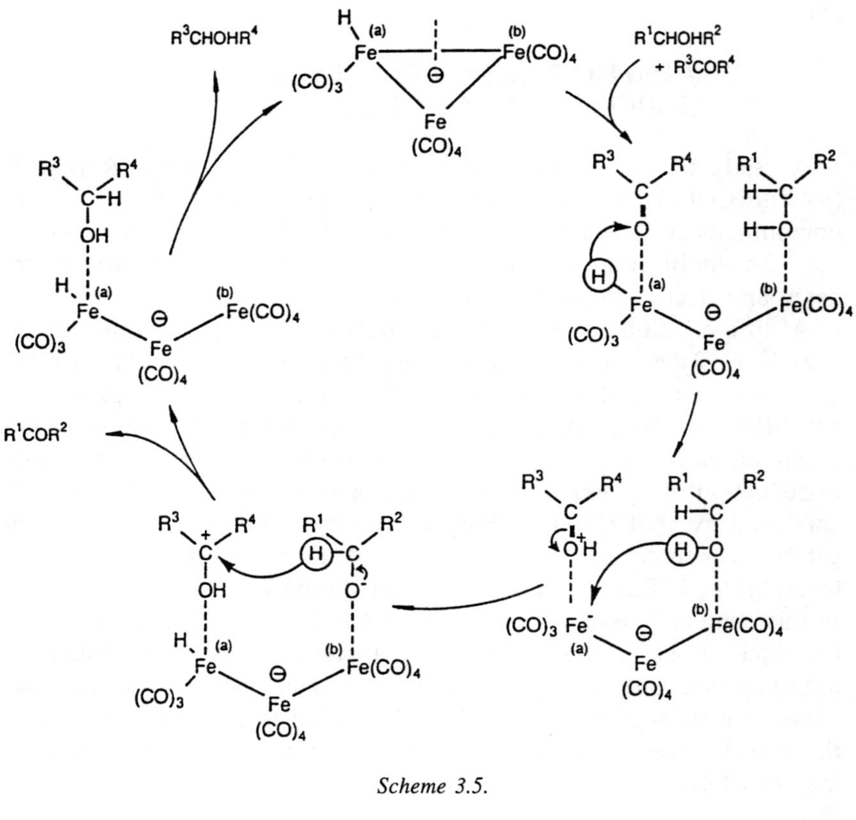

Scheme 3.5.

increased CO pressure. The rate is found to be inversely proportional to the concentrations of added phosphine. Similarly, with increased CO pressure a limiting inhibited rate was reached. These observations indicate that the rate-determining step is preceded by a rapid pre-equilibrium involving dissociation of $PnBu_3$ and probably also dissociation of CO.

A mechanism for hydrogen transfer from secondary alcohols (2-propanol and 1-phenylethanol) to ketones catalysed by $[Fe_3H(CO)_{11}]^-$ in the presence of NaOH was recently proposed (Scheme *3.5*) [89]. In the first step, $[Fe_3H(CO)_{11}]^-$ distorts by breaking an iron-iron bond, thus providing an additional vacancy for the donor and acceptor, which are coordinated to different iron atoms (a and b). The hydrogen atom on Fe(a) is transferred to the acceptor (coordinated to Fe(a)), followed by subsequent transfer of H^+ from the donor to Fe(a). In the last step, both the newly formed alcohol and ketone are dissociated with simultaneous closure of the bond between Fe(a) and Fe(b), thus regenerating $[Fe_3H(CO)_{11}]^-$.

3.3. Mechanisms of hydrogen transfer from alcohols

3.3.1. INTRODUCTION

We have already mentioned some of the most important catalysts for the reduction of unsaturated organic substrates by hydrogen transfer from secondary alcohols. A detailed discussion on the probable mechanisms for all the above mentioned reactions is not feasible at the moment because the data are scarce and incomplete. As a consequence, the mechanistic discussion will be limited to the following systems:

* Reduction of cycloalkenes catalysed by $[RhH(PPh_3)_4]$.
* Reduction of alkenes and ketones catalysed by cationic rhodium and iridium complexes.
* Reduction of ketones and α,β-unsaturated ketones catalysed by $[MHCl(CO)(PiPr_3)_2]$ (M = Ru,Os).
* Reduction of diphenylethyne catalysed by $[Ru(O_2CCF_3)_2(CO)(PPh_3)_2]$, in the presence of trifluoroethanoic acid.

3.3.2. REDUCTION OF CYCLOALKENES CATALYSED BY $[RhH(PPh_3)_4]$ [19]

It has been shown that $[RhH(PPh_3)_4]$ is an active catalyst for hydrogen transfer from 2-propanol to cycloheptene in toluene as solvent. A detailed kinetic study has led to equation *3.15*. Thus the rate-determining step of the reaction can be considered to be hydrogen transfer from the alcohol to a Rh(I) species.

$$\frac{-d[\text{Cycloheptene}]}{dt} = \frac{k_{15}[\text{Rh}]_{\text{Tot}}[(\text{CH}_3)_2\text{CHOH}]}{[\text{PPh}_3] + k_{15}} \qquad (3.15)$$

Given this result and other experimental data, the process of hydrogen transfer may be roughly divided into the following six steps: i) formation of unsaturated coordination sites on the rhodium (I) metal by dissociation of triphenylphosphine from $[RhH(PPh_3)_4]$; ii) coordination of 2-propanol; iii) coordination of cycloheptene; iv) hydrogen transfer from the hydrogen donor to the metal to form a rhodium (III) trihydride complex; v) transfer of a hydrogen atom to the alkene to form an alkyl complex; and vi) transfer of one more hydrogen atom to give cycloheptane and a rhodium (I) species. As plausible combinations of these steps, three reaction pathways may be considered:
1) i-ii-iv-iii-v-vi
2) i-iii-ii-iv-v-vi
3) i-iii-v-ii-iv-vi.

The assumption that step ii occurs prior to step iii in the hydrogen transfer rules out pathways *2* and *3*, and this assumption is supported by the following observations: a) If step iii proceeds prior to step ii, most of the rhodium complexes must exist in the form of alkene complexes or alkyl complexes, which have not been detected by spectroscopic studies. b) The initial rates of the reduction of some cycloalkenes with rather different coordinating ability and steric requirements are very similar.

Based on the considerations above, Imai *et al.* [19] proposed the cycle shown in Scheme *3.6* as the most reasonable mechanistic proposal for hydrogen transfer from 2-propanol to cycloalkenes catalysed by $[RhH(PPh_3)_4]$.

3.3.3. REDUCTION OF ALKENES AND KETONES CATALYSED BY CATIONIC RHODIUM AND IRIDIUM COMPLEXES

The catalytic activity of the cationic systems $[M(diene)L_2]^+$ (M = Rh, Ir) depends on the nature of the mono- or bidentate ligands. For the systems generated *in situ* by addition of group XV ligands to

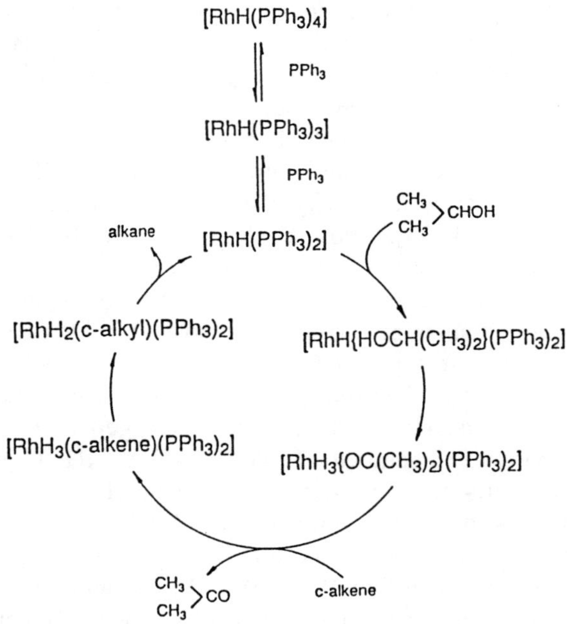

Scheme 3.6.

$[Rh(NBD)_2]^+$, the outcome of catalytic hydrogen transfer from 2-propanol to acetophenone and cyclohexene, in the presence of potassium hydroxide, can be rationalized in terms of the properties of the species obtained by treating $[Rh(NBD)_2]^+$ with the different ligands [29a].

i) Monodentate nitrile, arsine and stibine ligands

The (2:1 or 1:1 stoicheiometry) reaction of these ligands with $[Rh(NBD)_2]^+$ does not displace the diene, but leads to the formation of pentacoordinated species of the type $[Rh(NBD)_2L]^+$ [90], which are very poor catalysts.

ii) Monodentate phosphine and bidentate N or P ligands

The reaction of $[Rh(NBD)_2]^+$ with 2L (or L-L) gives rise to the displacement of one mole of diene and the formation of the catalyst precursors $[Rh(NBD)L_2]^+$ or $[Rh(NBD)(L\text{-}L)]^+$ [90–92]. The results of the hydrogen transfer reaction show that acetophenone is generally more efficiently reduced than cyclohexene.

iii) Amine or quinoline type monodentate nitrogen-donor ligands
In these systems the end product of the reaction is generally $[Rh(NBD)L_2]^+$, although species of the type $[Rh(NBD)L(solvent)]^+$ must also be taken into account [90]. These species give a higher conversion for cyclohexene than for acetophenone.

For iridium complexes of the formula $[Ir(COD)L_2]^+$ (L = nitrogen-donor ligand), the catalytic activity also depends on the nature of the nitrogen donor ligand [93]. Those containing nitrile ligands give rise to moderately active systems, whereas those with pyridine or quinoline-type ligands are generally catalysts of low activity. Complexes with bidentate nitrogen ligands give rise to the most active systems, particularly complexes of the type $[Ir(COD)(Me_xphen)]^+$, which have been extensively studied by Mestroni *et al.* [94–96]. Similar results have been obtained for the $[Ir(TFB)L_2]^+$ complexes generated *in situ* by addition of group XVI ligands to $[Ir(TFB)(1,4\text{-}Me_2C_6H_4)]^+$ [97]. In the case of diamine derivatives, primary amines generally give catalysts of higher activity than do tertiary amines [29, 93, 97].

Whereas hydrogenation with molecular hydrogen catalysed by cationic complexes of the type $[M(diene)L_2]^+$ (M = Rh,Ir) can be carried out at room temperature, transfer reactions require somewhat higher temperatures. This probably arises from the need to form an intermediate hydride derivative from the coordinated iso-propoxide generated in the basic media used (equation *3.16*) [29a].

$$M\text{-}OCHMe_2 \longrightarrow H\text{-}M\text{-}OCMe_2 \qquad (3.16)$$

Thus, $[IrH(diene)L_2]$ and acetone are obtained from the reaction of $[Ir(diene)L_2]^+$ (diene = COD, TFB, L = PPh_3, $AsPh_3$; L_2 = DPPE, DPPP) with potassium hydroxide in 2-propanol [34]. These complexes, which can also be obtained by addition of an equimolar amount of the ligand, L_2, to the dimers $[\{Ir(\mu\text{-}OR)(diene)\}_2]$ in alcohol, catalyse hydrogen transfer from 2-propanol to cyclohexanone. The species $[IrH(diene)L_2]$ are coordinatively saturated, and therefore activation is needed to initiate the catalytic reaction. This activation involves an isomerization or reduction of the coordinated diene, leading to the unsaturated complex $[IrH_xL_2]$, which can then coordinate the ketone and initiate the catalytic cycle (Scheme *3.7*) [34, 98].

Square-planar complexes with the general formula [Ir(OR)(COD)(PCy_3)] (R = Me, Et, *i*Pr) have recently been isolated, and the exchange of the alkoxy group by reaction with the alcohol which acts as solvent has been investigated [99, 100]. This exchange proceeds most probably

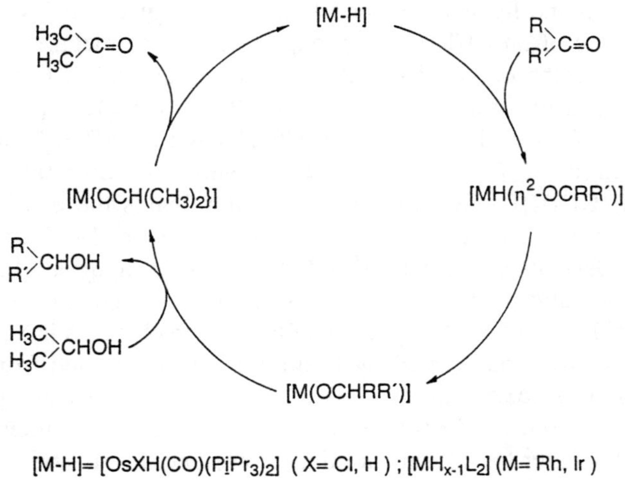

Scheme 3.7.

via a hydrogen bonded adduct, some examples of which have been characterized [101].

3.3.4. REDUCTION OF KETONES AND α,β-UNSATURATED KETONES CATALYSED BY [MHCl(CO)(P*i*Pr$_3$)$_2$] (M = Ru, Os)

It was suggested above that the hydrogen transfer from 2-propanol to ketones involves four steps (Scheme *3.7*): i) coordination of the ketone to the coordinatively unsaturated metal centre, ii) formation of an alkoxy metal intermediate by hydrogen migration from the metal to the ketone double bond, iii) exchange of the alkoxy group by reaction with the alcohol which acts as solvent, and iv) a β-elimination process. Under non-catalytic conditions, by using the five-coordinate osmium hydride complex [OsHCl(CO)(P*i*Pr$_3$)$_2$] as the starting material, experimental evidence has been obtained for the coordination, insertion and β-elimination steps [102].

The ^{1}H NMR spectrum of [OsHCl(CO)(P*i*Pr$_3$)$_2$] in C_6D_6 at room temperature shows a triplet at δ –31.9 with P-H coupling of 14Hz [103]. In acetone-d$_6$ this triplet disappears and a broad signal at δ –28.3 is observed. Upon lowering the temperature, a new compound is formed by insertion of the coordinated ketone into the Os-H bond. Thus, the ^{1}H NMR spectrum of the complex in acetone-d$_6$ at –60°C shows the signals of the phosphine ligands together with a new signal at δ 3.43 which is characteristic of an $OCH(CD_3)_2$ group linked to the metal.

These spectroscopic observations can be rationalized in terms of a rapid equilibrium between $[OsHCl(CO)(PiPr_3)_2]$, $[OsHCl(CO)(PiPr_3)_2\{\eta^2\text{-}(CD_3)_2CO\}]$ and $[Os\{OCH(CD_3)_2\}Cl(CO)(PiPr_3)_2]$ according to equation *3.17*. The coordinatively unsaturated species $[Os\{OCH(CD_3)_2\}Cl(CO)(PiPr_3)_2]$ which dominates the equilibrium at –60°C rearranges at higher temperatures to give the 18-electron intermediate $[OsHCl(CO)(PiPr_3)_2\{\eta^2\text{-}(CD_3)_2CO\}]$ by β-hydride elimination. These observations are very similar to those reported for the rhenium isopropoxide complex, $[Re_3(\mu\text{-}OiPr)_3(OiPr)_6]$, which exists in equilibrium with a monohydride formed by β-elimination of acetone from one of the terminal alkoxo ligands [104]. Insertion reactions are generally viewed as a concerted process involving a four centre intermediate; thus, for the insertion of the coordinated acetone-d_6 into the Os-H bond of $[OsHCl(CO)(PiPr_3)_2\{\eta^2\text{-}(CD_3)_2CO\}]$, η^2- bonding of the acetone to osmium is required.

$$\begin{aligned}&[OsHCl(CO)(PiPr_3)_2] + (CD_3)_2CO \rightleftharpoons\\&\quad [OsHCl(CO)(PiPr_3)_2\{\eta^2\text{-}(CD_3)_2CO\}] \rightleftharpoons\\&\quad [Os\{OCH(CD_3)_2\}Cl(CO)(PiPr_3)_2]\end{aligned} \qquad (3.17)$$

The complex $[OsHCl(CO)(PiPr_3)_2]$ catalyses hydrogen transfer from 2-propanol to cyclohexanone and acetophenone [54]. The NMR spectroscopic observations mentioned above support Scheme *3.7* as the most reasonable mechanistic proposal for these reactions. Interestingly, the addition of $Na[BH_4]$ leads to a significant increase in the catalytic activity. Under these conditions, $[OsHCl(CO)(PiPr_3)_2]$ reacts with $Na[BH_4]$ to give the tetrahydride, $[OsH_4(CO)(PiPr_3)_2]$ [55]. As $[OsH_4(CO)(PiPr_3)_2]$ is coordinatively saturated, activation most probably involves the loss of one dihydrogen molecule per molecule of $[OsH_4(CO)(PiPr_3)_2]$ to produce $[OsH_2(CO)(PiPr_3)_2]$ (equation *3.18*), which is structurally related to $[OsHCl(CO)(PiPr_3)_2]$ and is presumably the active catalyst.

$$[OsH_4(CO)(PiPr_3)_2] \rightleftharpoons [OsH_2(CO)(PiPr_3)_2] + H_2 \qquad (3.18)$$

The lability of $[OsH_2(CO)(PiPr_3)_2]$ prevents the detection of the supposed intermediates in the reaction catalysed by $[OsHCl(CO)(PiPr_3)_2]/Na[BH_4]$. However, a detailed kinetic study of the reduction of cyclohexanone has revealed that the reaction is first order in catalyst and in substrate [102]. The second order rate law strongly supports the suggestion that the catalytic cycle shown in Scheme *3.7* describes not only the mechanism for hydrogen transfer with [OsHCl(CO)(P*i*Pr$_3$)$_2$] as catalyst but also with [OsH$_2$(CO)(P*i*Pr$_3$)$_2$] as the active catalytic species.

The complex [OsHCl(CO)(P*i*Pr$_3$)$_2$] complex and its analogue [RuHCl(CO)(P*i*Pr$_3$)$_2$] also catalyse hydrogen transfer from 2-propanol to benzylideneacetone and benzylideneacetophenone. The reaction shows a preference for selective reduction to the saturated ketone. No unsaturated alcohol is observed during the reactions. In both cases, the addition of Na[BH$_4$] leads to a large increase in the catalytic activity of the system.

Scheme *3.8* shows a likely catalytic cycle for the reduction of α,β-unsaturated ketones under these conditions [56]. As [OsHCl(CO)(η^2-CH$_2$=CH-CO-CH$_3$)(P*i*Pr$_3$)$_2$] can be isolated [103], it is assumed that the α,β-unsaturated ketone coordinates to the metal *via* the C=C bond. The preferential coordination of the alkene bond implies it is reduced preferentially, *via* the hypothetical formation of an oxaallyl intermediate by hydrogen migration from the metal to the β-carbon atom. The driving force for the formation of this intermediate is the presumed stability of the π-oxaallyl structure, which is analogous to a π-allyl complex. It has been suggested that a transition metal π-oxaallyl species acts as an intermediate in certain reactions of iron [105], ruthenium [42, 44, 106] and cobalt [107]. Saegusa and coworkers [108] obtained

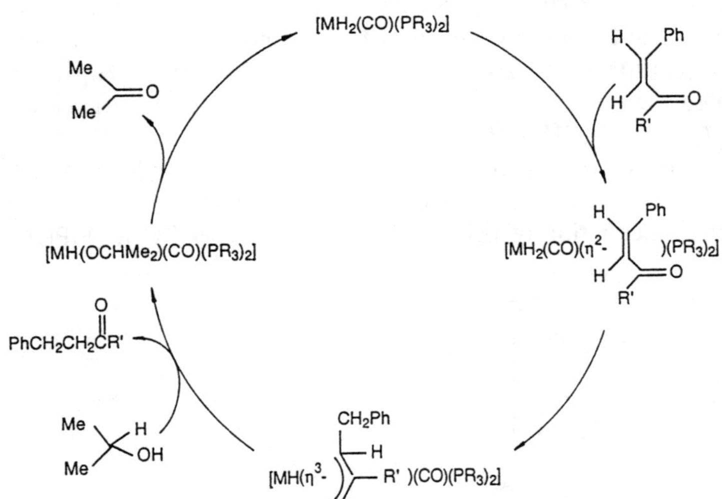

Scheme 3.8.

evidence for a palladium π-oxaallyl intermediate which is believed to undergo alkene insertion. If a β-hydrogen atom is present, β-hydride elimination occurs, leading to α,β-unsaturated carbonyl compounds. More recently, the photochemical formation of some π-oxaallyl molybdenum and tungsten complexes has been described [109].

The electrophilic attack of the 2-propanol proton on the π-oxaallyl ligand could lead to the formation of the saturated ketone and a metal-isopropoxide intermediate, which can give the coordinatively unsaturated metal dihydride by β-hydride elimination with subsequent formation of acetone.

On the other hand, the complexes [MHCl(CO)(P*i*Pr$_3$)$_2$] (M = Ru,Os) are poor catalysts for hydrogen transfer from 2-propanol to styrene in the presence of Na[BH$_4$], and competitive reduction experiments for mixtures of cyclohexanone:styrene (1:1) show preferential reduction of cyclohexanone [54]. Conjugation of the C=C and C=O bonds in α,β-unsaturated ketones may decrease both the coordinating ability of the carbonyl group and the susceptibility to hydride attack on the carbon atom of the carbonyl group compared to saturated carbonyl compounds [110]. This could explain the preferential reduction of α,β-unsaturated ketones by comparison with cyclohexanone:styrene (1:1) mixtures.

The difference in catalytic activity between [MH$_2$(CO)(P*i*Pr$_3$)$_2$] and [MHCl(CO)(P*i*Pr$_3$)$_2$] is probably due to the different prerequisites of the insertion step. In the intermediates of the type [MHCl(CO)(η^2-PhCH=CH-CO-R)(P*i*Pr$_3$)$_2$] (Scheme *3.8*), the intramolecular insertion of the alkene ligand into the M-H bond is not favoured, owing to the *trans*- position of these ligands (Figure *3.1a*) [103]. However, the expected coplanarity of the M(C=C)H system in the proposed [MH$_2$(CO)(η^2-PhCH=CH-CO-R)(P*i*Pr$_3$)$_2$] intermediate (Figure *3.1b*) could lead to easy insertion of the substrate into the M-H bond. The formation of the assumed π-oxaallyl intermediates is likely to involve a prior isomerization of [MHCl(CO)(η^2-PhCH=CH-CO-R)(P*i*Pr$_3$)$_2$].

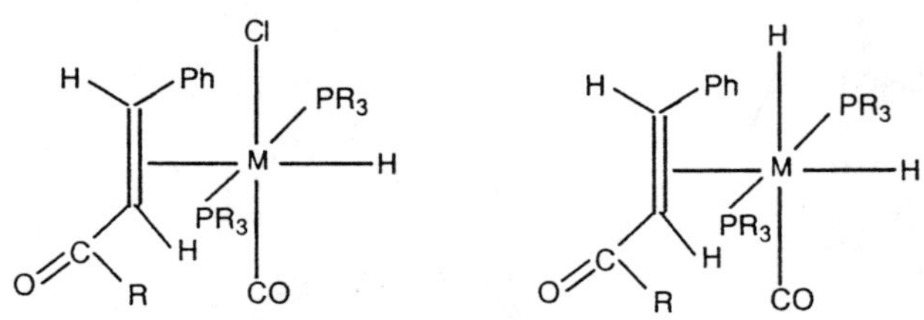

Fig. 3.1. Proposed intermediates [MHX(CO)(η^2-PhCH=CHCOR)(PR$_3$)$_2$]; (a) X=Cl; (b) X=H.

3.3.5. REDUCTION OF DIPHENYLETHYNE CATALYSED BY $[Ru(O_2CCF_3)_2(CO)(PPh_3)_2]$, IN THE PRESENCE OF TRIFLUOROETHANOIC ACID [60]

Diphenylethyne inserts into the Ru-H bond of the $[RuH(O_2CCF_3)(CO)(PPh_3)_2]$ to give the 1,2-diphenylalkenyl derivative $[Ru(PhC{=}CHPh)(O_2CCF_3)(CO)(PPh_3)_2]$. Cleavage of the metal alkenyl linkage with trifluoroethanoic acid under mild conditions gives the corresponding trifluoroethanoate complex $[Ru(O_2CCF_3)_2(CO)(PPh_3)_2]$ and *cis*-stilbene. The product $[Ru(O_2CCF_3)_2(CO)(PPh_3)_2]$, released during the acidolysis of the alkenyl complex, readily dehydrogenates primary alcohols to the corresponding aldehydes, free trifluoroethanoic acid and the hydride $[RuH(O_2CCF_3)(CO)(PPh_3)_2]$, thereby completing a cycle for the catalytic hydrogen transfer from alcohol to internal alkynes (Scheme *3.9*).

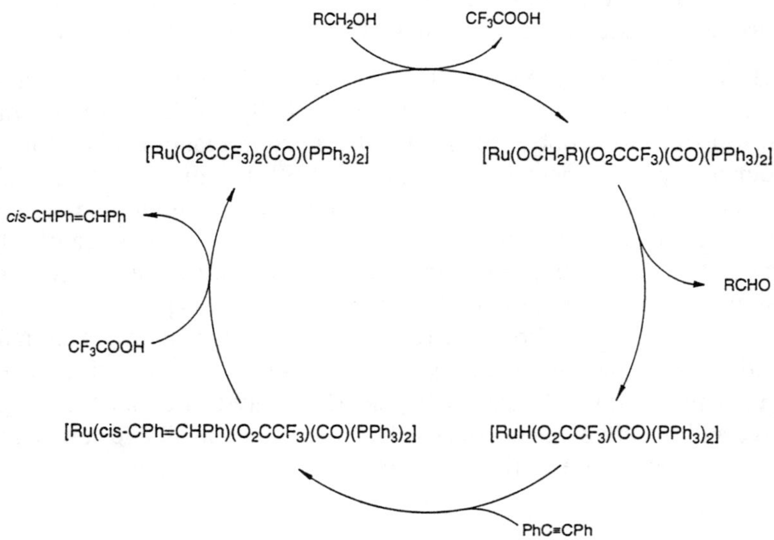

Scheme 3.9.

3.4. Hydrogen transfer and hydrogenation

To summarise, reduction reactions with molecular hydrogen are very effective, but in some cases the use of hydrogen transfer catalysts can produce interesting selectivity behaviour. Thus, the complexes

$[MHCl(CO)(PiPr_3)_2]$ (M = Ru,Os) catalyse the reduction of cyclohexanone:styrene mixtures (1:1). Under hydrogen transfer conditions, the catalysts preferentially reduce cyclohexanone; however, under hydrogenation conditions only the styrene is reduced [54].

Catalytic hydrogen transfer from 2-propanol to 1,5-cyclooctadiene, catalysed by $[Rh(NBD)(6\text{-}Mequin)(PPh_3)]^+$ occurs with a high degree of isomerization to 1,3-cyclooctadiene and 1,4-cyclooctadiene, selective formation of cyclooctene and subsequent appearance of Me_2CO. This complex is also active in the hydrogenation of 1,5-cyclooctadiene but its selectivity for the corresponding monoene is lower than that observed for transfer reactions [111].

In the hydrogenation of α,β-unsaturated ketones to allylic alcohols catalysed by iridium compounds it has been proposed [112] that the selectivity is mainly associated with the steric crowding around iridium metal center rather than the hydride character of the coordinated hydride. The factors determining the selectivity of the enone reduction *via* hydrogen transfer appear to be totally different. The selectivity of $[H_2\overline{Ir\{Ph_2P(o-C_6H_4N(Me)C}H_2)\}\{Ph_2P(o-C_6H_4NMe_2)\}]$, which is quite effective in promoting the formation of allyllic alcohols *via* hydrogen transfer, does not exceed 50% if the same reaction is performed under normal hydrogenation conditions [39]. On the other hand, catalysts such as $[IrH_3(PR_3)_3]$ with high selectivity in hydrogenation are not selective in hydrogen transfer reactions [65, 112]. Recently, the same authors have suggested that selectivity appears to be related to the basicity of the metal in the hydrogen transfer process [39].

It should be clear from the reported results that hydrogen transfer reactions must be viewed as a viable alternative to the classical reduction reactions with molecular hydrogen. The catalysts are sometimes more active and selective, the donors used are commonly available compounds and in general, the experimental technique is very simple.

References

1. G. Brieger and T.J. Nestrick, Chem. Rev., **74**, 567 (1974)
2. A.E. Shilov, *The activation of saturated hydrocarbons by transition metal complexes*, Riedel, Dordrecht, 1984
3. R.H. Crabtree, Chem. Rev., **85**, 245 (1985)
4. R.H. Crabtree, J.M. Mihelcic and J.M. Quirk, J. Am. Chem. Soc., **101**, 7738 (1979)
5. R.H. Crabtree, M.F. Mellea, J.M. Mihelcic and J.M. Quirk, J. Am. Chem. Soc., **104**, 107 (1982)

6. R.H. Crabtree, P.C. Demon, D. Eden, J.M. Mihelcic, C.P. Parnell, J.M. Quirk and G.E. Morris, J. Am. Chem. Soc., **104**, 6994 (1982)
7. R.H. Crabtree, M.F. Mellea and J.M. Quirck, J. Am. Chem. Soc., **106**, 2913 (1984)
8. a) D. Bandry, M. Ephritikine and H. Felkin, J. Chem. Soc., Chem. Commun., 1243 (1980). b) D. Bandry, M. Ephritikine, H. Felkin and R. Holmes-Smith, J. Chem. Soc., Chem. Commun., 788 (1983)
9. H. Felkin, T. Fillebeen-Khan, Y. Gauls, R. Holmes-Smith and J. Zahrzewski, Tetrahedron Lett., 1279 (1984)
10. M.G. Clerici, S. Di Gioacchino, F. Maspero, E. Perrotti and A. Zanobi, J. Organomet. Chem., **84**, 379 (1975)
11. H. Imai, T. Nishiguchi and K. Fukuzumi, J. Org. Chem., **41**, 2688 (1976)
12. T. Nishiguchi, K. Tachi and K. Fukuzumi, J. Org. Chem., **40**, 237 (1975)
13. T. Nishiguchi, K. Tachi and K. Fukuzumi, J. Org. Chem., **40**, 240 (1975)
14. T. Nishiguchi and K. Fukuzumi, J. Organomet. Chem., **80**, C42 (1974)
15. T. Nishiguchi, K. Tanaka and K. Fukuzumi, J. Org. Chem., **43**, 2968 (1978)
16. T. Nishiguchi, K. Tachi and K. Fukuzumi, J. Am. Chem. Soc., **94**, 8616 (1972)
17. T. Nishiguchi and K. Fukuzumi, J. Am. Chem. Soc., **96**, 1893 (1974)
18. C. Masters, A.A. Kiffen and J.P. Visser, J. Am. Chem. Soc., **98**, 1357 (1976)
19. H. Imai, T. Nishiguchi and K. Fukuzumi, J. Org. Chem., **39**, 1622 (1974)
20. H. Imai, T. Nishiguchi and K. Fukuzumi, J. Org. Chem., **41**, 665 (1976)
21. J.A. Osborn, F.H. Jardine, J.F. Young and G. Wilkinson, J. Chem. Soc., A, 1711 (1966)
22. a) M.C. Baird, C.J. Nyman and G. Wilkinson, J. Chem. Soc., A, 348 (1968). b) K. Ohno and J. Tsuji, J. Am. Chem. Soc., **90**, 99 (1968). c) R.E. Harmon, J.L. Parson, D.W. Cooke, S.K. Gupta and J. Shoolenborg, J. Org. Chem., **34**, 3648 (1969)
23. a) I.S. Kolmnikov, V.P. Kukolev and M.E. Vol'pin, Russ. Chem. Rev., Engl. Transl., **43**, 399 (1974). b) F.H.Jardine, Prog. Inorg. Chem., **28**, 64 (1981). c) G. Mestroni, A. Camus and G. Zassinovich; *Aspects of Homogeneous Catalysis*, R. Ugo, Ed., D. Reidel Publishing Company, Boston 1981, Vol 4, p. 71–98. d) B. Heil, L. Marbó and S. Toros, *Homogeneous Catalysis with metal phosphine complex*, L.H. Pignolet Ed., Plenum : New York, 1983, Vol 10, p. 317. e) F.H. Jardine, Prog. Inorg. Chem., **31**, 265 (1984)
24. E. Farnetti, F. Vinzi and G. Mestroni, J. Mol. Catal., **24**, 147 (1984)
25. T. A. Smith and P.M. Maitlis, J. Organomet. Chem., **289**, 385 (1985)
26. E. Malunowicz, S. Tyrlik and Z. Lasocki, J. Organomet. Chem., **72**, 269 (1974)
27. D. Beanpere, L. Nadjo, R. Uzan and P. Bauer, J. Mol. Catal., **18**, 73 (1983) and **20**, 185 (1983)
28. G. Mestroni, G. Zassinovich, E. Alessio and M. Tornatore, J. Mol. Catal., **49**, 175 (1989)
29. a) R. Uson, L. A. Oro, R. Sariego and M.A. Esteruelas, J. Organomet. Chem., **214**, 399 (1981). b) R. Uson, L.A. Oro, M.A. Ciriano and F.J. Lahoz, J. Organomet. Chem., **240**, 429 (1982). c) L.A. Oro, M.P. Lamata and M. Valderrama, Transition Met. Chem., **8**, 48 (1983)
30. R. Sariego, I. Carkovic, M. Martinez and M. Valderrama, J. Mol. Catal., **35**, 161 (1986).
31. R. Sariego, M. Martinez, I. Carkovic, R. Contreras and S.A. Moya, J. Mol. Catal., **51**, 67 (1989)
32. M. Visintin, R. Spogliarich, J. Kaspar and M. Graziani, J. Mol. Catal., **24**, 277 (1984)
33. M. Visintin, R. Spogliarich, J. Kaspar and M. Graziani, J. Mol. Catal., **32**, 349 (1985)

34. M.J.Fernandez, M.A. Esteruelas, M. Covarrubias and L.A. Oro, J. Organomet. Chem., **316**, 343 (1986)
35. A.S. Goldman and J. Halpern, J. Am. Chem. Soc., **109**, 7537 (1987)
36. R. Zanella, F. Canziani, R.Ros and M. Graziani, J. Oragnomet. Chem., **67**, 449 (1974)
37. X. Lu, Y. Lin and D. Ma, Pure, Appl. Chem., **60**, 1299 (1988)
38. E. Farnetti, G. Nardin and M. Graziani, J. Chem. Soc. Chem. Commun., 1264 (1988)
39. E. Farnetti, J. Kaspar and M. Graziani, J. Mol. Catal., **63**, 5 (1990)
40. V.Z. Sharf, L.K. Freidlin and V.N. Krutii, Bull. Acad. Sci. USSR, Div. Chem. Sci., 666 (1973)
41. Y. Sasson and G.L. Rempel, Tetrahedron Lett., 3221 (1974)
42. Y. Sasson and J. Blum, J. Org. Chem., **40**, 1887 (1975)
43. Y. Sasson and G.L. Rempel, Tetrahedron Lett., 4133 (1974)
44. M. Dedieu and Y.L. Pascal, J. Mol. Catal., **9**, 71 (1980)
45. S.L. Regen and G. M. Whitesides, J. Org. Chem., **37**, 1832 (1972)
46. Y. Sasson, P. Albin and J. Blum, Tetrahedron Lett., 833 (1974)
47. L.K. Freidlin, V.Z. Sharf, V.N. Krutii and S.I. Shcherbakova, J. Org. Chem. USSR, **8**, 986 (1972)
48. C. O'Connor, J.D. Gilbert and G. Wilkinson, J. Chem. Soc. A, 84 (1969)
49. R. Graser and H. Steigerwald, J. Organomet. Chem., **193**, C67 (1980)
50. S.M. Pillai, S. Vancheesan, J. Rajaram and J.C. Kuriacose, Indian J. Chem. A, **19**, 1074 (1980)
51. B.L. Shaw and R.E. Stainbank, J. Chem. Soc. Dalton Trans., 2108 (1972)
52. H.D. Empsall, E.M. Hyde, E. Mentzer, B.L. Shaw and M.F. Uttley, J. Chem. Soc., Dalton Trans., 2069 (1976)
53. G. G. Hlatky and R.H. Crabtree, Coord. Chem. Rev., **65**, 1 (1985)
54. M.A. Esteruelas, E. Sola, L.A. Oro, H. Werner and U. Meyer, J. Mol. Catal., **45**, 1 (1988)
55. H. Werner, M.A. Esteruelas, U. Meyer and B. Wrackmeyer, Chem. Ber., **120**, 11 (1987)
56. M.A. Esteruelas, E. Sola, L.A. Oro, H. Werner and U. Meyer, J. Mol. Catal., **53**, 43 (1989)
57. H. Werner, U. Meyer, M.A. Esteruelas, E. Sola and L. A. Oro, J. Organomet. Chem., **366**, 187 (1989)
58. M. Aracama, M.A. Esteruelas, F.J. Lahoz, J.A. Lopez, U.Meyer, L.A. Oro and H. Werner, Inorg. Chem., **30**, 288 (1991)
59. A. Dobson, D.S. Moore, S.D. Robinson, M.B. Hursthose and L. New, J.Organomet. Chem., **177**, C8 (1979)
60. B.Graser and H. Steigerwald, J.Organomet. Chem., **193**, C67 (1980)
61. G. Speier and L. Marko, J. Organomet. Chem., **210**, 253 (1981)
62. Y. Lin and Y. Zhou, J. Organomet. Chem., **381**, 135 (1990)
63. R. Spogliarich, G. Zassinovich, J. Kaspar and M. Graziani, J. Mol. Catal., **16**, 359 (1982)
64. R. Spogliarich, E. Farnetti, J. Kaspar, M. Graziani and E. Cesaroti, J. Mol. Catal., **50**, 19 (1989)
65. R. Spogliarich, J. Kaspar, M. Graziani and F. Morandini, J. Organomet. Chem., **306**, 407 (1986)
66. P. Kvintovics, J. Bakos and B. Heil, J. Mol. Catal., **32**, 111 (1985)

67. H.W. Kranse and A.K. Bhatnagar, J. Organomet. Chem., **302**, 265 (1986)
68. R. Chauvin, J. Mol. Catal., **62**, 147 (1990)
69. M. Bianchi, U. Matteoli, P. Frediani, G. Menchi and F. Piacenti, J. Organomet. Chem., **236**, 375 (1982)
70. G. Zassinovich, C. del Bianco and G. Mestroni, J. Organomet. Chem., **222**, 323 (1981)
71. S. De Martín, G. Zassinovich and G. Mestroni, Inorg. Chim. Acta, **174**, 9 (1990)
72. G. Zassinovich and F. Grisoni, J. Organomet, Chem., **247**, C24 (1983)
73. G. Zassinovich, R. Bettella, G. Mestroni, N. Bresciani-Pahor, S. Geremia and L. Randaccio, J. Organomet. Chem., **370**, 187 (1989)
74. G.Zassinovich and G. Mestroni, J. Mol. Catal., **42**, 81 (1987)
75. P. Kvintovics, B.R. James and B. Heil, J. Chem. Soc., Chem. Commun., 1810 (1986)
76. P. Kvintovics and B. Heil, J. Organomet. Chem., **361**, 117 (1989)
77. C. Botteghi, G. Chelucci, G. Chessa, G. Delogu, S. Gladiali and F. Soccolini, J. Organomet. Chem., **304**, 217 (1986)
78. a) A. M. Lopez, Ph.D. Thesis, University of Zaragoza, 1991. b) M.A. Esteruelas, M.P. García, A.M. L6pez and L.A. Oro, Organometallics, **11**, 702 (1992)
79. M.P.Garcia, A.M. Lopez, M.A. Esteruelas, F. J. Lahoz and L.A. Oro, J. Organomet. Chem., **388**, 365 (1990)
80. L.A. Oro, D. Carmona, J. Reyes, C. Foces-Foces and F. H. Cano, J. Chem. Soc., Dalton Trans., 31 (1986)
81. a) R. Uson, L.A. Oro, D. Carmona and M.A. Esteruelas, J. Organomet. Chem., **263**, 109 (1984) b) L.A. Oro, D. Carmona, M.A. Esteruelas, C. Foces-Foces and F. H. Cano, J. Organomet. Chem., **307**, 83 (1986)
82. M. P. Garcia, A. Portilla, L.A. Oro, C. Foces-Foces and F. H. Cano, J. Organomet. Chem., **322**, 111 (1987)
83. M. P. Garcia, J.L. Millan, M.A. Esteruelas and L.A. Oro, Polyhedron, **6**, 1427 (1987)
84. Y. Blum and Y. Shvo, J. Organomet. Chem., **263**, 93 (1984)
85. A. Basu, S. Bhaduri and K. Sharma, J. Organomet. Chem., **319**, 407 (1987)
86. M. Bianchi, U. Maffeoli, P. Freniani, G. Menchi and F. Piacenti, J. Organomet. Chem., **240**, 59 (1982) and **236**, 375 (1982)
87. M. Bianchi, U. Maffeoli, G. Menci, P. Frediani, S. Pratesi, F. Piacenti and C. Botteghi, J. Organomet. Chem., **198**, 73 (1980)
88. S. Bhaduri and K. Sharma, J. Chem. Soc., Chem. Commun., 173 (1988)
89. K. Jothimony and S. Vancheesan, J. Mol. Catal., **52**, 301 (1989)
90. B. Denise and G. Pannetier, J. Organomet. Chem., **161**, 171 (1978)
91. R. Uson, L.A. Oro, M.A.Garralda, M.C. Claver and P. Lahuerta, Transition Met. Chem., **4**, 55 (1979)
92. R. Uson, L.A. Oro, J. Artigas and R. Sariego, J. Organomet. Chem., **179**, 65 (1979)
93. R. Uson, L.A. Oro, D. Carmona and M.A. Esteruelas, Inorg. Chim. Acta, **73**, 275 (1983)
94. A. Camus, G. Mestroni and G. Zassinovich, J. Mol. Catal., **6**, 231 (1979)
95. A. Camus, G. Mestroni and G. Zassinovich, J. Organomet. Chem., **184**, C10 (1980)
96. G. Mestroni, G. Zassinovich, A. Camus and F. Martinelli, J. Organomet. Chem., **198**, 87 (1980)
97. R. Uson, L.A. Oro, D. Carmona, M.A. Esteruelas, C. Foces-Foces, F. H. Cano and S. Garcia-Blanco, J. Organomet. Chem., **254**, 249 (1983)

98. M. Covarrubias, Ph. D. Thesis, University of Zaragoza, 1988
99. M.J. Fernandez, M.A. Esteruelas, M. Covarrubias, L.A. Oro, M.C. Apreda, C. Foces-Foces and F. H. Cano, Organometallics, **8**, 1158 (1989)
100. M.A. Esteruelas, O. Lahuerta and L. A. Oro, manuscript in preparation
101. a) S. E. Kegley, C. J. Schaverien, J. H. Freudenberger and R.G. Bergman, J. Am. Chem. Soc., **109**, 6563 (1987). b) D. Braga, P. Sabatino, C. Di Bugno, P. Leoni and M. Pascuali, J. Organomet. Chem., **334**, C46 (1987). c) Y. Kim, K. Osakada, A. Takenada and A. Yamamoto, J. Am. Chem. Soc., **112**, 1096 (1990)
102. M.A. Esteruelas, C. Valero, L.A. Oro, U. Meyer and H. Werner, Inorg. Chem., **30**, 1159 (1991)
103. M.A. Esteruelas and H. Werner, J. Organomet. Chem., **303**, 221 (1986)
104. D. M. Hoffman, D. Lappas and D.A. Wierda, J. Am. Chem. Soc., **111**, 1531 (1989)
105. H. Alper and E. C. H. Keung, J. Org. Chem., **37**, 2566 (1972)
106. R. H. Prince and K.A. Raspin, J. Chem. Soc. (A), 612 (1969)
107. R.W. Goetz and M. Orchin, J. Am. Chem. Soc., **85**, 2782 (1963)
108. a) Y.Ito, T. Hirao and T. Saegusa, J. Org. Chem., **43**, 1011 (1978). b) Y. Ito, H. Aoyama, T. Hirao, A. Mochizuki and T. Saegusa, J. Am. Chem. Soc., **101**, 494 (1979)
109. a) E. R. Burkhardt, J. J. Doney, R. G. Bergman and C. H. Heathcock, J. Am. Chem. Soc., **109**, 2022 (1987). b) E. R. Burkhardt, J. J. Doney, G.A. Slough, J.M. Stack, C.H. Heathcock and R. G. Bergman, Pure Appl. Chem., **60**, 1 (1988)
110. H. Fujitsu, E. Matsumura, K. Takeshita and I. Mochita, J. Chem. Soc., Perkin Trans., **1**, 2650 (1981)
111. R. Uson, L.A. Oro and M.A. Esteruelas, Transition. Met. Chem., **7**, 242 (1982)
112. E. Farnetti, J. Kaspar, R. Spogliarich and M. Graziani, J. Chem. Soc., Dalton Trans., 947 (1988)

CHAPTER 4

HOMOGENEOUS HYDROGENATION IN ORGANIC SYNTHESIS

4.1. Introduction

Homogeneous hydrogenation reactions have been used widely in organic synthesis, both of the industrial and laboratory scale. [1–5]. Early workers used the reactions in the most obvious sense, to remove unwanted carbon-carbon double bonds, but recent years have seen much more interesting developments, particularly in the areas of diastereoselective and enantioselective reductions, and in stereoselective reduction of carbonyl groups. In research applications the platinum group metals have provided most of the catalysts, but in considering larger scale applications of the reactions it will be necessary in future to be aware of cost, particularly in view of the very large fluctuations in the price of rhodium over recent years.

4.2. Reduction of simple alkenes

The range of complexes which catalyse the reduction of simple, unfunctionalised alkenes is very large. The most important from a synthetic point of view are those from which we can expect good rates under mild conditions, which show good chemoselectivity and are relatively free of side reactions. Wilkinson's catalyst remains the most widely used in this field, and has seen wide applications, perhaps because it is simple to synthesise, and commercially available [6].

The rates of reduction of alkenes in the presence of Wilkinson's catalyst are largely dependent on steric factors, and addition of hydrogen is from the less hindered face of the double bond. Reactions (4.1) and (4.2) show two examples in which the least sterically hindered double bond in a molecule was reduced. Reaction (4.1) was used in the synthesis of **R**-cryptone [7], and reaction (4.2) in the preparation of cyathins [8].

P.A. Chaloner, M.A. Esteruelas, F. Joó and L.A. Oro, Homogeneous Hydrogenation 119–181.

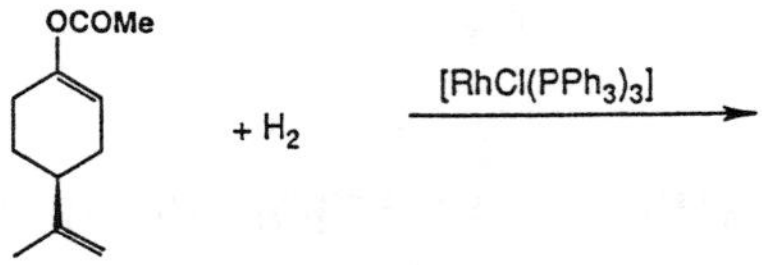

(4.1)

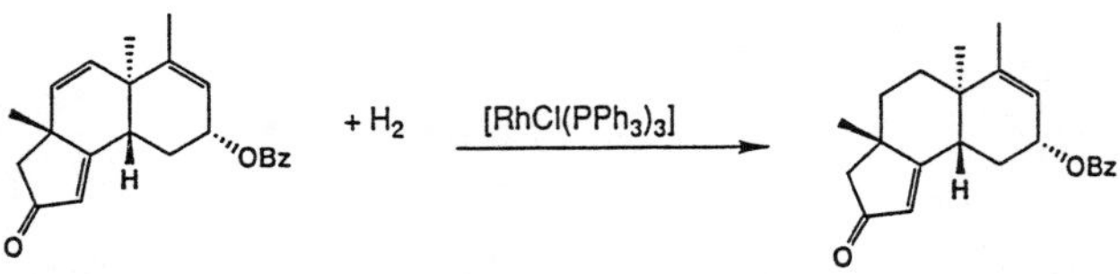

(4.2)

Moreover, additions are rigourously *cis*. The mechanism of the reaction involves an alkylrhodium hydride which is short-lived, so that β-hydride elimination reactions are discouraged, and most additions are site specific with little isomerisation of the staring material. This has been usefully applied in some deuteriations and tritiations. The tritiation of **4-1** was both site and stereospecific [9]. When the substrates are simple alkenes, any small amount of isomerised product can usually be reduced, and thus has no synthetic consequences. Arene, carbonyl, hydroxy, halo, azo, ether, ester, and carboxy groups are tolerated. Although thiophenes inhibit the reaction at atmospheric pressure, thiophene containing alkenes have been reduced in the presence of Wilkinson's catalyst at pressures of 3–4 atmospheres. Aldehydes and allyl alcohols are often decarbonylated with deactivation of the catalyst.

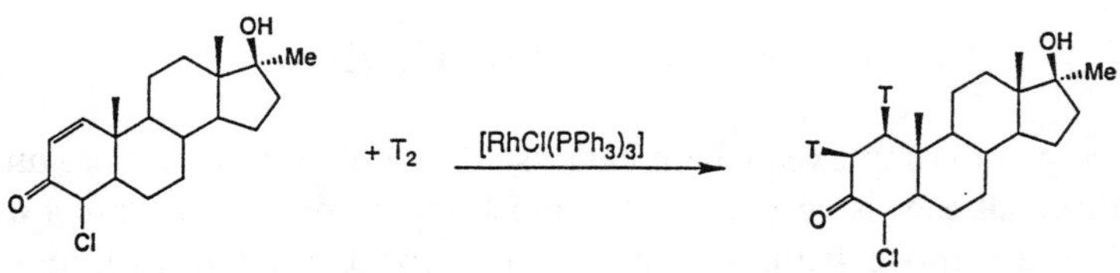

(4-1)

The main disdavantage of the use of Wilkinson's catalyst (and related rhodium complexes), as the above examples demonstrate, is that tri- and tetrasubstituted alkenes are not in general reduced. Thus 1,2-dimethyl-1, 4-cyclohexadiene is reduced to 1,2-dimethylcyclohexene [10]. This problem can be largely overcome by the use of iridium complexes, such as $[Ir(COD)(PCy_3)(py)]^+$, prepared by Crabtree [11]. Using this and related catalysts, 1-methylcyclohexene was readily reduced, and even tetrasubstituted alkenes were slowly hydrogenated. Reaction (4.3) shows an example which was used in the synthesis of the avian toxin talaromycin B.

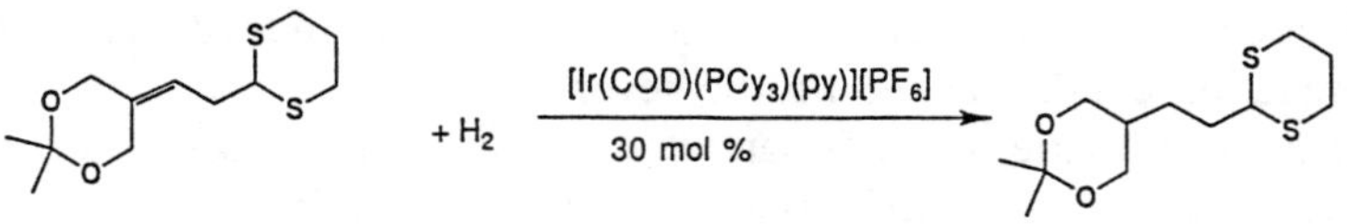

(4.3)

It is perhaps worthy of note that $[RuH_2(PPh_3)_4]$ and $[RuH_4(PPh_3)_3]$ are selective catalysts for the reduction of 1-alkenes only, although the reaction has not found wide applications in synthesis.

4.3. Reduction of functionalised alkenes

This section will consider only rather simple versions of this reaction, with examples involving stereoselectivity being considered later. In general conjugated alkenes are reduced more readily than unconjugated ones. Thus using Wilkinson's catalyst, styrene is reduced 2–6 time more rapidly than 1-hexene, and $[Co(CN)_5]^{3-}$ is active as a catalyst only for the reduction of conjugated alkenes.

It is usually easier to reduce carbon-carbon rather than carbon-oxygen double bonds, so the reduction of enals to saturated aldehydes might be expected to be straightforward. However, using Wilkinsons's catalyst, the main reaction is decarbonylation. Increased hydrogen pressure enhances hydrogenation at the expense of decarbonylation, but also promotes reduction of the carbonyl group of the saturated aldehyde. The complex $[Rh(bipy)(solvent)_2]^+$ is reported to give saturated aldehydes with moderate selectivity, and using water gas as the reductant and $[Rh_6(CO)_{16}]$ as the catalyst allowed reduction of 3-phenylpropenal to 3-phenylpropanal with 100% selectivity [12]. The biphasic system prepared *in situ* from $[\{RhCl(COD)\}_2]$ and TPPTS gave excellent selectivity for saturated aldehydes, without decarbonylation [13].

$[Co(CN)_5]^{3-}$ is known for its specificity towards conjugated double bonds, but it is not an especially efficient catalyst for the reduction of unsaturated aldehydes; for the reduction of citral, **4-2**, $[Co(CN)_3(en)]^-$ proved to be the best catalyst, giving **4-3** with good selectivity [14]. The use of $[Co_2(CO)_8]$, in the presence of CO, as the catalyst, might

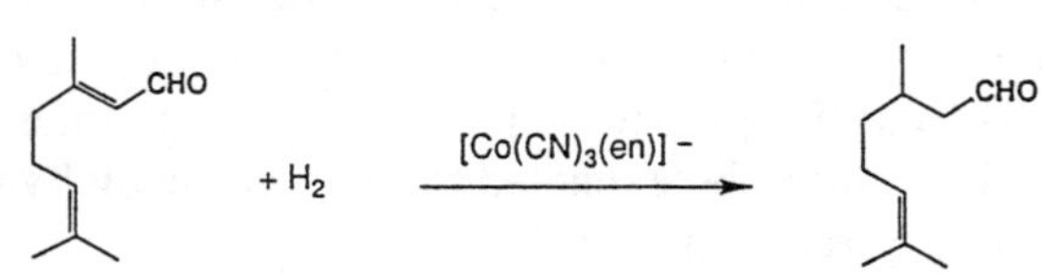

(4-2) (4-3)

be expected to supress decarbonylation, and this is indeed the case, though reaction conditions tend to be severe.

Moderate selectivities towards saturated aldehydes have been observed using catalysts such as $[RuHCl(CO)(PPh_3)_3]$, but allylic alcohols and saturated alcohols were generally also produced [15].

α,β-Unsaturated ketones offer no ready pathway for decarbonylation, and the range of catalysts available for their hydrogenation is consequently much greater. Thus, simple enones may be reduced to ketones in the presence of $[Rh(diene)L_2]^+$, $[Co(CN)_5H]^{3-}$, $[Co_2(CO)_8]$, $[PtCl_2(PPh_3)_2/SnCl_2$ or $[\{CuH(PPh_3)\}_6]$. As previously noted, many cobalt complexes catalyse only the reduction of conjugated double bonds; an example is shown in reaction (4.4) [16]. Some catalysts prepared *in situ* from $[\{Ir(COD)(OMe)\}_2]$ and phosphine have been used in hydrogenation of PhCH=CHCOMe. The outcome of the reaction depends on the nature and the amount of the added phosphine. When $PMePh_2$ was added in a 10-fold excess, the catalytically active species was thought to be $[IrH_3(PMePh_2)_3]$, and the reduction was entirely at the carbonyl group to give PhCH=CHCH(OH)Me. With smaller amounts of added phosphine the catalytic species was $[IrH_5(PR_3)_2]$ and reduction of the carbon-carbon double bond predominated [17].

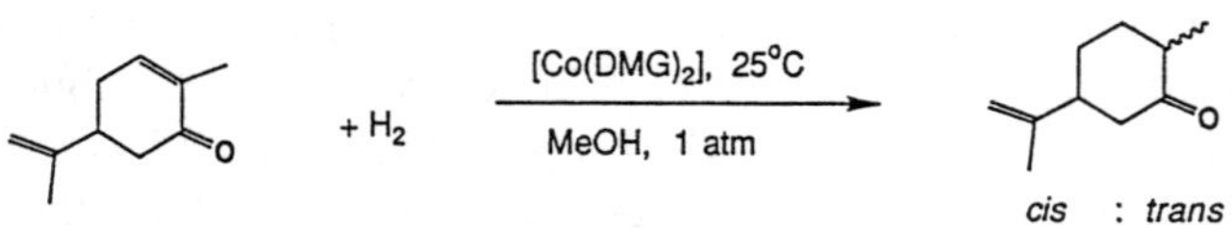

(4.4)

cis : *trans* = 3 : 7

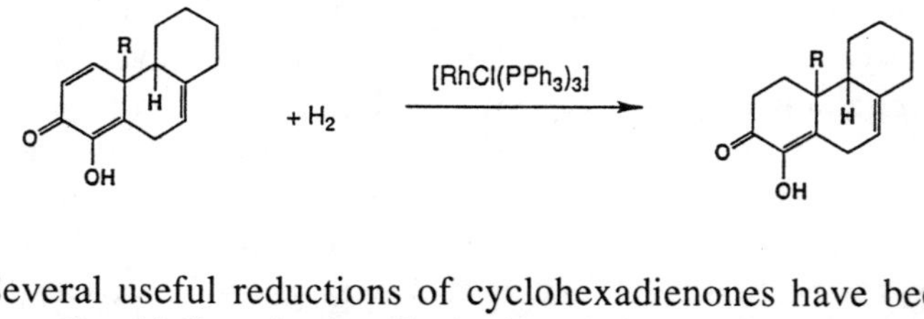

(4.5)

Several useful reductions of cyclohexadienones have been reported. In reaction (4.5) only the disubstituted double bond was reduced, the product being used in a synthesis of the podocarpane skeleton [18]. The reduction of (–)-artemisin, **4-4**, gave **4-5**, which was converted to yomogin and 1-deoxyivangustin [19]. The reaction is tolerant of functional groups which might be expected to undergo ready hydrogenolyis; thus **4-6** was cleanly reduced to **4-7** [20].

Selectivities in enone reduction have found wide application in steroid

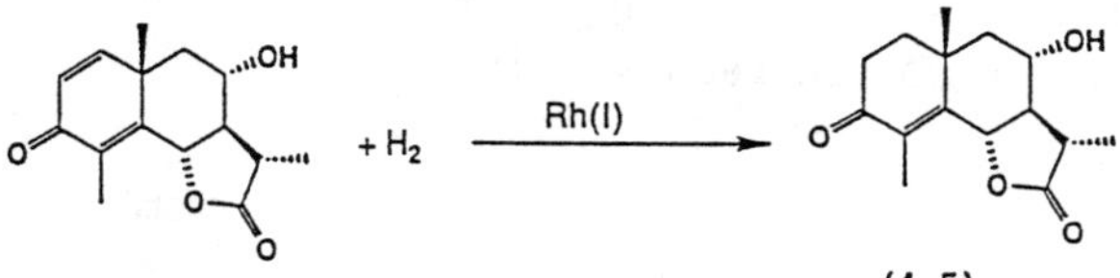

(4-4) (4-5)

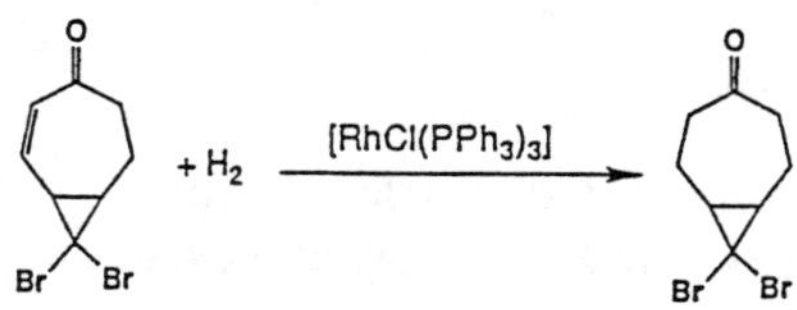

(4-6) (4-7)

chemistry. For example, androsta-1,4-diene-3,17-dione, **4-8**, was reduced to **4-9** in the presence of [RhCl(PPh$_3$)$_3$] or [RuClH(PPh$_3$)$_3$] and its related deuteriation occurred stereospecifically from the bottom face [21]. Using the more reactive [Ir(COD)(PCy$_3$)(py)][PF$_6$] as the catalyst, both double bonds of **4-8** could be reduced to give **4-10**; with this catalyst dienones reacted more rapidly than enones [22].

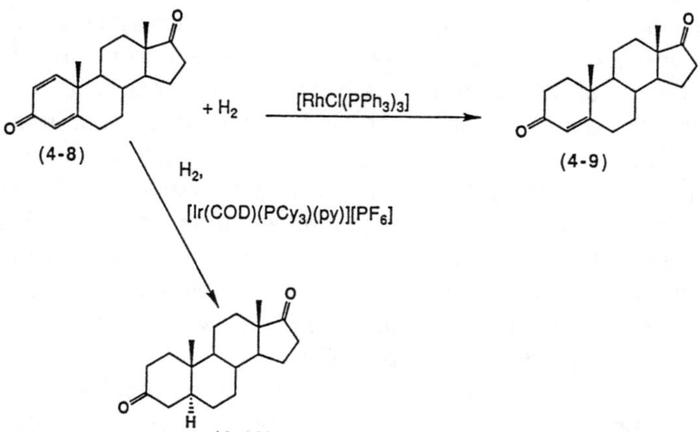

(4-10)

α,β-Unsaturated acids and esters are readily reduced using a variety of catalysts, but genuine synthetic applications, rather than demonstrations of reactivity, have been few. Deuteriation of maleic and fumaric acid derivatives has been used to prove rigorous *cis*-addition of hydrogen. Reduction of PhCOOC(Me)=CHCOOMe with D_2 in the presence of [RhCl(PPh$_3$)$_3$] gave the **R,R**- and **S,S**-products from the **Z**-isomer and the **R,S**- and **S,R** products from the *E*-isomer with > 99.8% selectivity

[23]. Reduction of **4-11** is interesting in that the thiophene does not inhibit the reaction or poison the catalyst [24].

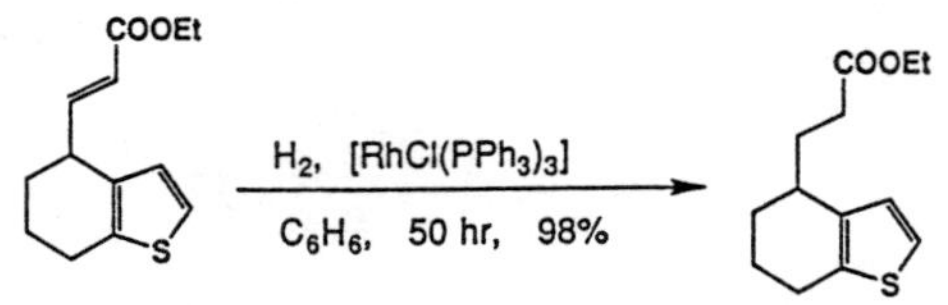

(4-11)

Reduction of unsaturated nitriles or nitro compounds is similar; reduction generally occurs exclusively at the carbon-carbon double bonds, but applications have been limited.

4.4. Reduction of dienes

There exist relatively few reports of homogeneous hydrogenation of allenes, but there is no reason to suppose that it is difficult; the usual product is a *cis*-alkene, and the less substituted double bond reacts proferentially [25].

In reactions of non-conjugated dienes, catalysts may be divided into two groups, those capable of catalysing isomerisation to give a conjugated diene before or during reduction, and those which are not isomerisation catalysts. One important commercial reaction in this class is the selective reduction of cyclododecatriene (produced by butadiene cyclotrimerisation) to cyclododecene. On ozonolysis this yields the long chain diacid, an intermediate is the production of high quality polyamides. The ability to act as an isomerisation catalyst is important in this case, since the double bonds are not initially conjugated. Among the catalysts most selective for the monoene were $[Co_2(CO)_6(PR_3)_2]$, $[RuCl_2(CO)_2(PPh_3)_3]/PPh_3$ and $[NiI_2(PPh_3)_2]$.

1,4-Cyclohexadienes are produced by the Birch reduction of arenes. The most useful catalysts for reduction of the less hindered double bond are those where isomerisation is not a competing reaction, as shown in reactions (4.6) and (4.7) [26,27].

The selective reduction of unsaturated fats is an important commercial problem, involving reactions of both saturated and unsaturated dienes. Soybean and linseed oils consist of mixtures of mono- di- and triene carboyxlate esters. Linolenates (9,12,15-octadecatrienoates) have an unpleasant flavour, and it is desireable to reduce them to linoleates (*cis,cis*-9,12-octadecadienoates) for food use. Control of the propor-

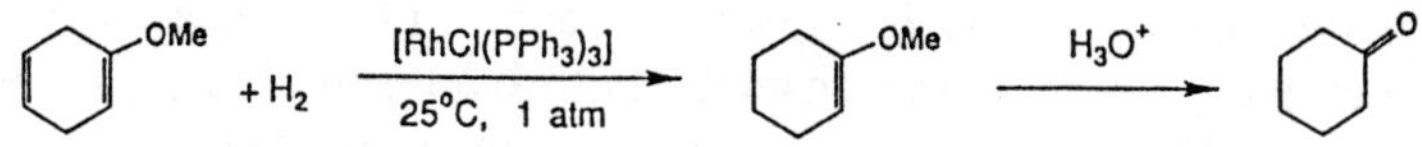

(4.6)

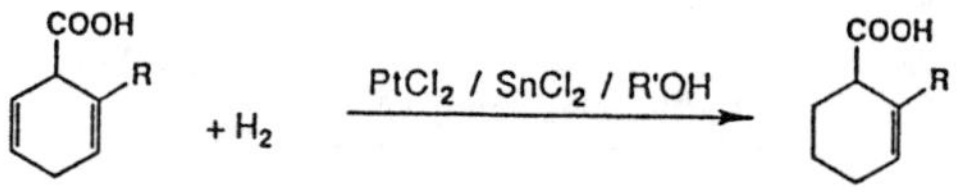

(4.7)

tions of saturated and unsaturated fats may also be desireable on health grounds. The standard model for fat reduction is methyl sorbate (methyl hexadienoate). This may be reduced to the 2- ($[Co(CN)_5]^{3-}$, 1 atm H_2, or $NiCl_2/Na[BH_4]$), 3- ($[Cr(arene)(CO)_3]$, which gives mainly *cis*-product) or 4-enoate ($[Co(CN)_5]^{3-}$, 50 atm H_2), with reasonable selectivity [28–30]. With more complex fats, selectivities tend to be lower. Some catalysts are selective for conjugated dienes, and some are capable of causing isomerisation. Recent years have seen few advances, and research into homogeneous catalysts for this reaction currently seems to be at a low level.

The aims of studies on studies on conjugated dienes are usually to generate a monoene either by 1,4-addition of hydrogen, or by selective reduction of one or other of the double bonds. Arene chromium tricarbonyl complexes are selective catalysts for the reduction of 1,3-dienes by 1,4-hydrogen addition. They are also relatively selective for *trans,trans*-dienes which can attain a cisoid conformation at chromium prior to reduction. The usual product is the *cis*-monoene. Non-conjugated double bonds are not reduced, and a range of functional groups are tolerated. Reaction (4.8) was the final step in a synthesis of a juvenile hormone [31] and reaction (4.9) was used in a carbacyclin synthesis [32].

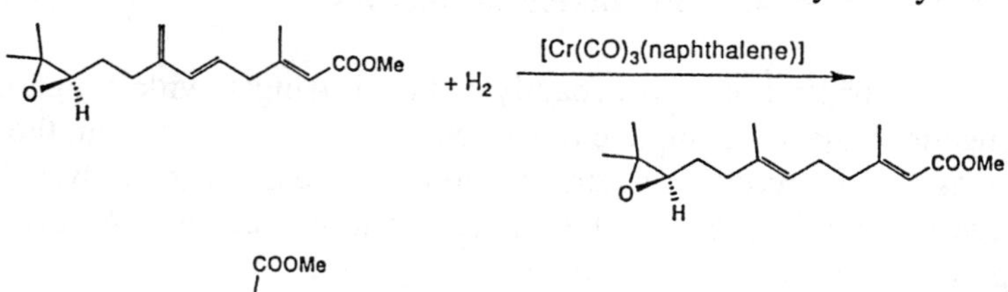

(4.8)

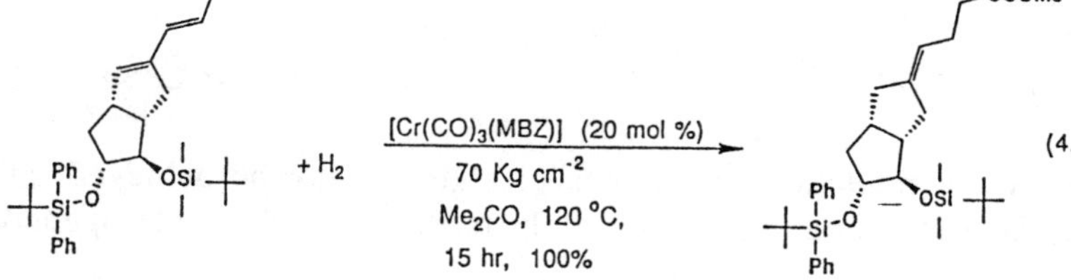

(4.9)

The other type of selectivity is shown in reaction (4.10) of a isocarbacyclin precursor, though in this instance both overreduction (16%) and 1,4-reduction (24%) were serious competing reactions [33].

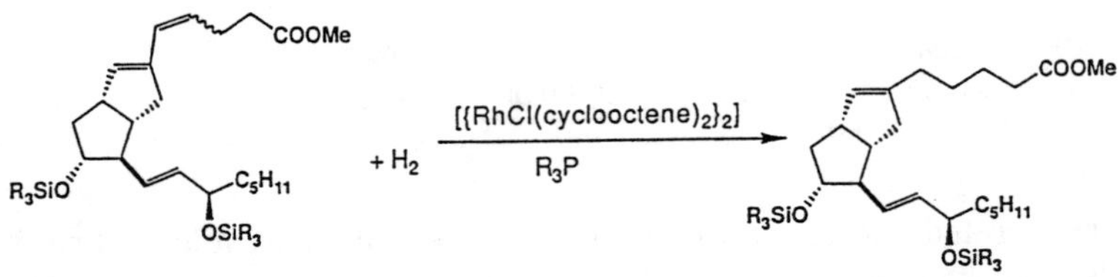

(4.10)

Using cobalt complexes as catalysts, reduction of dienes and dienones by either 1,2- or 1,4-addition of hydrogen can be accomplished by an appropriate choice of reaction conditions. Thus, with $[Co(CN)_5]^{3-}$, butadiene yields mainly 1-butene for $[CN]^-$:Co > 5 and *trans*-2-butene for $[CN]^-$:Co < 5, the reaction mechanism involving partition between a σ-butenyl and a π-butenyl cobalt intermediate [34]. The complex $[Co_2(CO)_6(PBu_3)_2]$ was used to catalyse terpene reduction (reaction (4.11)), but isomerisation was a major competing process, and selectivity rather limited [35].

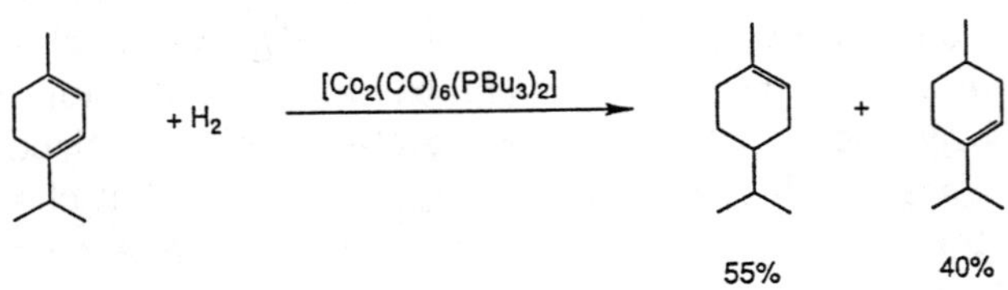

(4.11)

4.5. Reduction of alkynes

Carbon-carbon triple bonds are readily reduced using a wide range of homogeneous catalysts. Complete reduction to alkanes is facile, but there are few reasons for preferring a homogeneous to a heterogeneous system. One interesting application involved a deuteriation (reaction (4.12)) to give a product useful for a biosynthetic study [36].

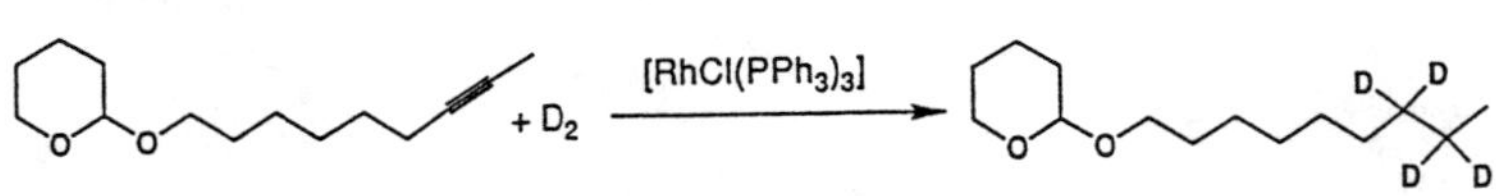

(4.12)

Selective reduction of alkynes to alkenes may be achieved quite readily, using homogeneous catalysts, though the heterogeneous

Lindlaar catalyst remains the most popular for practical applications. However, PhC≡CCOOEt is reduced to give mainly **Z**-PhCH=CHCOOEt in the presence of $[Rh(NBD)(PPhMe_2)_3]^+$; the Lindlaar catalyst gave mainly $PhCH_2CH_2COOEt$ with this substrate. This proved to be a generally useful system for internal alkynes, but was deactivated by 1-alkynes [37]. 1-Alkynes may be selectively reduced to 1-alkenes in the presence of $[Cr(arene)(CO)_3]$, $PdCl_2$/diamine/$Na[BH_4]$, $PdCl_2$/DMF, $[RuH(DPPB)_2][PF_6]$, and many related species. One interesting example of the functional groups tolerated in shown in reaction (4.13); there was no racemisation of the chiral sulfoxide [38].

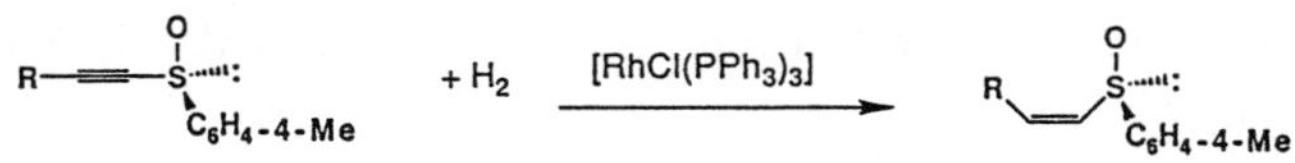

(4.13)

4.6. Reduction of arenes

The reduction of arenes generally requires rather severe conditions, and heterogeneous catalysts operating at elevated temperatures and pressures have dominated the field. The design of the early homogeneous catalysts was more concerned with activity than selectivity, but advances in the understanding of metal arene chemistry have led to some improvements. Reduction of benzene may be accomplished in the presence of $[Co(\eta^3\text{-}C_3H_5)\{P(OMe)_3\}_3]$ at one atm. hydrogen pressure and 25°C. Cyclohexane is the sole product, and cyclohexene and cycohexadiene are not observed during the reaction. As well as operating under very mild conditions, the catalyst has the advantage of essentially complete stereoselectivity. Deuteriation of benzene yielded all *cis*-$C_6H_6D_6$ without H/D exchange. *cis*-Dimethylcyclohexanes are produced from xylene, and **4-12** from anthracene. The use of the catalyst is limited by the fact that it is destroyed by protic substituents, and reaction rates are reduced by substitution. Ruthenium complexes such as **4-13** are quite effective catalysts, but cyclohexenes are a major by-product, and stereoselection

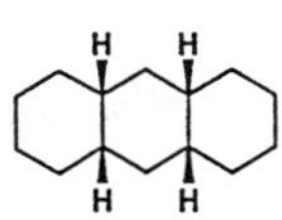

(4-12)

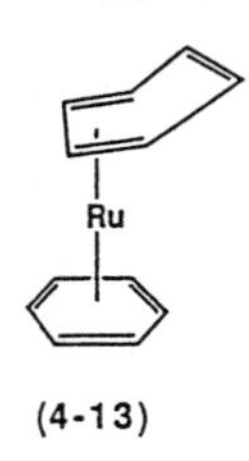

(4-13)

is incomplete. Recently the niobium complex, $[Nb(OC_6H_3\text{-}2,6\text{-}Me_2)(CH_2C_6H_4\text{-}4\text{-}Me)_3]$ was used to catalyse reduction of PPh_3 to PCy_3; no cyclohexenyl or cyclohexadienyl intermediates were detected, but the exact nature of the catalyst system is unknown [39].

More useful applications have been noted in the reduction of polycyclic aromatic compounds. $[Co_2(CO)_8]$ catalysed reaction (4.14) and a number of related processes, but the conditions are rather severe [40]. Anthracene was reduced to **4-14** with unexpectedly high selectivity in the presence of $[Rh(anthracene)(DPPE)]^+$ [41] or $K[RuH_2(PPh_3)_2(Ph_2PC_6H_4)]$. With the latter catalyst isolated arene rings were not reduced [42]. Very selective reduction of polycyclic aromatic compounds has been reported to occur in the presence of $RhCl_3$/Aliquat 336 [43].

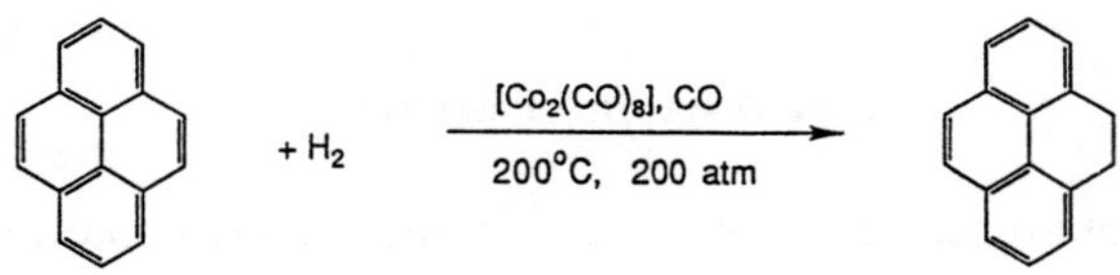

(4.14)

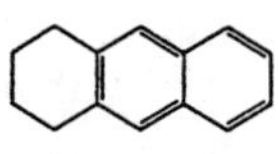

(4-14)

Heterocyclic rings are much more readily reduced that carbocycles, and this too has led to a number of usefully selective reductions using a range of catalysts. Reaction (4.15) shows an example involving quinoline; a careful deuterium labelling study indicated that the reduction of the carbon-nitrogen double bond was reversible [44]. Other catalysts which accomplish essentially the same reaction include $[RhCl(PPh_3)_3]$, $[RuClH(PPh_3)_3]$ $[Fe(CO)_5]/K[OH]$, or $[Rh_6(CO)_{16}]$ with water gas. Benzothiophenes were selectively reduced in the heterocyclic ring in the presence of $[RuCl_2(PPh_3)_3]$, $[RhCl(PPh_3)_3]$ or $[Ir(COD)(PPh_3)_2][PF_6]$.

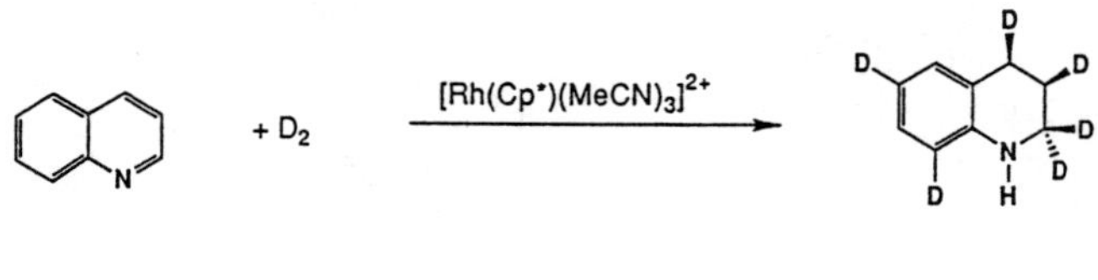

(4.15)

4.7. Reduction of carbonyl groups

Catalytic addition of molecular hydrogen is rarely the method of choice for carbonyl reduction in small-scale synthetic sequences. Alternatives are provided by hydrosilylation, hydride reduction, and transfer hydrogenation. In some case the product alcohol is a poison for the catalyst. A few special cases are, however, useful.

Aldehydes may be reduced to primary alcohols at 1 atm and 25°C in the presence of $[Rh(diene)(PR_3)_2][ClO_4]$ where R_3P is a basic phosphine. The lifetime of the catalyst is short, with irreversible deactivation due to decarbonylation. Slightly more forcing consitions are necessary using $[RuClH(PPh_3)_3]$ as the catalyst, but decarbonylation is less serious, and does not inactivate the catalyst to further reaction. Using this system glucose was reduced to **D**-glucitol, without decarbonylation [45]. $[Et_4N][Cr_2(CO)_{10}(\mu\text{-}H)]$ has been used as a catalyst for reduction of aldehydes (50 atm, 100°C) [46].

Reduction of α,β-unsaturated aldehydes to give allylic alcohols has been reported (reaction (4.16)) but for practical applications hydrosilylation or hydride reduction are generally superior [47]. One interesting application involves the use of $RuCl_3$/TPPTS as catalyst; selectivity for carbonyl reduction was 96% [13]. Selective transfer hydrogenation of aldehydes has been accomplished using $[RuCl_2(TPPMS)_2]$ [48] or $[RhCl(PTA)_3]$ [49] as catalysts, and [HCOO]Na as the hydrogen source.

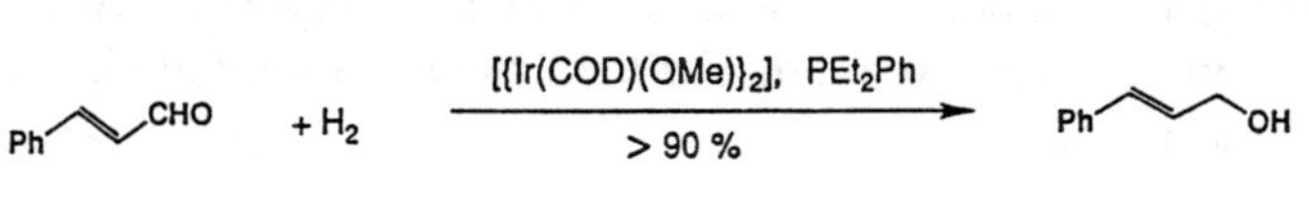

(4.16)

Homogeneous bydrogenation of ketones may be achieved using $[Rh(diene)(PR_3)_2]^+$ as the catalyst, and the reaction mechanism is reasonably well understood. Reduction of 4-*tert*-butylcyclohexanone yields two diastereomeric products; their proportions vary widely, depending on whether the ketone becomes coordinated to the catalyst as an *n*-donor or a π-donor [50].

More rapid reductions, under milder conditions, have been achieved by transfer hydrogenation, using catalysts such as $[Ir(C_2H_4)_2Cl(3,4,7,8\text{-}Me_4phen)]$; reduction of 4-*tert*-butylcyclohexanone using this system gave the *trans*-product with 97% selectivity, and the catalyst was long-lived [51]. In the transfer hydrogenation of 5α-androstane-3,17-dione, **4-15**, by 2-propanol in the presence of $H_2[IrCl_6]$ and phosphorous acid, the axial alcohol, **4-16**, was formed in up to 95% selectivity [52]. A similar system was used in the reduction of **4-17**, which gave the 2β-hydroxy compound

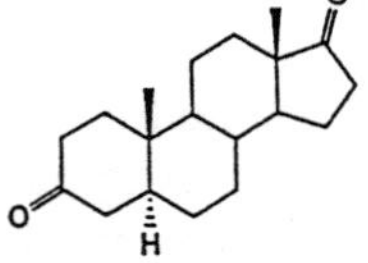
(4-15)

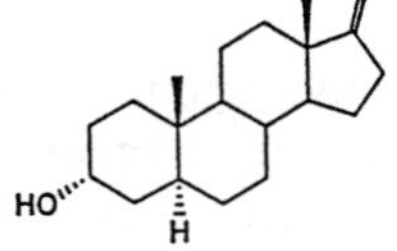
(4-16)

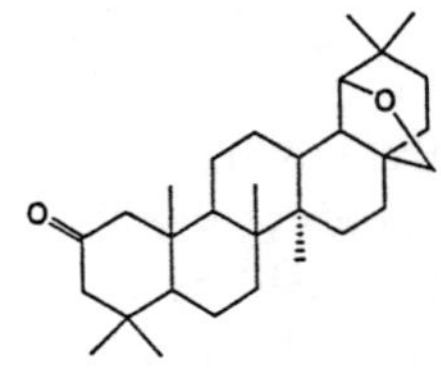
(4-17)

stereospecifically [53]. Enones are usually reduced to allylic alcohols by hydride reducing agents. However, up to 93% selectivity for carbonyl reduction was also noted for the transfer hydrogenation of **Z**-PhCH=CHCOMe using 2-propanol as the reductant and $[\{Ir(COD)(OMe)\}_2]/Ph_2P(C_6H_4\text{-}2\text{-}NMe_2)$ as the catalyst system [54]. The catalyst $[RuCl_2(PPh_3)_3]$ has been used for the transfer hydrogenation of the carbonyl group in ketones and enones, with 2-propanol as the donor. Turnover was relatively slow, but there were no competing aldol reactions [55].

Ketoesters and diketones are more difficult to reduce than simple ketones using hydride reducing agents. The reverse is true for hydrogenation, where the additional unsaturated group assists coordination. Thus ketoesters are reduced using a number of catalysts which are inactive for simple ketones. Wilkinson's catalyst was used for reaction (4.17). Reaction (4.18) shows a reductive cyclisation of a ketoacid, and lactones were also obtained by reduction of cyclic anhydrides (reaction (4.19)). Selectivities were excellent and the products found applications in lignan synthesis [56].

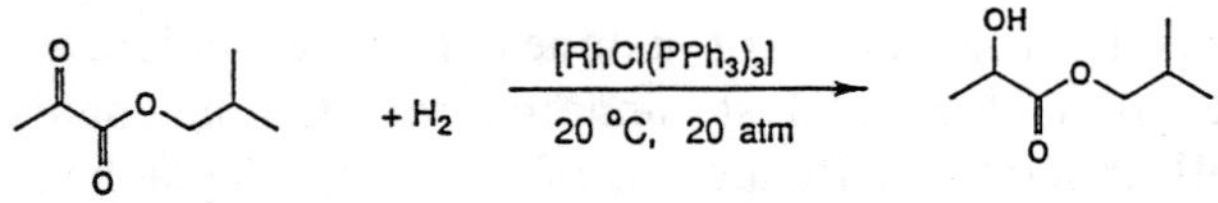

(4.17)

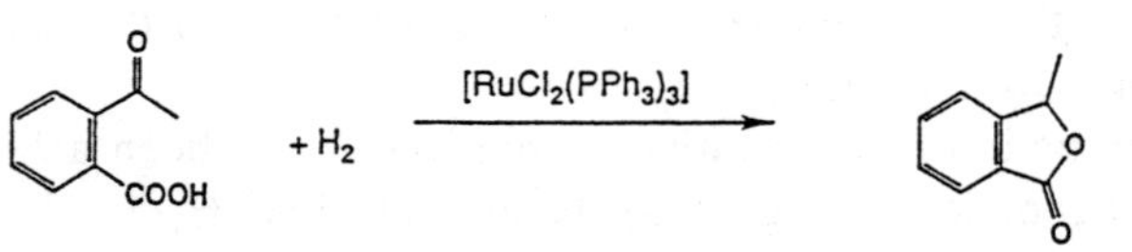

(4.18)

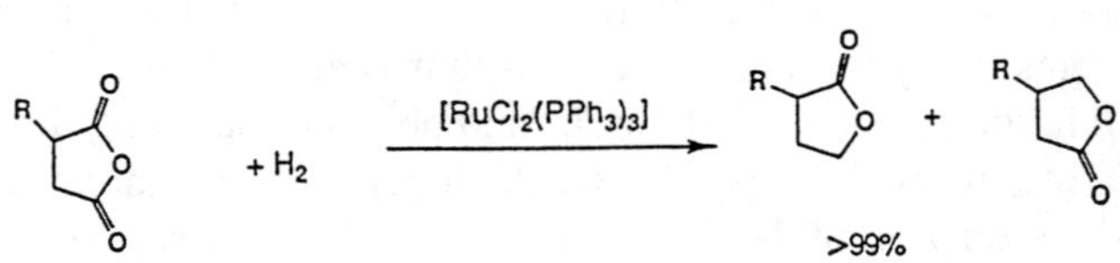

(4.19)

4.8. Reduction of carbon-nitrogen double bonds

The reduction of carbon-nitrogen double bonds has been sought as a synthetic route to substituted amines. Although catalysts such as $[Pd_5(PPh)_2]$, $[Fe(CO)_5]$, $[Co_2(CO)_8]$, $[Mo(CO)_6]/Na[OMe]$ or $[RhCl_2(DMF)(py)_2][BF_4]$ lead to facile reduction, they have not been widely used. Recently, the use of Li[Ir(chelating biphosphine)I_4] has been described (reaction (4.20)); this catalyst tolerates a wide range of functional groups including alkene, ketone, ester, nitrile and nitro [57]. Transfer hydrogenation of imines by 2-propanol has been catalysed by $[Ru_3(CO)_{12}]$ [58] or $[RuCl_2(PPh_3)_3]$ [59]. There are few restrictions on the nature of the substituents, but rates decrease with steric hindrance. Reductive amination is a closely related process, in which an imine is made and reduced *in situ* (reactions (4.21) and (4.22)) [60].

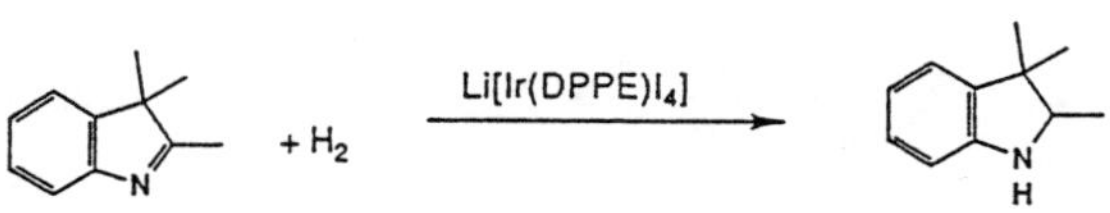

(4.20)

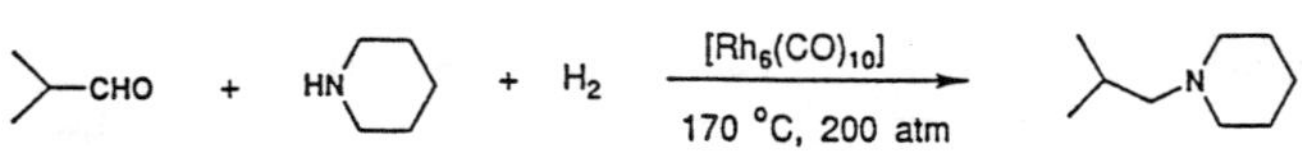

(4.21)

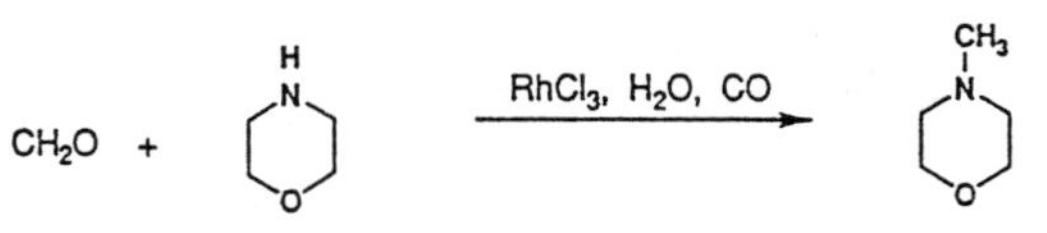

(4.22)

4.9. Reduction of other nitrogen containing functional groups

Hydrogenation of nitriles to primary amines over heterogeneous catalysts may be complicated by the formation of large amounts of tertiary and secondary amines, formed by reaction of the primary amine with imine intermediates. Similar problems tend to be encountered in homogeneous systems. Thus, catalysis by $[Fe(CO)_5]$, $[Ni(CO)_4]$ or $[Co_2(CO)_8]$ gave mixtures of primary, secondary and tertiary amines under severe conditions. The use of $K_2[RuH_4(PPh_2)(PPh_3)_2]$ or $K[RuH_2(C_6H_4PPh_2)(PPh_3)_2]C_{10}H_8.Et_2O$ gave up to 98% selection for ethylamine from MeCN, but the conditions needed were still severe [61].

The reduction of $ArNO_2$ is a problem of considerable industrial importance. The pathway usually discussed involves the intermediacy of ArNO, ArNHOH, ArN(O)=NAr, ArN=NAr and ArNHNHAr. In a few cases these intermediates may be isolated in reasonable selectivities [62], but they have not been widely synthetically, Catalysts for complete reduction to the aniline include $[Co(CN)_5]^{3-}$, $[Ru_3(CO)_{12}]$, $[RhH(CO)_4]$, $[Co(DMG)_2]$, $[RuCl_2(PPh_3)_3]$, $[Rh_6(CO)_{12}]$, *trans*-$[PdCl_2(py)_2]$ or $[PtCl_2(PPh_3)_2]/SnCl_2$ (using water gas as the reductant). A long-standing problem has been the reduction of one nitro group selectively in the presence of another. The selective methods depend on the fact that nitroso compounds are reduced more rapidly than nitro derivatives (reactions (4.23) and (4.24). The reduction of **4-18** could be accomplished at either the nitro or the carbonyl group using the same catalyst $[RuCl_2(PPh_3)_3]$, by changing from direct to transfer hydrogenation [63].

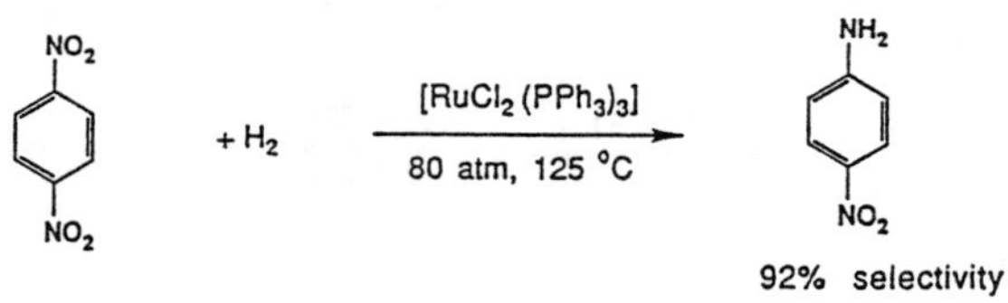

(4.23)

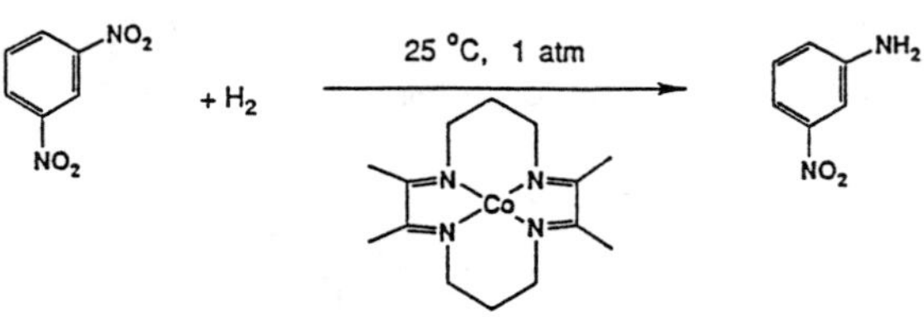

(4.24)

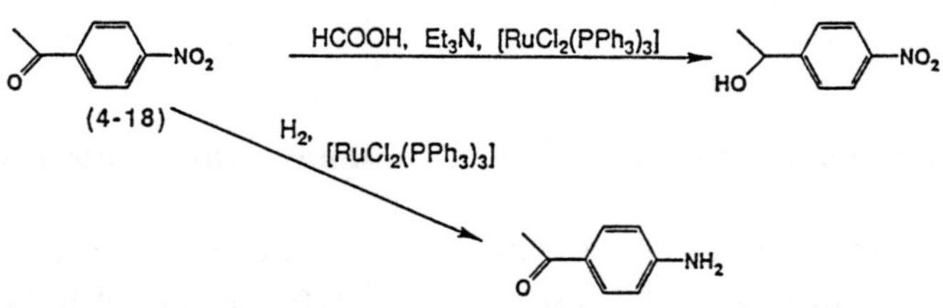

Reductions of aliphatic nitro compounds have been less widely reported. Nitrocyclohexylamine was partially reduced to the oxime in the presence of Cu_2Cl_2/en, presumably *via* tautomerisation of the intermediate nitroso compound [64]. Hydrogenation of **4-19** gave **4-20**, a lysine precursor. In the presence of a range of chiral amines enantiomer excesses up to 13% were observed; although this could arise from hydrogenation of an intermediate oxime, the mechanism was not discussed in detail [65].

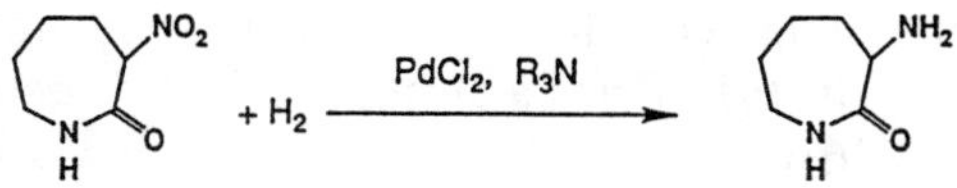

(4-19) (4-20)

4.10. Diastereoselectivity in alkene reduction

In sections 4.1 and 4.2 a number of reactions in which hydrogen was added to the less hindered side of a molecule were noted. These are of course diastereoselective reactions. However, over the last few years the most important group of such processes has been that in which the diastereoselectivity has been directed by the chelation of a neighbouring polar functional group. Heterogeneous reactions of this type are also known [66, 67]. The requirement for a successful reaction is a polar functional group in proximity to the double bond to be reduced, which remains coordinated to the metal during the catalytic cycle, and thus directs the sterochemical course of hydrogen addition. Hydroxyl is the most frequently used group, but esters, amides and carbamates also have useful applications [68, 69].

The first significant report in this area was of the hydrogenation of **4-21** in the presence of Wilkinson's catalyst. When R = H hydrogenation was unsuccesful at 100 psi and 50°C, but with the potassium salt the reaction gave the *cis*-product, **4-22**, exclusively, *via* a transition state supposed to involve coordination of the alkoxide to rhodium. Since then a number of groups have explored the reduction of variously substituted cyclohexenols. The reducation of 2-substituted-2-cyclohexene-1-ols in the presence of Wilkinson's catalyst gave mainly *trans*-products (reaction (4.25)); this was first interpreted in terms of the thermodynamic stabilities of the products or the intermediate

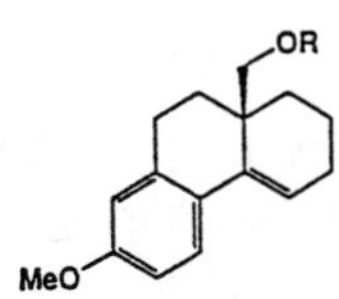

(4-21)

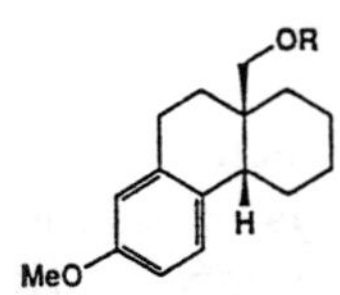

(4-22)

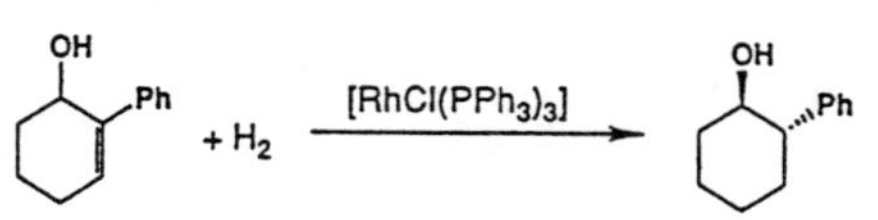

(4.25)

metal alkyls, but metal coordination may also be important. Using $[Ir(COD)(PCy_3)(py)][PF_6]$ as the catalyst gave 96% selectivity in the best cases (reaction (4.26)). The catalyst loading needed was low, and the contrast with Wilkinson's catalyst was attributed to the latter's greater saturation, which discourages hydroxy binding. With 3-cyclohexene-1-ols such as **4-23**, stereoselection for reduction from the same face as the hydroxyl group was excellent, and assumed to proceed *via* an intermediate such as **4-24**. Whilst this species could not be observed during the reaction, **4-26** was observed by ^{1}H NMR spectroscopy at 0°C during the iridium catalysed reduction of **4-25** [70].

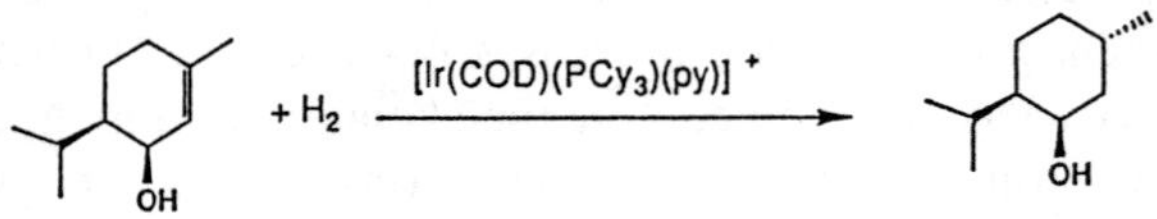

(4.26)

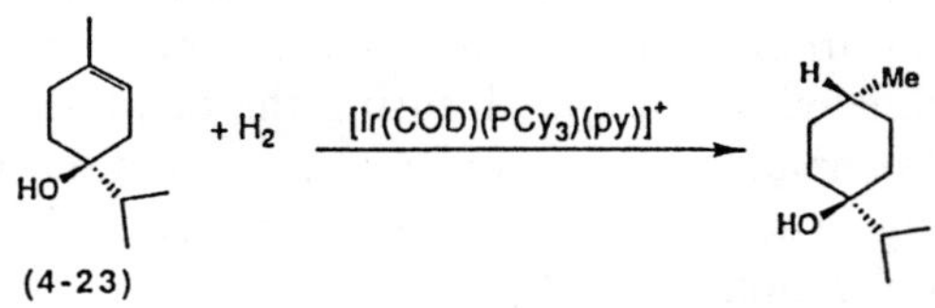

(4-23)

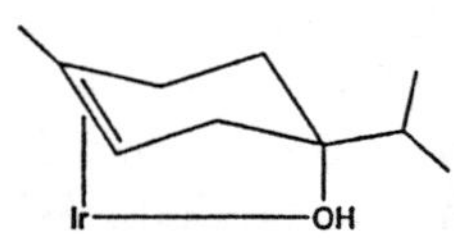

(4-24)

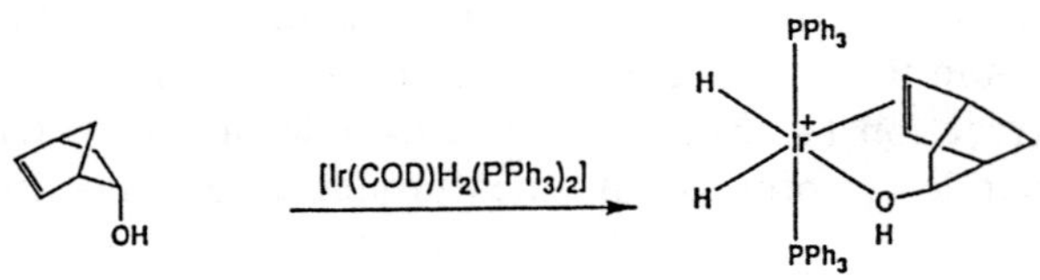

(4-25)

(4-26)

Successful reactions have also been observed when the double bond to be reduced is exocyclic to the six-membered ring. Reduction of **4-27** in the presence of $[Rh(DPPB)(NBD)]^+$ proceeded with > 98%

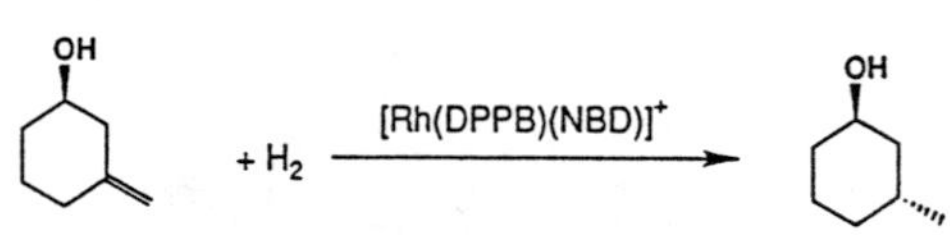

(4-27)

stereoselectivity, *via* an intermediate such as **4-28**. However, for **4-29** in which the metal cannot simultaneously coordinate the alcohol and the double bond, stereoselection was low, and the major product was 2-methylcyclohexanone, produced by isomerisation [71]. Reduction of **4-30** was accomplished with fair selectivity using $[Rh(DPPB)(NBD)]^+$ as catalyst, but the reaction was slow and incomplete. Using $[Ir(COD)(PCy_3)(py)]^+$ the reaction was faster, but the stereoselectivity lower. By contrast, reduction of **4-31** was fast and stereoselective, suggesting that 5-membered chelate rings are preferred to six-membered ones. A deuterium labelling study indicated that double bond migration was extremely facile in these systems, suggesting that they may be unsuitable for site specific reductions [72]. Under these conditions acids were incompletely reduced, implying that carboxylate complexes inhibit the reaction. Hydrogenation of the Birch reduction product, **4-32**, was extremely slow, in contrast to the reaction using Wilkinson's catalyst. An interesting application involved the reduction of **4-33**. Whilst heterogeneous catalysts gave reduction mainly to **4-34** (R = H), the use of homogeneous species resulted in the formation of **4-35** (R = Bz), with essentially complete selectivity when the catalyst was $[Rh(DPPB)(NBD)]^+$ [73]. More remote alcohols may also be useful in exerting a directing effect; the selectivity in reaction (4.27) was 96% [74].

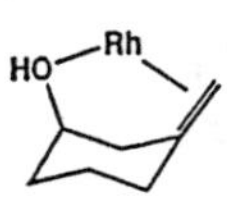

(4-28)

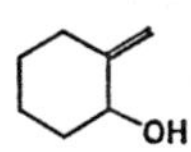

(4-29)

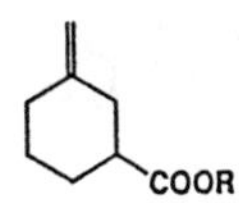

(4-30)

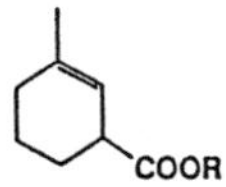

(4-31)

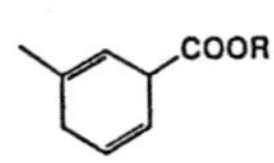

(4-32)

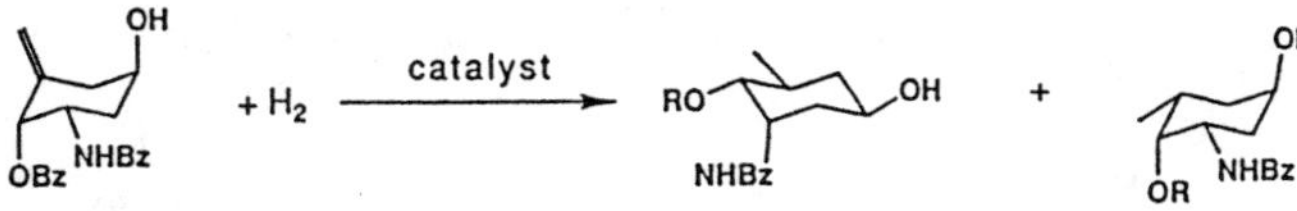

(4-33) (4-34) (4-35)

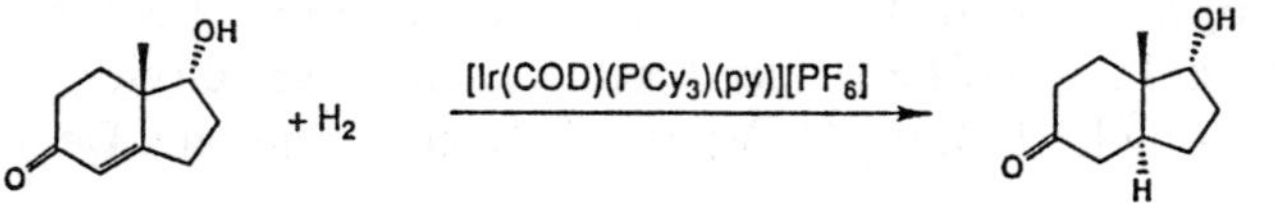

(4.27)

Reductions of cyclopentenes have been similarly directed by hydroxyl groups. Deuteriation of **4-36** gave a product with a *trans:cis* ratio of > 37:1 [75]. Reactions (4.28) and (4.29) were used respectively in syntheses of a gastroprotective substance [76] and **D**-mycosine [77]. Examples of reductions of bonds exocylic to 5-membered rings are shown in reactions (4.30) [78] and (4.31) [79]; the latter was used in a synthesis of (+)-mikrolin.

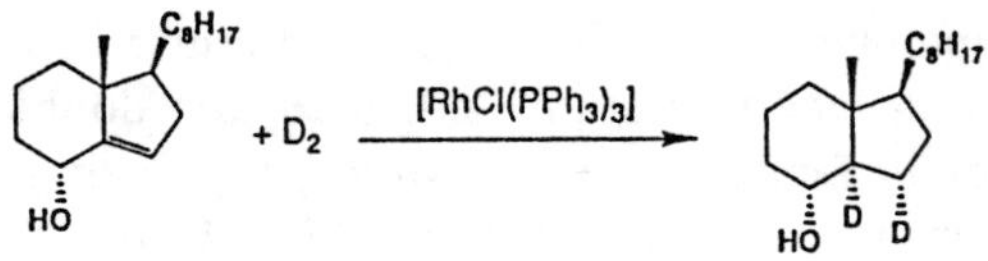

(4-36)

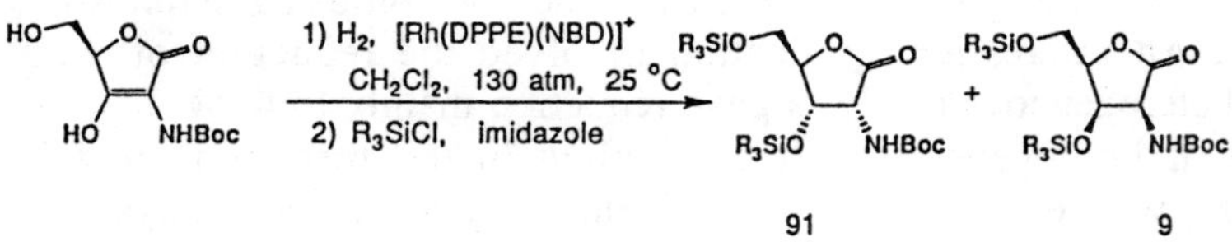

(4.28)

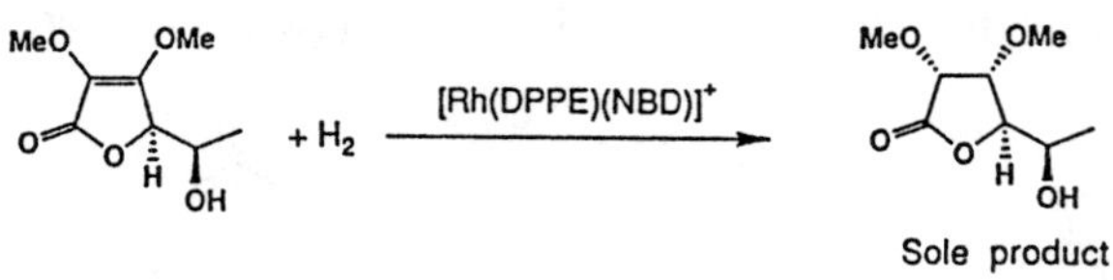

(4.29)

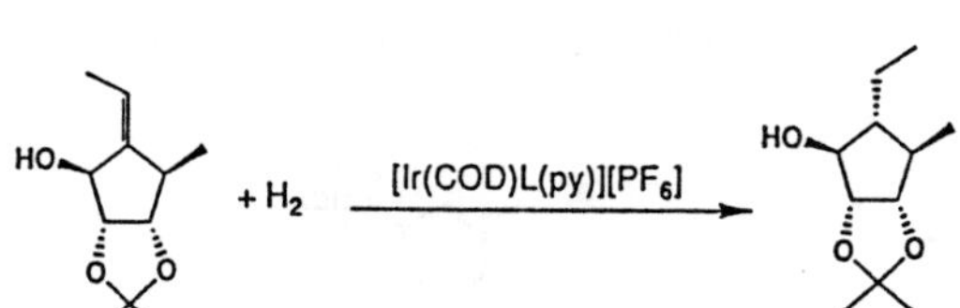

(4.30)

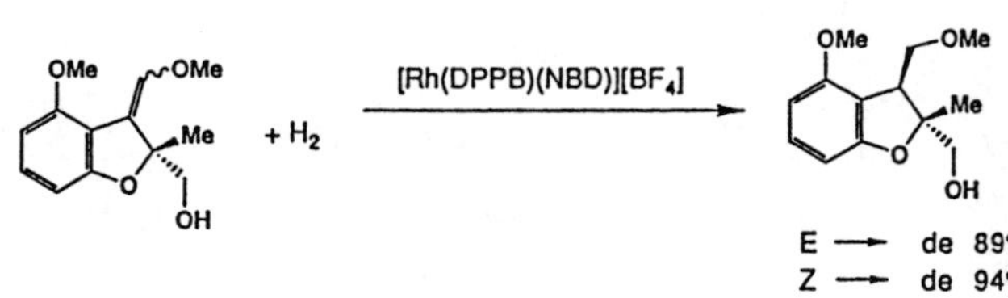

(4.31)

The hydrogenation of **4-37** was slow using most homogeneous catalysts, but **4-38** could be reduced with excellent selectivity towards **4-39** (99.4% using [Ir(COD)(PCy_3)(py)][PF_6]), at a rapid rate. Deuterium

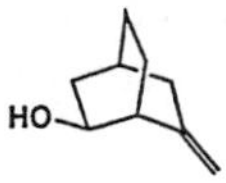

(4-37)

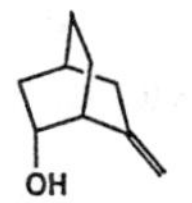

(4-38)

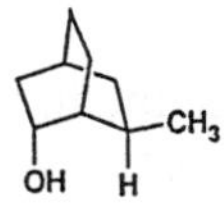

(4-39)

labelling studies indicated that there was some isomerisation to the endocyclic alkene prior to reduction [80].

Results with directing groups other than alcohols have been more variable. Ethers coordinate less well to the metal centres than do alcohols, and ethanoates prefer to bind at the more basic carbonyl oxygen atom, which is usually less well oriented for chelate formation. Esters and amides are, however, good directing groups, as shown in reactions (4.32) and (4.33); the ester must be in quite close proximity to the double bond for reduction to proceed with high stereoselection, but this requirement is less critical with the related amides [81]. This was usefully applied in the reduction of **4-40** in an approach to the synthesis of pumiliotoxins; in this instance reduction using a heterogeneous catalyst gave mainly the wrong isomer [82]. Hydrogenation of **4-41** was used

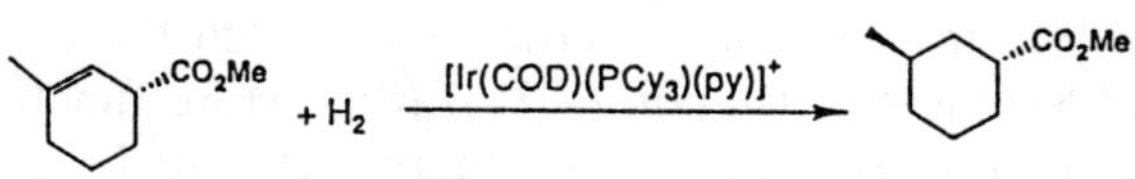

(4.32)

99.9%

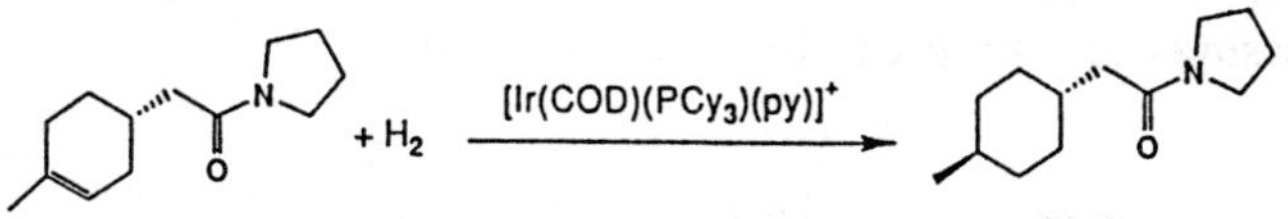

(4.33)

99 %

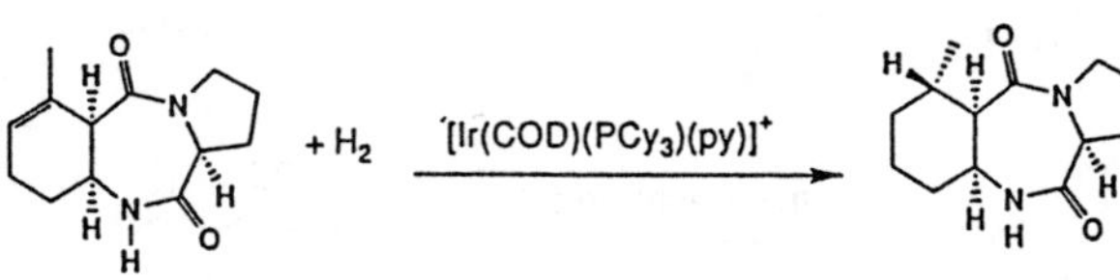

(4-40)

99 %

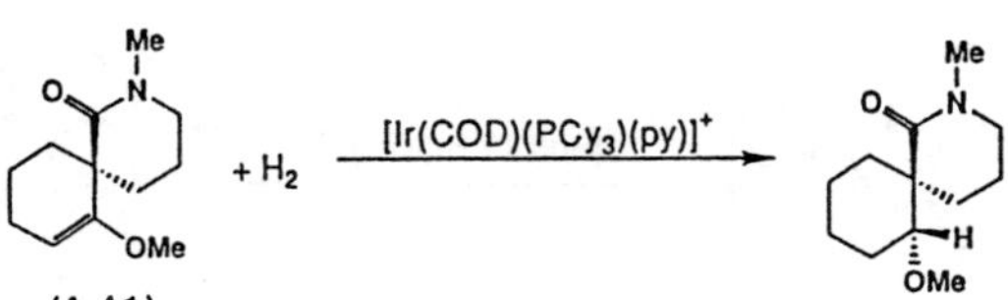

(4-41)

in an approach to the synthesis of sibirine, nitramine and isonitramine [83], whilst reduction of **4-42** was used in an approach to the clerodan diterpenoids [84].

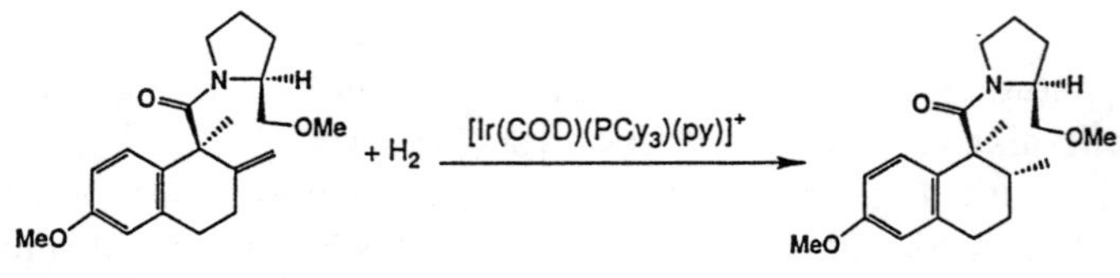

(4-42)

The outcome of the hydrogenation of cyclic substrates is usually readily predictable. With acyclic compounds this is less easy, but there have been some notable successes, and with growing experience the stereoselectivity observed can be rationalised. Allylic alcohols have proved to be excellent substrates, particularly using cationic rhodium or iridium catalysts. Thus reduction of **4-43** using [Rh(DPPB)(NBD)]$^+$ as the catalyst gave **4-44** as the major product. The overall yield was reduced as a result of rhodium catalysed isomerisation to the saturated ketone. The reaction methodology is now well enough known to have reached Organic Syntheses (reaction (4.34)) [85]. More functionalised substrates have sometimes proved to be more difficult; in the reductions of the allylic and the homoallylic alcohols **4-45** and **4-46**, selectivity towards reduction rather than isomerisation was increased by the use of high pressures and high catalyst loadings [86].

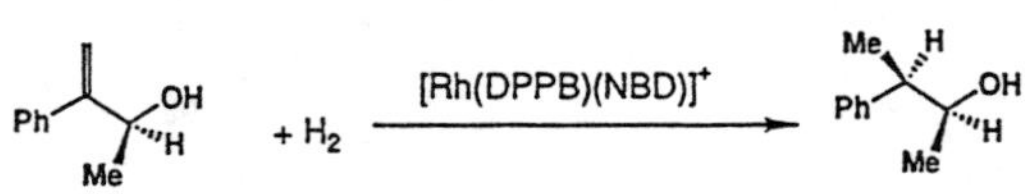

(S-4-43) (R,S-4-44)

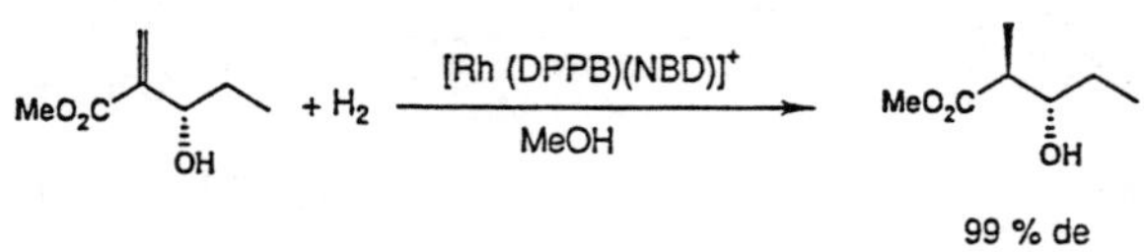

99 % de

(4.34)

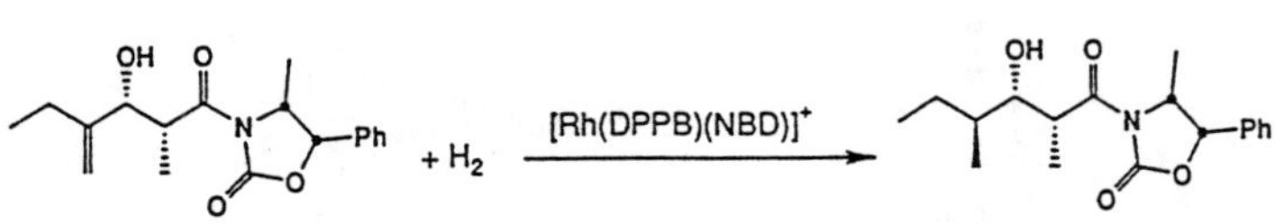

(4-45) 93 %

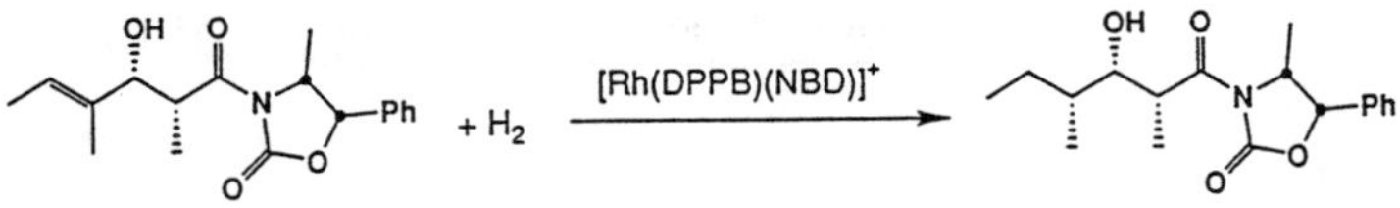

(4-46)

For acyclic allylic alcohols not bearing either very bulky groups, or other coordination sites, common principles have emerged. The reaction is almost invariably *anti*-selective. This is rationalised by Brown in terms of non-bonded interactions in the diastereomeric chelate complexes such as **4-47a** and **4-47b**. If correct, this suggests that the stereoselection is based entirely on the substrate, with only peripheral contributions from any influence of the phosphine ligands. That this is essentially the case will be shown later, when reactions using chiral phosphines are considered; both enantiomers of the substrate give an *anti*-dominated product from reduction in the presence of a complex of a chiral phosphine [68].

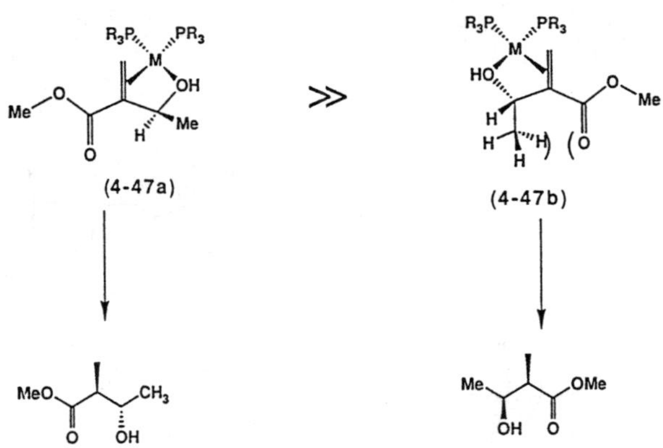

(4-47a) (4-47b)

Asymmetric induction under the influence of a more remote directing group has been achieved using homoallylic alcohols as substrates, and again there are now sufficient examples in the literature that stereoselectivity may be predicted with some confidence. Early examples were provided by reactions (4.35) and (4.36), although the selectivity is unexceptional in these cases. Evans' group have now reported a wide range of such reactions, mostly directed towards syntheses of macrolide and polyether antibiotics. Some examples are given in reactions (4.37), (4.38) and (4.39) [87,88]. Mechanistic considerations are complex; Brown has proposed that the selectivity derives from the generation of the intermediate, **4-48a** or **4-48b**, which has the less 1,3-allylic strain, and

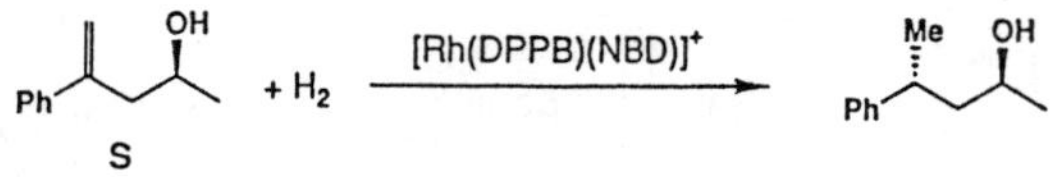

(4.35)

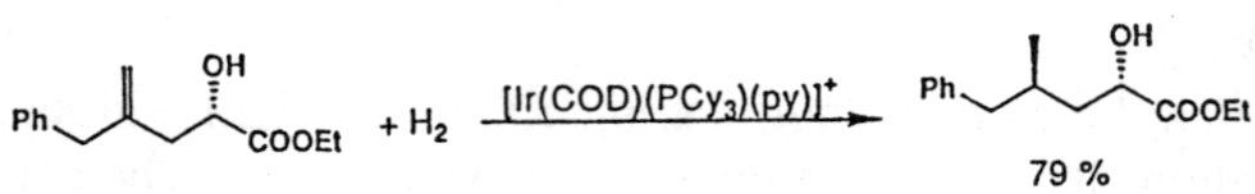

(4.36)

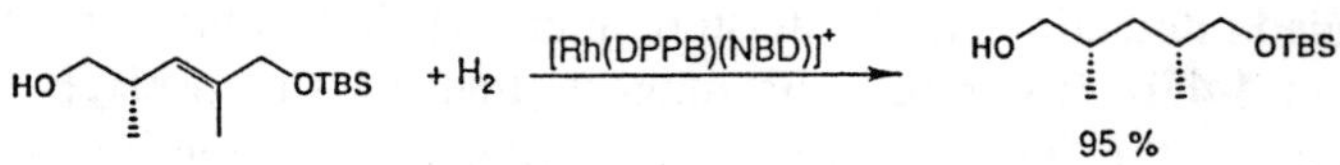

(4.37)

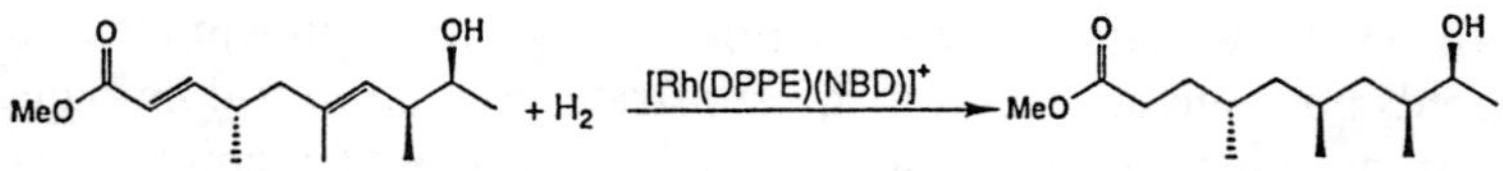

(4.38)

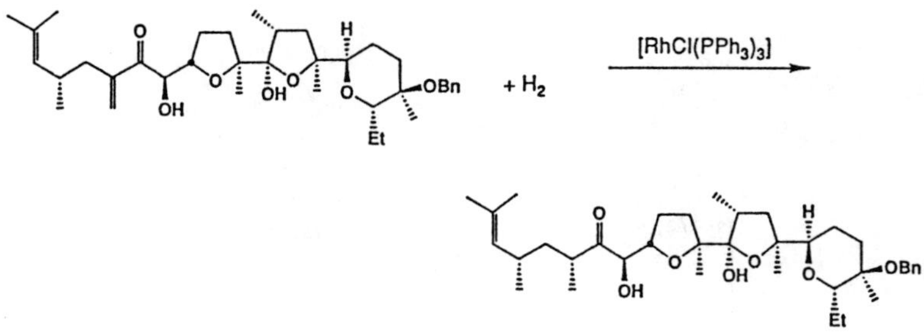

(4.39)

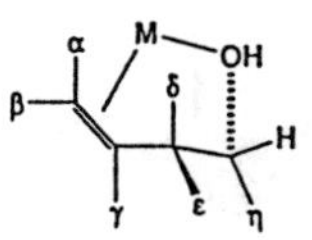

(4-48a)

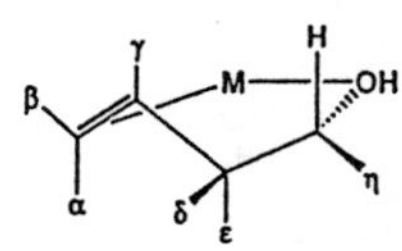

(4-48b)

this correctly predicts the result in the examples described by Evans. In examples such as that in reaction (4.40) the possibility of changes in conformation due to hydrogen bonding to the nearby carbonyl group must also be considered [89].

Examples of useful reductions involving other directing groups

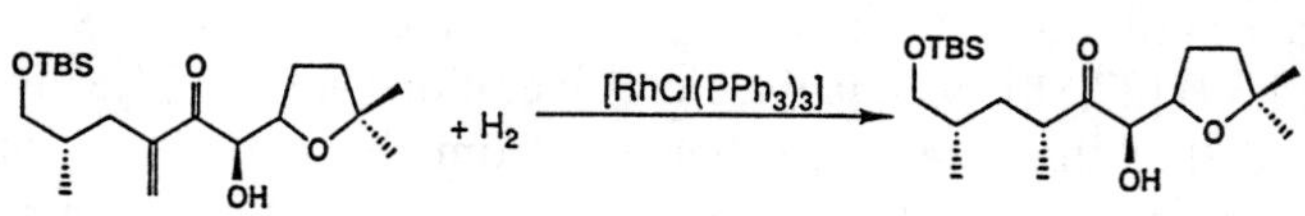

(4.40)

have been less common. Hydrogenation of **4-49** in the presence of $[Rh(DPPB)(NBD)]^+$ gave **4-50** with excellent selectivity [90]; using Pd/C as the catalyst gave predominantly the *syn*-product. Interestingly, reaction of **4-49** with $[Rh(acac)(C_2H_4)_2]$ gave **4-51**, characterised in an X-ray diffraction study; the NH group was not metal coordinated, and the product corresponds to the disfavoured isomer on the directed hydrogenation pathway [91]. The related substrate, **4-52**, was reduced to **4-53** with 99.1% selectivity, the product being converted to a chiral β-lactam. Unusually, the best catalyst was $[Ru(OCOCF_3)_2(PPh_3)_2]$, although > 95% selection was also obtained using $[Rh(COD)(DPPE)][ClO_4]$ [92].

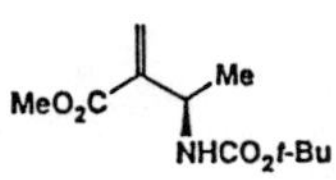

(4-49)

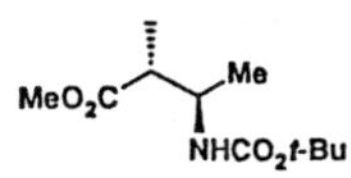

(4-50)

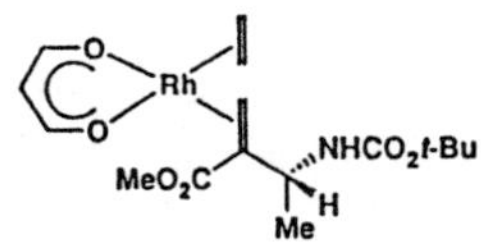

(4-51)

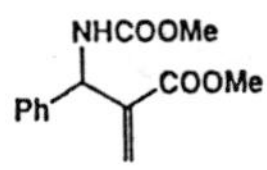

(4-52)

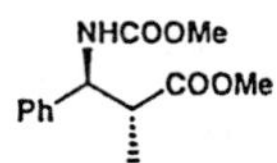

(4-53)

The reductions of a series of homoallylic alcohols which also bear an amide functionality has been studied by Brown's group (reactions (4.41) and (4.42)). These seem to follow a simple steric model, with little influence from the amide [93]. However, in the reactions of **4-54** and

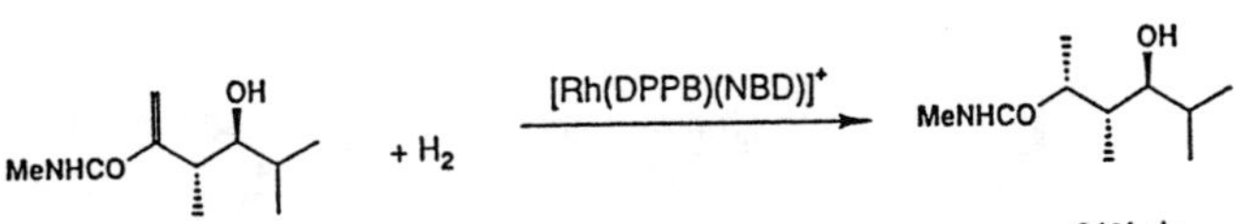

91% de

(4.41)

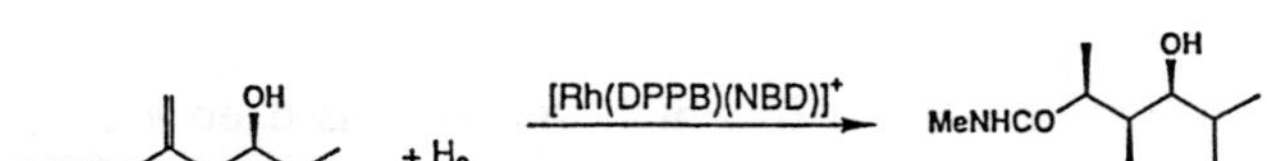

67% de

(4.42)

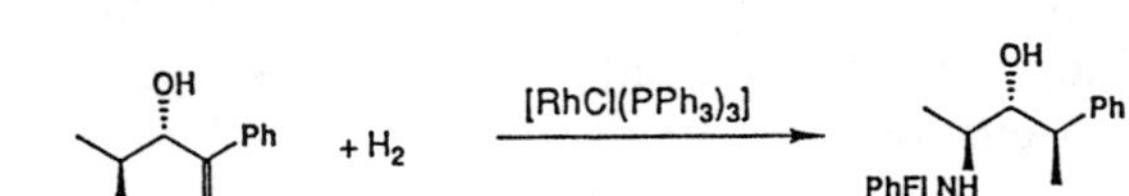

single isomer

(4-54)

4-55 (PhF1 = 9-phenylfluorene-9-yl), the stereochemical outcome depended strongly on the stereoisomer used, arguing either for some involvement of the nitrogen function, or that there is one of the possible transition states in which the coordination of the alcohol is inhibited by the size of the substituent at nitrogen [94]. The reduction of the sulphone bearing allylic alcohol, **4-56** in the presence of [Rh(DPPB)(NBD)]$^+$ gave good *anti*-selectivity for **4-57**. However, the sulphoxide, **4-58**, gave **4-59** in 99% diasterioisomer excess, and **4-60** gave **4-61**, indicating that the diastereoselectivity in this case is controlled entirely by the chiral sulphoxide, and not the allylic alcohol [95].

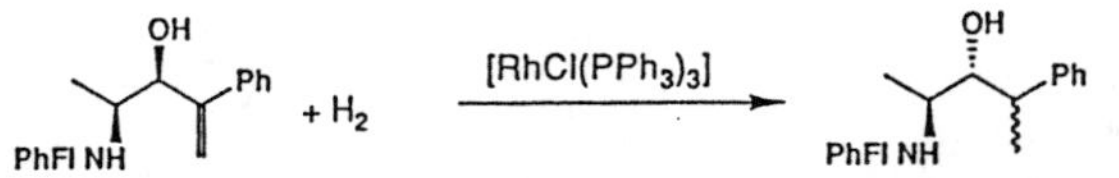

(4-55)

2:1 mixture of diastereoisomers

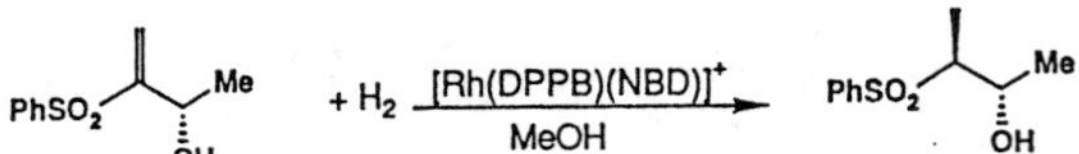

(4-56)

99% de

(4-57)

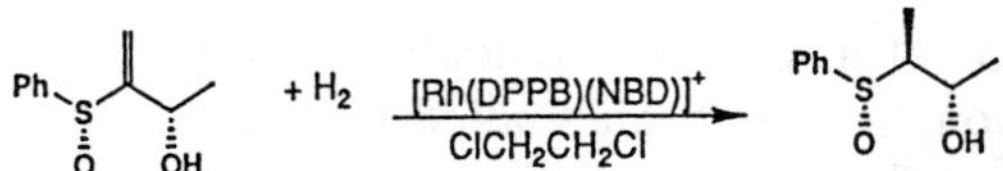

99% de

(4-58)

(4-59)

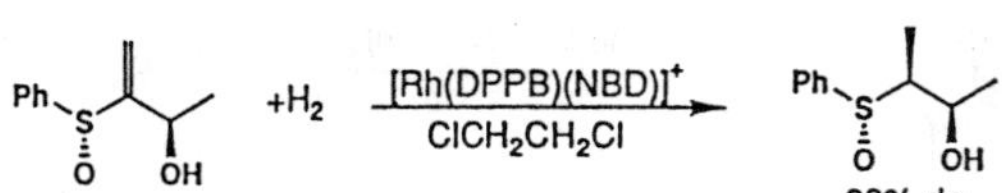

98% de

(4-60)

(4-61)

A number of hydrogenations in which the aim has been diastereoselective reduction have been carried out using chiral rhodium or iridium complexes as the catalysts. Some of these have given spectacularly good results, and in the best cases also have the potential for kinetic resolution. They will be discussed in detail in the section on enantioselective hydrogenation.

4.11. Enantioselective hydrogenation

Enantioselective hydrogenation reactions were first reported in the 1940's and involved heterogeneous catalysts either impregnated into chiral natural materials such as silk, or which were modified with simple chiral molecules. Since then there has been steady work in the area of chirally modified heterogeneous catalysts, mainly from groups in Japan and the Soviet Union. Modifiers are generally amino acids or hydroxy acids. Although there have been great improvements in this technique, the range of useful substrates remains generally limited to β-dicarbonyl compounds.

The discovery of Wilkinson's catalyst led quickly to the development of enantioselective varsions of the catalyst. A wide range of rhodium complexes of chiral phosphines was produced in the 1970s, and detailed work on the reaction mechanism made it one of the best understood in the field of homogeneous catalysis. Efforts slowed in the 1980s, because the range of substrates for which usefully selective reduction could be achieved seemed limited. More recently, interest has been revived, largely as a result of the introduction of ruthenium complexes of binaphthyl derived phosphines as catalysts; selectivities are high, and a much wider range of substrates, including ketones, give good results. Other systems have recently proved to be useful for the asymmetric reduction of imimes, and non-polar alkenes.

This field has been extensively reviewed [96–100].

4.11.1. HYDROGENATION OF DEHYDROAMINO ACID DERIVATIVES AND RELATED SUBSTRATES

4.11.1.1. *Reaction in the presence of rhodium complexes of chiral phosphines*

The enantioselective reduction of dehydroamino acid substrates in the presence of rhodium complexes of chiral phosphines and biphosphines has been a conspicuous success. The only sizeable commercial catalytic asymmetric synthesis involving a transition metal complex, the preparation of **L**-DOPA for the treatement of Parkinson's disease, falls into this category. A typical reaction is the reduction of acetamidocinnamic acid (**Z**-2-ethanoylamino-3-phenylpropenoic acid), **4-62**; some of the

(4-62)

better results for this substrate are shown in Tables *4.1–4.3*. Figure *4.1* shows the structures of the chiral phosphines which have been used.

TABLE 4.1
Hydrogenation of α-acetamidocinnamic acid, **4-62**, in the presence of $[Rh(diene)LL]^+$. where LL forms a five-membered chelate ring

Phosphine	Conditions[a]	ee%	Configuration
R,R-DIPAMP, (**4-63**)	25°C, 3.5 atm, 88% Me_2CHOH	92.8	**S**
R-PROPHOS, (**4-64**)	25°C, 1 atm, THF	91	**S**
S,S-CHIRAPHOS, (**4-65**)	25°C, 1 atm, EtOH	89	**R**
NORPHOS, (**4-66**)	25°C, 1 atm, EtOH, 10 h	95	**S**
PHELLANPHOS, (**4-67**)	25°C, 1 atm, EtOH	94	**S**
(**4-68**)	22°C, 57-45 atm, MeOH, 4 h	99	**S**
(**4-69**)	25°C, 1 atm, MeOH	89	**S**
(**4-70**)	20°C, 20 atm, EtOH:C_6H_6 = 2.1	99	**S**
(**4-71**)	25°C, 1 atm	85	b
(**4-72**)	25°C, 2 atm, MeOH[c]	99	**S**

[a] Chemical yields were essentially 100%
[b] Not given [101]
[c] Methyl α-acetamidocinnamate was the substrate [102]

TABLE 4.2
Hydrogenation of α-acetamidocinnamic acid, **4-62**, in the presence of $[Rh(diene)LL]^+$, where LL forms a seven-membered chelate ring [103, 104]

Phosphine	Conditions[a]	ee%	Configuration
S,S-DIOP, (**4-73**)	25°C, 1 atm, EtOH[b]	83.5	**S**
(**4-74**)	25°C, 1 atm, EtOH/C_6H_6	88	**R**
BPPM, (**4-75**)	20°C, 50 atm, Et_3N, EtOH, 20 h[b]	91	**R**
(**4-76**)	20°C, 1 atm, Et_3N, MeOH	92	**R**
(**4-77**)	50°C, 20 atm, Et_3N, EtOH	98	**R**
(**4-78**)	20°C, 1 atm, EtOH	73	**S**
(**4-79**)	25°C, 1 atm, MeOH	85	**R**
R-BINAP, (**4-80**)	25°C, 3–4 atm, EtOH, 48 h	84	**R**

[a] Yields essentially 100% in all cases
[b] Catalyst prepared *in situ* from $[\{RhCl(diene)\}_2]$

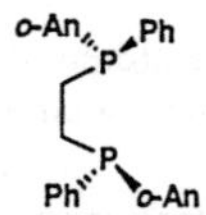

R,R-DIPAMP
(4-63)

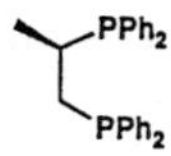

R-PROPHOS
(4-64)

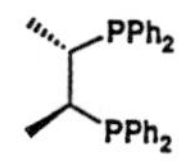

S,S-CHIRAPHOS
(4-65)

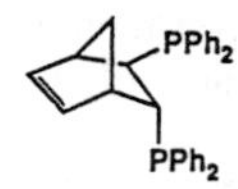

NORPHOS
(4-66)

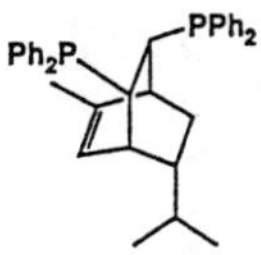

PHELLANPHOS
(4-67)

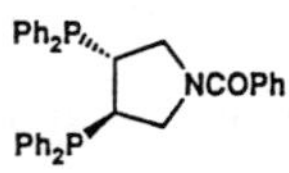

(4-68)

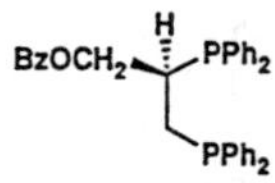

(4-69)

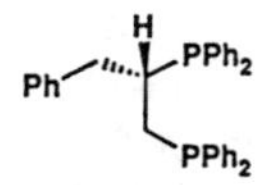

(4-70)

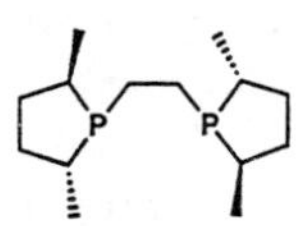

(4-71)

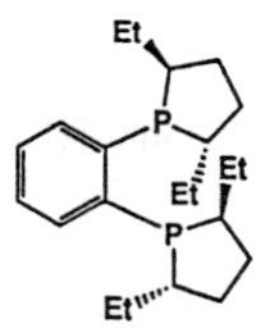

(4-72)

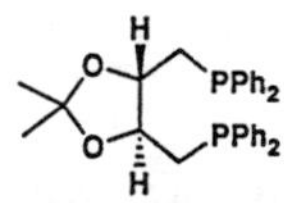

S,S-DIOP
(4-73)

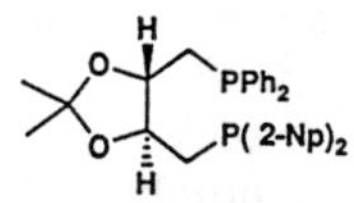

(4-74)

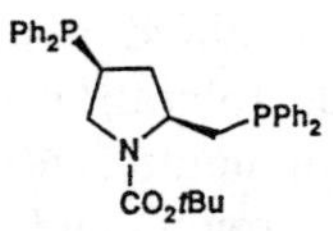

BPPM
(4-75)

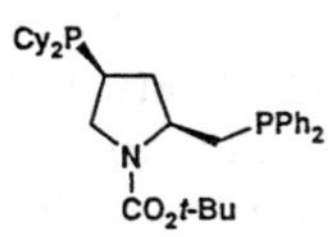

BCPM
(4-76)

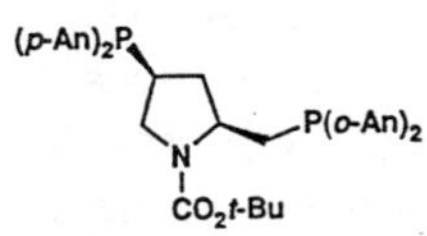

(4-77)

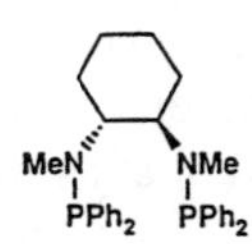

(4-78)

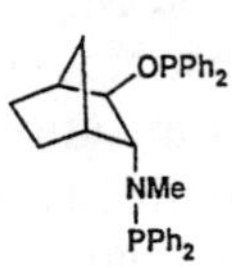

(4-79)

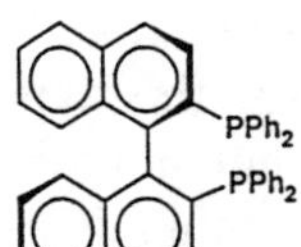

R-BINAP
(4-80)

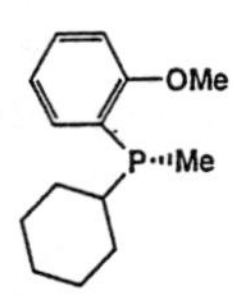

R-CAMP
(4-81)

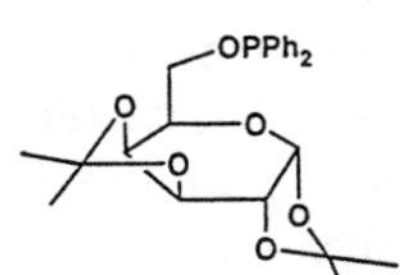

(4-82)

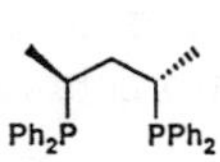

(4-83)

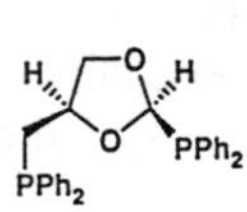

DIOXOP
(4-84)

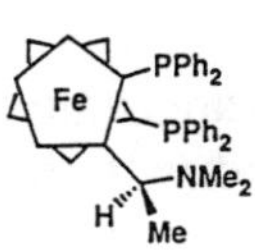

S,R-BPPFA
(4-85)

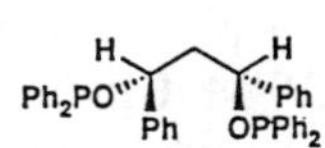

R,R-BDPPODP
(4-86)

Fig. 4.1. Chiral phosphines used in asymmetric hydrogenation.

TABLE 4.3
Hydrogenation of α-acetamidocinnamic acid, **4-62**, in the presence of $[Rh(diene)LL]^+$, where LL are monophosphines, or biphosphines which forms a six-membered or large chelate ring

Phosphine	Conditions[a]	ee%	Configuration
R-CAMP, (**4-81**)	25°C, 0.7 atm, Me_2CHOH	86	**R**
(**4-82**)	25°C, 1 atm, Et_3N, $EtOH/C_6H_6$[b]	67	**R**
(**4-83**)	30°C, 1 atm, MeOH	96	**R**
DIOXOP, (**4-84**)	20°C, 1 atm, Et_3N, MeOH	84	**S**
S,**R**-BPPFA, (**4-85**)	25°C, 5 atm, MeOH[b]	93	**S**
R,**R**-BDPODP, (**4-86**)	30°C, 1 atm, MeOH:C_6H_6 = 3:1[c]	96	**R**

[a] Essentially 100% chemical yields in all cases
[b] Catalyst prepared *in situ* from $[\{RhCl(diene)\}_2]$
[c] Catalyst prepared *in situ* from $[\{Rh(NBD)\}_2][BF_4]$ and $[\{CuCl($**R**,**R**-BDPODP$)\}_n]$

A number of points are worthy of note. Most of the reactions may be performed under ambient conditions, and in some instances elevated pressures are deleterious. The **Z**-isomers of dehydroamino acids react more rapidly than the **E**-isomers, and are generally reduced with better optical yield. Some satisfactory results have been obtained using **E**-isomers with benzene as the solvent. There is wide toleration of different substituents at the β-position (leading to different amino acids), to changing the substituent on the amide (though bulky groups tend to lead to slow reaction), and to conversion of the acid group to an ester. Biphosphines which form 5-membered chelate rings generally give better results than those which form the more flexible 7-membered ring.

In practical terms hydrogenation of this type of substrate in the presence of rhodium complexes is not a problem; optimisation of catalysts and conditions may proceed within fairly closely defined limits. There have been a number of surveys of the available ligands and substrates [96], containing many valuable data from the patent literature, but most workers will find their needs served by well-studied phosphines. Quite a number of these (DIOP, PROPHOS, CHIRAPHOS, BPPM, BINAP *etc.*) are commercially available, but costs tend to be high, even when compared with the high costs of the rhodium precursors.

4.11.1.2. *Mechanism of the reaction*

Enantioselective hydrogenation catalysed by rhodium complexes has been thoroughly studied mechanistically. Some controversy remains as to the precise details of the reaction kinetics, but the outline is well understood as a result of the work of the groups of Brown, Bosnich and Halpern. The mechanism of the reaction involving DIPAMP complexes is outlined in Figure *4.2*.

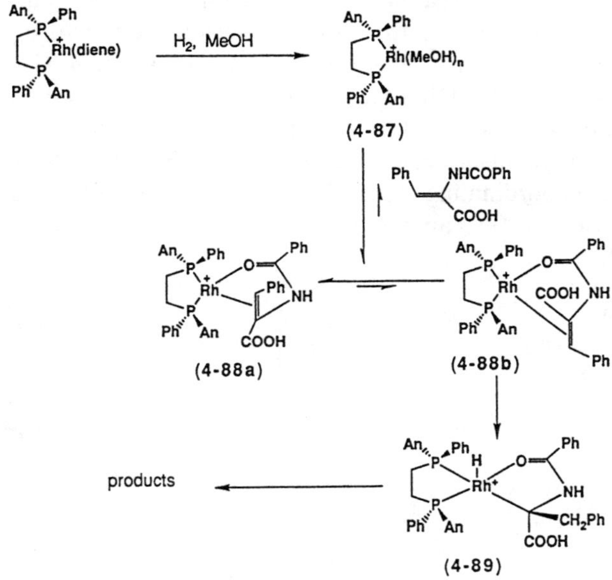

Fig. 4.2. Chiral phosphines used in asymmetric hydrogenation.

The structure of the solvate, **4-87**, was deduced initially by ^{31}P NMR spectroscopy. Dimerisation of the solvate to give arene-bridged species occurs in solvents of low coordinating power [105]. The structures of the enamide complexes in solution were elucidated by ^{13}C-labelling experiments, and in the solid state by X-ray crystallography, for derivatives of DPPE, CHIRAPHOS or DIPAMP. Circular dichroism data showed that the diastereoisomer isolated by crystallisation for CHIRAPHOS was the same as the major species in solution. Exchange between the diastereomeric enamide complexes **4-88a** and **4-88b** was investigated by dynamic NMR spectroscopy using the DANTE technique [106]. It was shown that one mode for exchange involved alkene dissociation only, and that this was a faster path than the one involving complete dissociation and exchange. Reaction of the minor diastereoisomer with hydrogen is much more rapid than that of the major diastereoisomer, and it is through this species that the main catalytic flux passes [107].

The alkyl hydride, **4-89**, is moderately stable at –50°C, and has been characterised by NMR spectroscopy. Related experiments on DIPAMP iridium complexes gave alkyl hydrides stable at room temperature [108]. The reaction of the minor isomer was also appreciably faster than that of the major diastereoisomer. No dihydride enamide complexes have so far been observed in the rhodium serious, but hydrogenation of **4-90** for less than 1 min at –70°C gave a mixture of **4-91a** and **4-91b** in the ratio 4:1 (PP = DIPAMP). Above –45°C these were converted to a mixture of **4-92a** and **4-92b** (one of which predominates to the extent of 95%), which insert at –25°C to give **4-93a** and **4-93b** (the OMe group in the coordinating sphere is part of the DIPAMP ligand). Further rearrangement at 0°C involved reductive elimination and C–H activation [109].

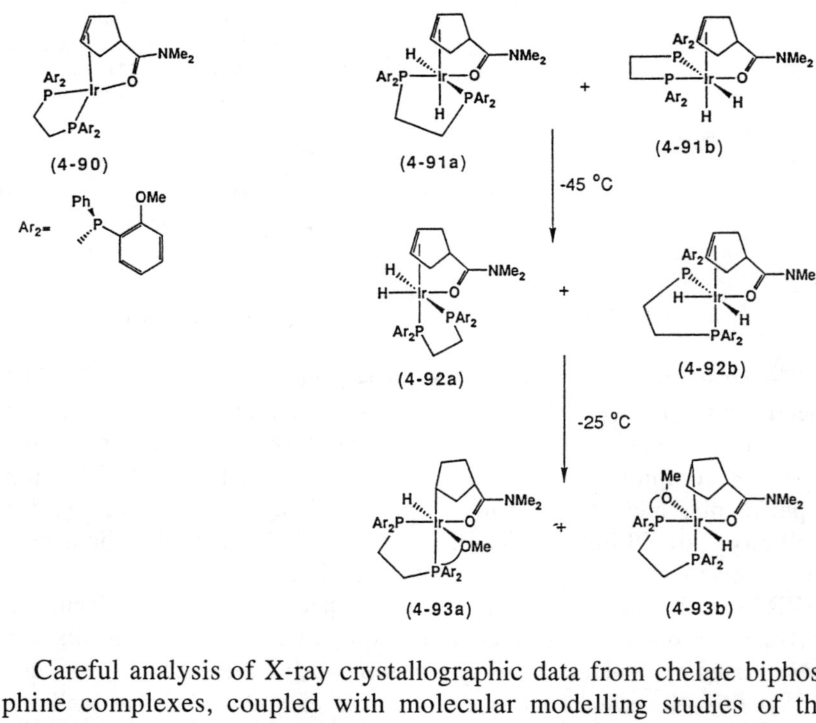

Careful analysis of X-ray crystallographic data from chelate biphosphine complexes, coupled with molecular modelling studies of the addition of hydrogen to the enamide complexes, suggests that although the minor enamide experiences repulsive interactions in the early stages of the approach of the hydrogen molecule, these are much smaller than those noted for the major enamide in the later stages of hydrogen addition [110]. Bosnich and his coworkers have undertaken detailed calculations on the addition of hydrogen to the major and minor diastereoisomers of CHIRAPHOS enamide complexes; they concluded that of the eight possible modes of hydrogen addition, six were impracticable on energy grounds, with one mode permitted for each diastereoisomer. It should be useful to apply this method systematically to a range of catalyst systems [111].

It is of some note that only derivatives of a relatively small number of biphosphines which form 5-membered chelate rings have been studied in detail. In general only the enamide derivatives (if those) or other types of phosphines have been investigated. Since it is well known that binding tends to be weaker with derivatives of 7-membered ring chelate phosphines, and the chelate ring is considerably more flexible, there is no guarantee that Figure *4.2* will be a good model in all cases.

4.11.1.3. *Hydrogenation of dehydroamino acid derivative using complexes of other metals and non-phosphine ligands*

Ruthenium derivatives of BINAP, **4-80**, such as [Ru(BINAP)$(OCOR)_2$], have been used in efficient enantioselective reduction of dehydroamino acid derivatives. The reaction conditions needed are somewhat more severe than for the comparable rhodium complexes, and the opposite enantiomer is obtained from that from the related rhodium catalysts [112].

There have been attempts to combine the high activity of the Crabtree catalyst, $[Ir(COD)(PCy_3)(py)]^+$, with a chiral phosphine. Although $[Ir(COD)L(PhCN)]^+$ (L = NMDPP or PAMP) proved to be adequate catalysts for the reduction of **4-94** under ambient conditions, optical yields were disappointing [113]. Studies of iridium complexes have, however, played a significant role in understanding of the modes of binding of various substrates to metals. Reaction of $[\{Ir(C_2H_4)_2Cl\}_2]$ with **4-95** (R = (–)-menthyl) gave a slowly interconverting mixture of diastereosiomers, **4-96** and **4-97**, in an approximate ratio of 1:2. Pure **4-96** could be crystallised, and reacted rapidly with **S,S**-CHIRAPHOS to give the stable complex **4-98**. When the enantiomer of **4-95** was used, the related reaction was very much slower and gave rise only to a metastable compound. Thus if a mixture of phosphine enantiomers is used to react with **4-97**, only one enantiomer is rapidly consumed, providing a rapid process for phosphine resolution. The unreacted enantiomer may then be treated with $[Rh(NBD)_2]^+$ to give a rhodium complex suitable for practical catalysis of hydrogenation [114].

There has been some interest in hydrogenations using palladium(II) complexes of chiral amines as the catalysts. Thus benzamidocinnamic

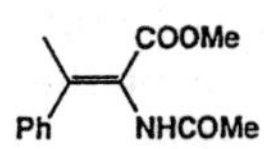

(4-94)

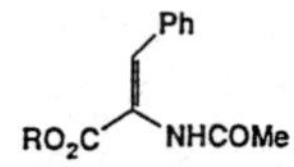

(4-95)

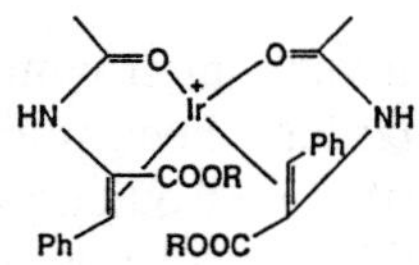

(4-96)

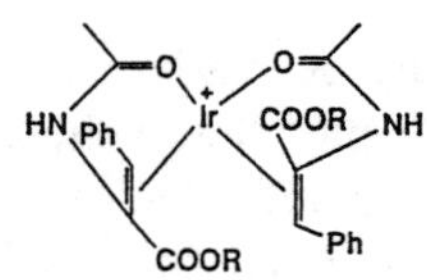

(4-97)

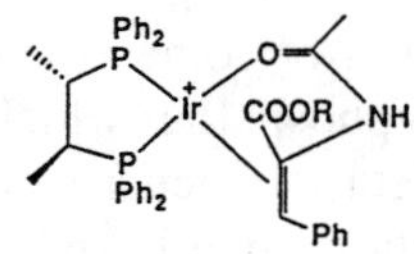

(4-98)

acid (**Z**-2-Benzoylamino-3-phenyl-2-propenoic acid) was reduced in low optical yield by a complex formed *in situ* between $PdCl_2$ and **S**-PhCH(Me)NH_2. Chiral diamines, amino acids, and Schiff bases have also been used as modifiers, and some reactions use $RhCl_3$, but optical yields have generally been disappointing [115, 116].

There have been a range of catalysts reported which comprise [Co(DMG)$_2$] with a chiral modifier, either a phosphine or an amino acid or peptide derivative [117]. Few of these give optical yields which are competitive in modern terms, but one example is worthy of note. The hydantoin **4-99**, was reduced in 79% enantiomer excess in the presence of [Co(DMG)$_2$]/PPh_3/**4-100**. Although the optical yield is not outstanding, all the reagents are relatively inexpensive [118].

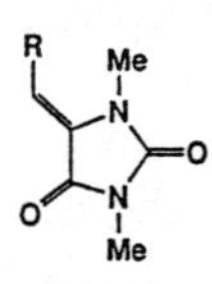

(4-99)

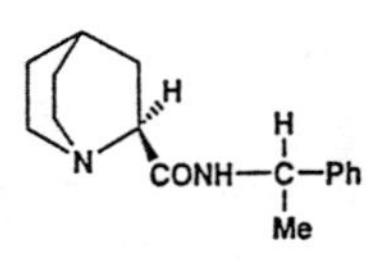

(4-100)

4.11.1.4. *Applications*

Apart from the manufacture of L-DOPA, these reactions have had relatively few practical applications, since most natural chiral amino acids may be obtained readily by microbiological methods. Although an asymmetric hydrogenation route to aspartame might prove interesting, little has been seen in the open literature.

Such applications as have been reported have mainly been in the synthesis of unusual or **D**-amino acids, particularly within small peptides. Dehydrodipeptides are good substrates for asymmetric hydrogenation

(Figure *4.3*) [119, 120]. The direction of the enantioselectivity is generally determined by the catalyst, with only a small contribution from the directing effect of the existing chiral centre. It is of note that most of the standard protecting groups used in peptide synthesis survive homogeneous hydrogenations of this type. Reactions (4.43) and (4.44) [121] show related examples used in the synthesis of an enkephalin analogue. Deuteriation and tritiation of small peptides is often useful in studying their transport and metabolism. The diastereoisomer ratio of the products from reaction (4.45) was established by ^{3}H NMR spectroscopy [122].

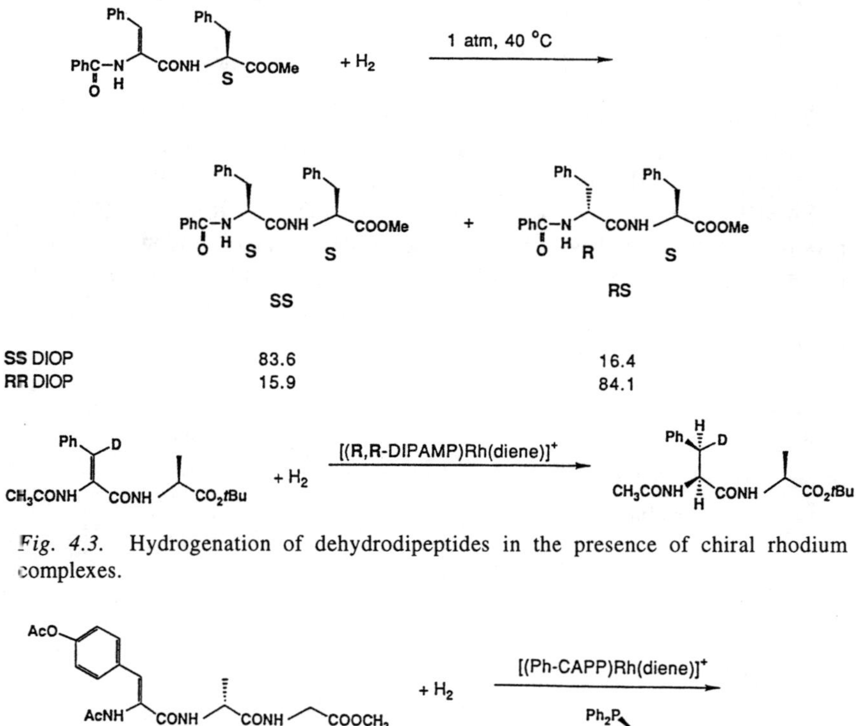

Fig. 4.3. Hydrogenation of dehydrodipeptides in the presence of chiral rhodium complexes.

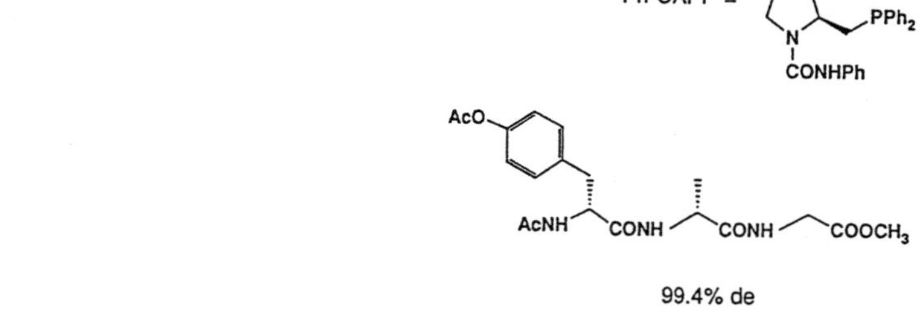

(4.43)

(4.44)

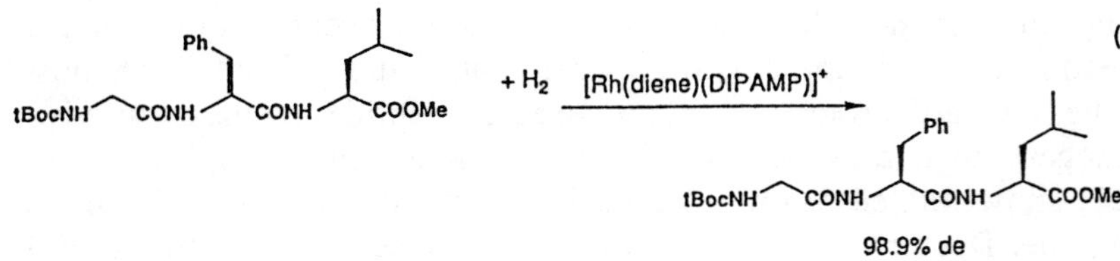

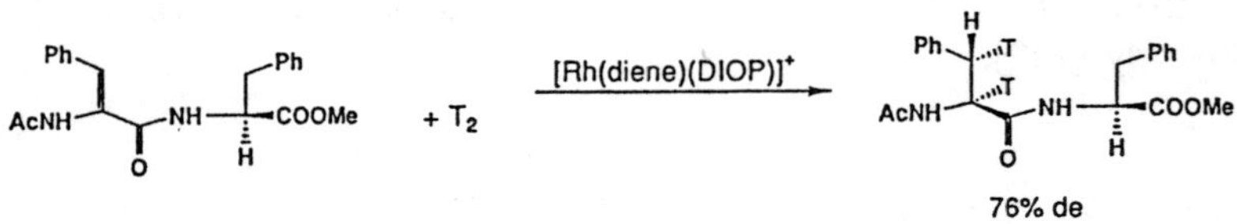

(4.45)

The reduction of **4-101** in the presence of a rhodium DIPAMP catalyst was a key step in the total synthesis of 4949-III, a natural inhibitor of aminopeptidase B from Ehrlich Ascites Carcinoma cells [123]. A DIPAMP based catalyst was also used in the reaction of **4-102**; the product was obtained with 99% diastereomeric purity, and was used in routes to chlamydocin and dihydrochlamydocin [124]. Additionally reactions the reactions of Figure *4.4* were used in a synthesis of biphenomycin [125].

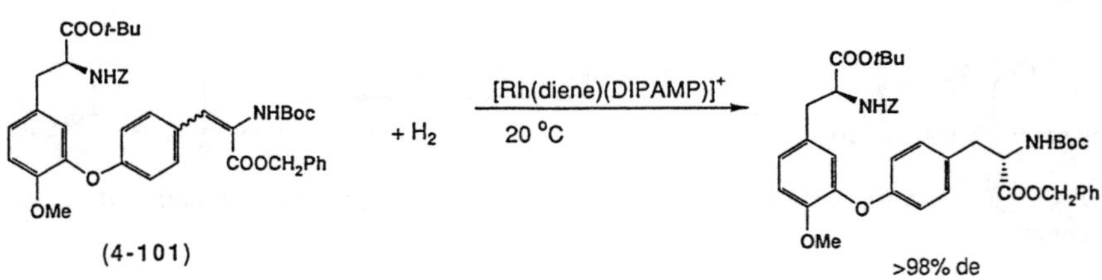

(4-101)

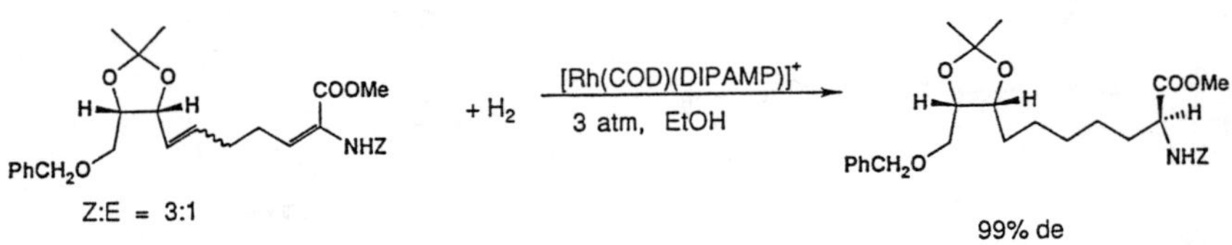

Z:E = 3:1

(4-102)

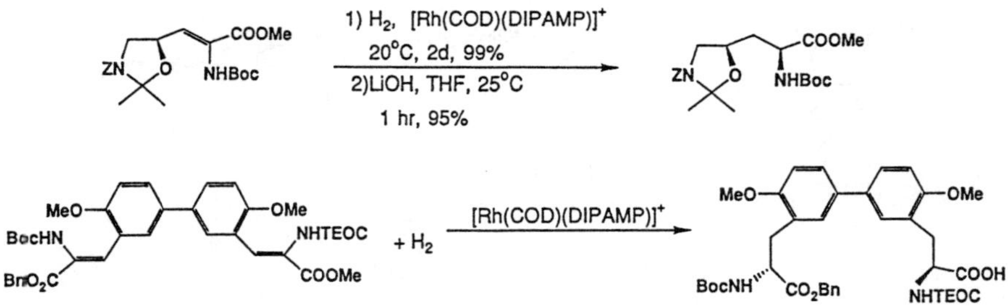

Fig. 4.4. Enantioselective hydrogenation reactions in the synthesis of biphenomycin.

4.11.2. REDUCTION OF OTHER ALKENES

4.11.2.1. *Alkenes with three binding sites*

Considerable effort has been expended on the testing of rhodium complexes of chiral phosphines as catalysts for the reduction of other substrates. Some useful results have been obtained for compounds with three potential metal binding sites (Tables *4.4* and *4.5*). Unfortunately

TABLE 4.4
Hydrogenation of itaconic acid, $CH_2=C(CH_2COOH)COOH$, in the presence of rhodium complexes of chiral phosphines

Phosphine		Added Et_3N	ee%	Configuration
R,**R**-DIPAMP,	(**4-63**)		77	**R**
NORPHOS,	(**4-66**)		63	**S**
	(**4-71**)[a]		91	
BPPM,	(**4-75**)	yes	71	**S**
BPPM,	(**4-75**)[b]		> 97	**S**
BCPM,	(**4-76**)	yes	92	**S**
R-BINAP,	(**4-80**)[c]	yes	90	**S**
	(**4-82**)[d]	yes	49	**S**
DIOXOP,	(**4-84**)		27	**R**[e]
	(**4-103**)	yes	68	**S**
	(**4-104**)		60	**R**
	(**4-105**)[f]		66	**R**
	(**4-106**)		85	**R**
BICHEP,	(**4-107**)[b]		> 99	**R**

[a] Substrate was dimethyl itaconate, chirality of product was not given
[b] Catalyst prepared from BPPM + $RhCl_3$ or $Rh_2(OCOMe)_4$ + **S**-PhCH(Me)NH_2. The hydrogen source was HCOOH [126]
[c] Catalyst was $[Ru_2($**R**-BINAP$)_2Cl_4H_2]$
[d] Catalyst prepared *in situ* from $[\{RhCl(diene)\}_2]$
[e] 1 atm pressure; at 20 atm pressure the optical yield was 22% **S**
[f] 0°C
[g] Substrate was dimethyl itaconate, pressure 5 atm [127]

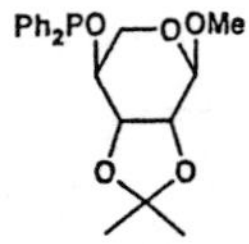

(4-103)

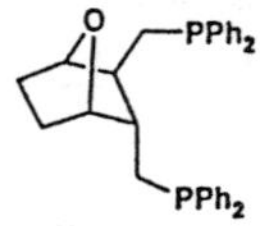

(4-104)

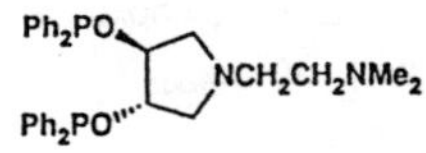

(4-105)

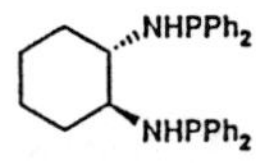

(4-106)

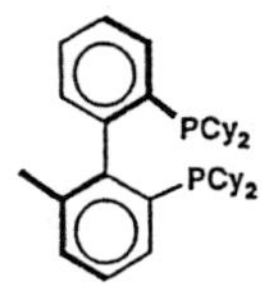

BICHEP
(4-107)

TABLE 4.5
Asymmetric hydrogenation of alkenes bearing two polar functional groups

Substrate	Phosphine	ee%	Configuration
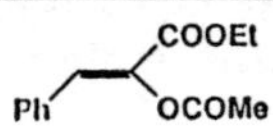	**R,R**-DIPAMP (**4-63**)	95	a
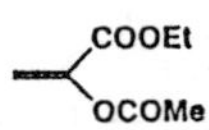	**R**-PROPHOS (**4-64**)	81	**S**
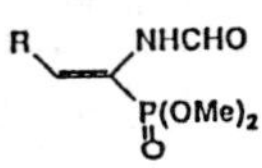	**S,S**-DIOP (**4-73**)	64	**S**
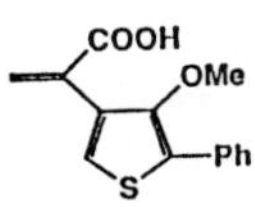	**R,R**-DIOP[b]	88	(+)
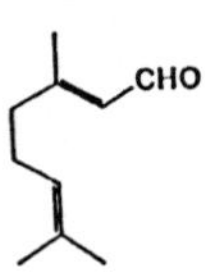	(**4-108**)[c]	48	**S**

[a] Not specified
[b] Et_3N added
[c] Catalyst prepared from $[Rh(CO)H(PPh_3)_3]$ and (**4-108**)

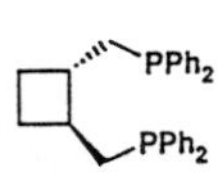

(4-108)

rather few of these reactions have produced products of much practical importance. It is perhaps worthy of note that phosphines which form seven-membered chelate rings are somewhat more tolerant in terms of substrate acceptability than the more rigid 5-membered ring complexes. Some binding studies exist, notable for itaconates [128] (itaconic acid, 3-carboxy-3-butenoic acid), but rather few generalisations can be made. Hydrogenation of the related diacid. **4-109**, was accomplished in 84% **S** enantiomer excess using $[Ru_2(R\text{-}BINAP)_2Cl_4]/Et_3N$ as the catalyst system. One recrystallisation gave optically pure material, used as the starting material in a lignan synthesis [129].

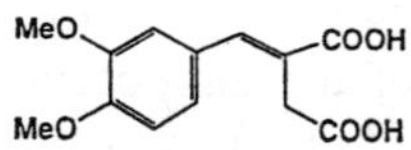

(4-109)

Two specific examples are perhaps worthy of note. Deuteriation of **4-110** using a rhodium complex of **R**-PROPHOS as the catalyst gave, after hydrolysis and recrystallisation, chiral methyl chiral lactic acid in 100% enantiomer excess. Since PROPHOS is itself prepared from lactic acid, this reaction sequence allows in principle for infinite catalyst generation [130].

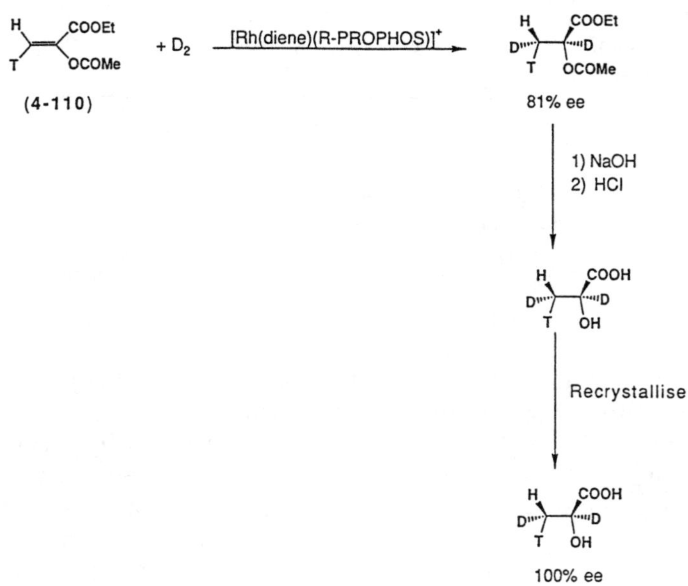

4.11.2.2. *Alkenes with a single polar binding site*

Early work on the reduction of alkenes with a single polar binding site (such as α,β-unsaturated carboxylic acids) had rather mixed success. Most of the catalysts used at this time were the rhodium complexes which had proved so successful in the reduction of dehydroamino acids. The pattern of greater substrate tolerance with phosphines which formed seven-membered chelate rings continues, as the data in Table *4.6* show. 2-Phenylpropenoic acid (atropic acid) has been a popular substrate, since the product is a model for a range of well-known anti-inflammatory compounds. So far as carboxylic acid derivatives are concerned, only one development in rhodium chemistry seems to hold real future promise. This involves derivatives of the aminobiphosphine, **4-111**. The remote amine functionality is thought to form a salt with carboxylic acid derivatives, and thus to direct coordination to the rhodium cation. This has allowed the reduction of **4-112** with better than 97% optical yield [131].

TABLE 4.6
Hydrogenation of 2-phenyl-propenoic acid, CH_2=C(Ph)COOH, in the presence of cationic rhodium complexes

Phosphine	Added Et_3N	ee%	Configuration	
PHELLANPHOS, (**4-67**)		12	S	
R,**R**-DIOP[a]		yes	63	S
DIOXOP, (**4-84**)	yes	58	S	
(**4-69**)	yes	0		

[a] Catalyst prepared *in situ* from [{RhCl(diene)}$_2$]

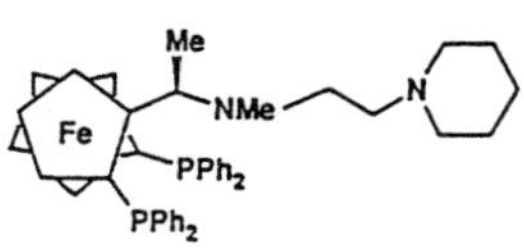

(4-111)

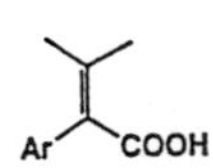

(4-112)

There is no doubt, however, that the most important advance derives from Noyori's work on the use of ruthenium complexes of BINAP, **4-80** [132]. This biphosphine is notable in that it is fully substituted with aryl groups, which exert a profound steric influence, and enhance the Lewis acidity of its complexes, the latter being particularly important in reactions involving substrates of relatively poor binding ability. Furthermore, aryl phosphines are more stable than alkyl derivatives.

We have already noted its usefulness in both rhodium and ruthenium complexes able to catalyse enantioselective reduction of dehydroamino acids.

Both α,β- and β,γ-unsaturated amino acids are hydrogenated in the presence of [Ru(BINAP)(OCOR)$_2$] with good optical yields. Examples are provided by reactions (4.46) [133] and (4.47), the latter giving a good synthesis of the anti-inflammatory compound, Naproxen. The sense and the extent of the optical induction are quite dependent on reaction conditions, particularly hydrogen pressure. The active catalyst may also be derived from [Ru(arene)(BINAP)Cl]$^+$; using the **S**-enantiomer of this catalyst E-$CH_3CH{=}C(CH_3)COOH$ was reduced in 89% **S** optical yield [134].

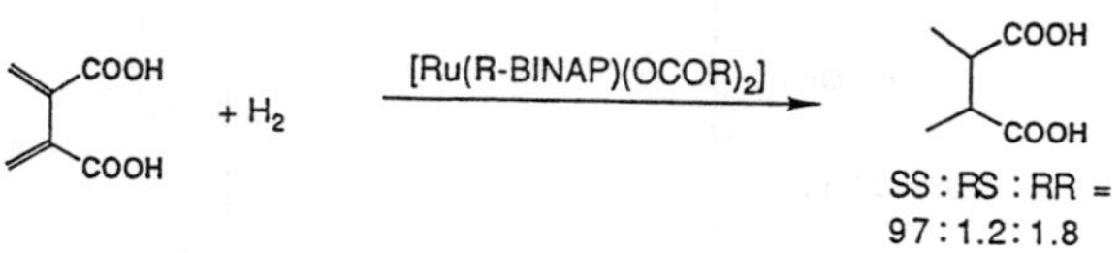

(4.46)

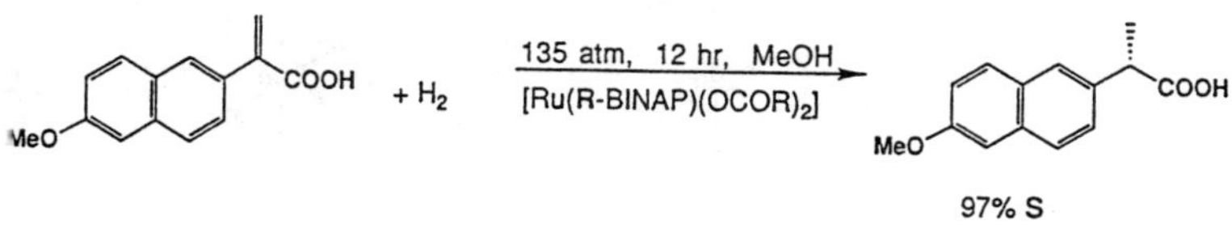

(4.47)

There has been a careful study of the deuteriation of **E**-$CH_3CH{=}C(CH_3)COOH$ to give **4-113**. It was shown that the deuterium at the α-position derives from molecular deuterium, but that the label at the β-position is solvent derived (MeOD). As the hydrogen pressure is increased, more of the added atom at the β-position is derived from molecular deuterium. This strongly suggests a monohydride mechanism for the reduction, shown in Figure *4.5* [135]. This is supported by kinetic measurements and an X-ray diffraction study of [Ru(BINAP)(MeCH=C(Me)CO$_2$)$_2$] [136]. This may well be related to earlier work concerning the preparation of **4-114a** and **4-114b**, both of which catalyse reduction of itaconic acid, and both of which react reversibly with H_2 to give the same molecular hydrogen complex, **4-115** [137].

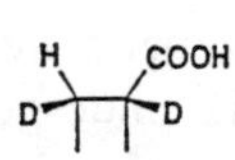

(4-113)

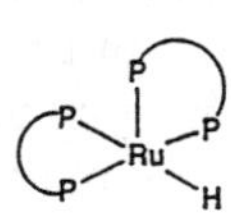

(4-114a)

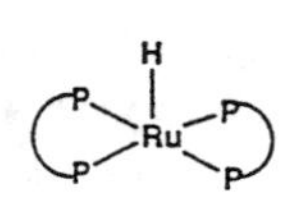

(4-114b)

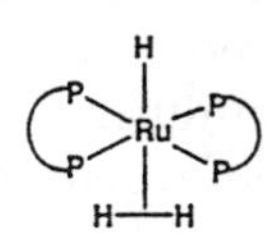

(4-115)

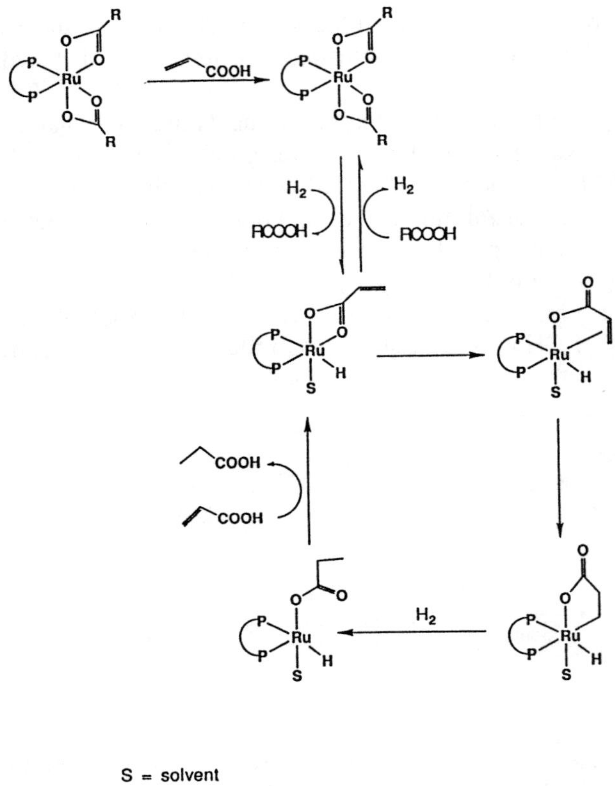

Fig. 4.5. Proposed mechanism for hydrogenation of α,β-unsaturated acids in the presence of ruthenium BINAP complexes.

There has not been comparable success in the enantioselective reduction of α,β-unsaturated ketones, and reports in the literature remain scattered, possibly because few of the products prepared to date have any very high synthetic value. The reduction of piperitenone, **4-116**, is one of the few to have been studied in detail. One study used $[Rh(CAMP)_2(COD)][BF_4]$ as the catalyst; the best enantiomer excess obtained for pulegone, **4-117** was 33%, with 74% chemical selectivity [138]. Using [RuClH(**R,R**-*trans*-**4-108**)$_2$] as the catalyst resulted, somewhat surprisingly, in substantial initial reduction of the more substitued double bond to give **4-118**, with up to 39% enantiomer excess, Subsequent reaction was complicated by the fact that **4-118** undergoes kinetic resolution in the presence of the catalyst [139]. Other studies have used isophorone as the substrate; catalysts included $[Co_2(CO)_6(NMDPP)_2]$

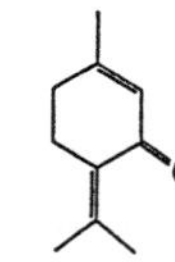

(4-116)

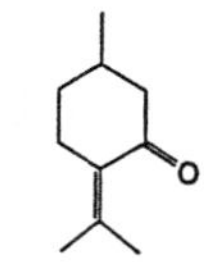

(4-117)

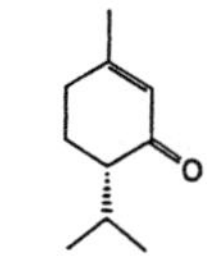

(4-118)

(NMDPP = neomenthyldiphenylphosphine, 16% **S**) [140], and [RuClH{(–)-1,2-*trans*-**4-108**}$_2$] (62% **R**) [141].

Simple enamides were popular substrates in early work on asymmetric hydrogenation catalysed by rhodium DIOP complexes, but this work has now been entirely eclipsed by the more recent studies involving BINAP complexes. In the presence of ruthenium BINAP complexes the 1-alkylidenetetrahydroisoquinoline, **4-119**, is reduced to **4-120** under 1–4 atm. hydrogen pressure. The use of **R**-BINAP complexes leads to the formation of **R**-products, and the reaction has been used in enantioselective syntheses of terahydropapaverine, **4-121**, laudanosine, **4-122** and salsolidine, **4-123**. Enantioselective hydrogenation, in combination with Grewe type annulation led to the preparation of optically active morphine, **4-124**, and analogues. It seems likely that this approach will in future be widely applied in the synthesis of the isoquinoline alkaloids [132].

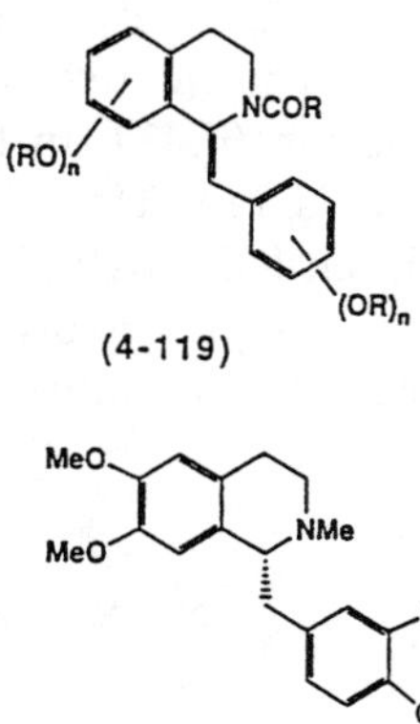

(4-119)

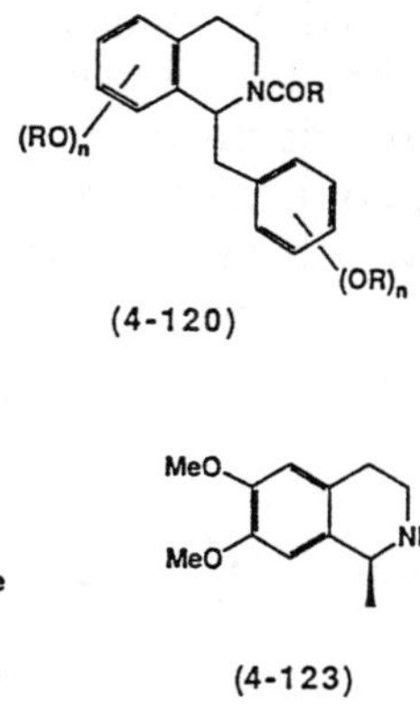

(4-120)

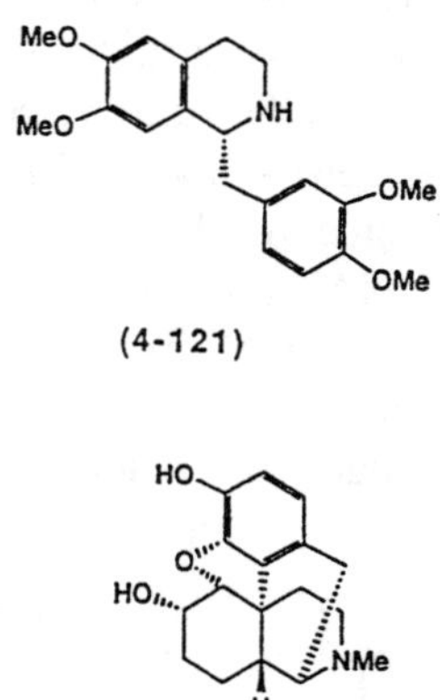

(4-121)

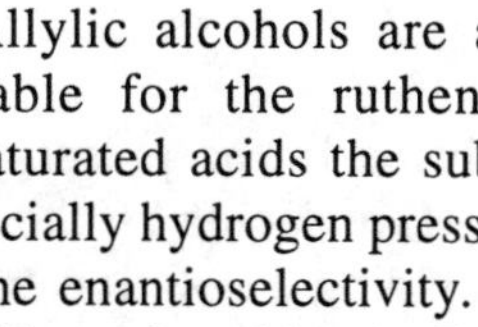

(4-122)

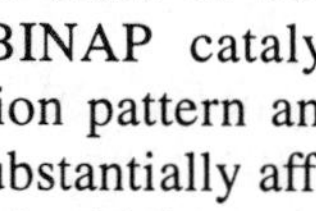

(4-123)

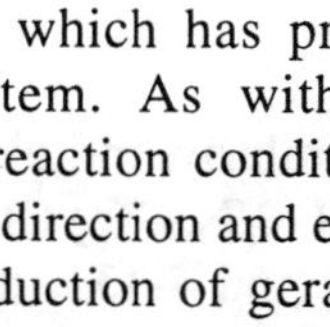

(4-124)

Allylic alcohols are another class of substrate which has proved suitable for the ruthenium BINAP catalyst system. As with the unsaturated acids the substitution pattern and the reaction conditions, especially hydrogen pressure, substantially affect the direction and extent of the enantioselectivity. Thus the high pressure reduction of geraniol, **4-125**, catalysed by the **R**-BINAP ruthenium complex, gave **S**-citronellal, **4-126** in 96–99% optical yield. Much poorer results were obtained using

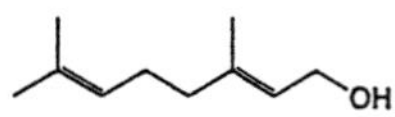

(4-125)

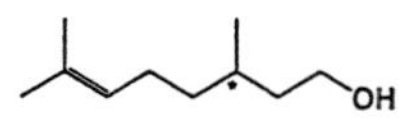

(4-126)

rhodium BINAP complexes as the catalysts. By contrast, nerol, **4-127**, was reduced to **R**-citronellal in the presence of the **R**-BINAP ruthenium complex, again with an excellent optical yields. The process has been applied in the synthesis of **S**-dolichol from polyprenols, and the side chain of vitamin E, **4-128** [132].

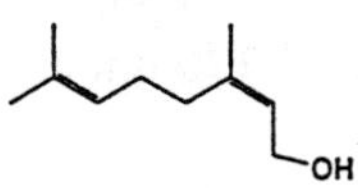

(4-127)

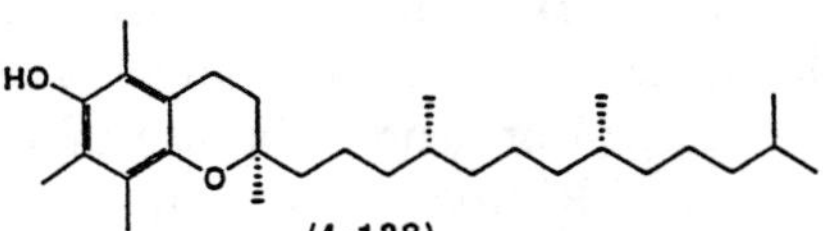

(4-128)

A number of the most interesting enantioselective reductions of allylic alcohols have in fact also been diastereoselective reactions. When a reduction which has an inherently high diastereoselectivity using an achiral catalyst is performed in the presence of a chiral complex, we expect to see double stereoselection. The preference of the catalyst may reinforce the preference of the molecule, or it may act in the opposite direction. A useful illustration is provided by reaction (4.48). In the presence of the achiral catalyst diastereoselection is fair. Using an **R**-BINAP rhodium derivative it is much improved, but **S**-BINAP gives a poorer result. We should note that in this, as in most examples involving allylic or homoallylic alcohols, the preference of the substrate dominates the reaction [87, 142].

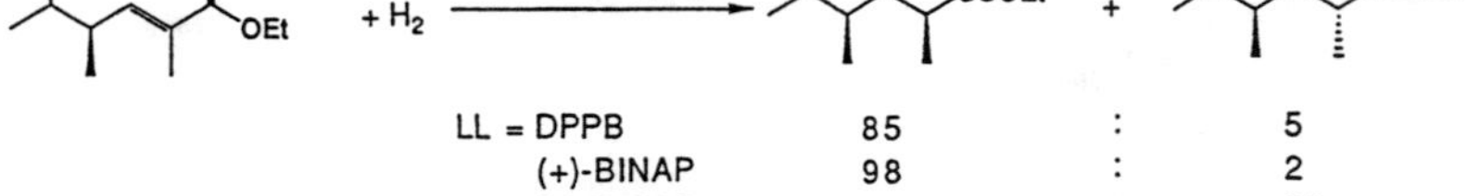

(4.48)

LL = DPPB	85	:	5
(+)-BINAP	98	:	2
(-)-BINAP	67	:	33

Similarly, the allylic alcohol **4-129** was reduced in the presence of [Ru(**R**-BINAP′)$_2$(OCOR)$_2$] (BINAP' has 4-MeC$_6$H$_4$ replacing Ph), to give **4-130** and **4-131** in the ratio 99.9:0.1, thus providing a solution to the 1-β-methylcarbapenem problem. The same reaction in the presence of the **S**-BINAP' derivative resulted in only modest stereoselection, 22:78.

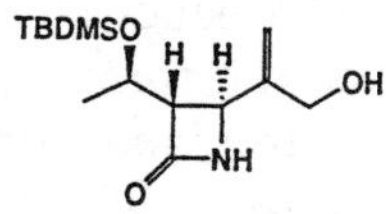

(4-129)

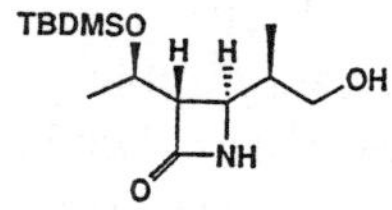

(4-130)

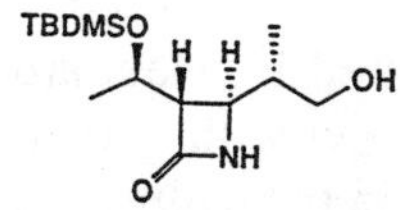

(4-131)

4.11.2.3. *Reduction of non-polar alkenes*

The remaining problem in the reduction of carbon-carbon double bonds is that of the reduction of strictly non-polar substrates such as PhC(Et)=CH_2. A number of studies have used well established catalysts; these vary widely in quality, and few patterns can be discerned. The best results have been reported using rhodium derivatives of PHELLANPHOS (31% **S**), **4-132** (37% **R**) [143], or **4-83** (54% **S**) [144], or $PtCl_2$/$SnCl_2$/DIOP (36% **S**). The most promising development involves the catalyst [$TiCl_2$(**4-133**)$_2$] which catalysed the reduction of PhC(Et)=CH_2 with up to 96% optical yields [145]. Related catalysts have proved very effective in stereoselective polymerisation of simple alkenes, suggesting that appropriate and reproducible catalyst substrate interactions are readily established.

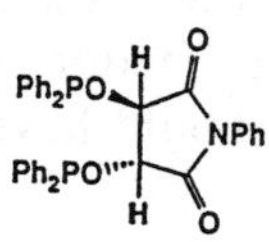

(4-132)

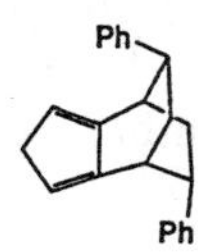

(4-133)

4.11.2.4. *Kinetic resolution*

We noted earlier that α-(hydroxyalkyl)acrylates, **4-134**, were good substrates for diastereoslective hydrogenation in the presence of achiral catalysts, leading to the *anti*-product in all cases. If the starting material is racemic, and it is reduced with a chiral catalyst we might expect that one enantiomer would react more rapidly than the other. This indeed proved to be the case when a rhodium DIPAMP catalyst was used. Thus **4-135** was reduced in THF at 0°C and the reaction was allowed to

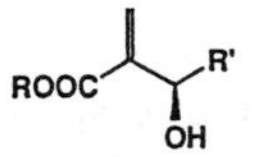

(4-134)

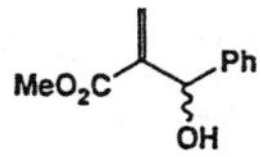

(4-135)

proceed to 70% completion. Starting material and product could be separated by standard chromatographic procedures, the former being recovered with > 90% optical purity. Double asymmetric induction was noted in the hydrogenation of **L**- or **D**-menthyl α-(hydroxyethyl)acrylate, the **L**-enantiomer being reduced with the greater selectivity. A similar result was achieved for **4-136** using [Ru(**S**-BINAP)$_2$ (OCOMe)$_2$] as the catalyst. Recovery of the starting material was modest, but it was of very high optical purity [146].

The kinetic resolution of the simpler allylic alcohols, **4-137** and **4-138** has also been studied using ruthenium BINAP catalysts. k_R/k_S for **4-137** was up to 76, under optimised conditions, and **R-4-138** in an important building block in prostaglandin synthesis [147].

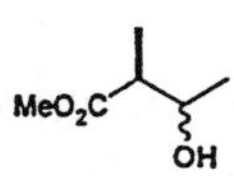

(4-136)

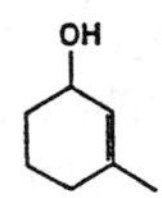

(4-137)

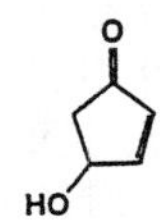

(4-138)

Other groups have also proven to be useful in directing kinetic resolution. Thus **4-139** reacts slowly with hydrogen in the presence of [Rh(diene)(DIPAMP)]$^+$, but its enantiomer reacts more rapidly, with k_R/k_S = 10–16. Results with **4-140** were still better, suggesting that this process should be rather general, and should have wide applicability [148].

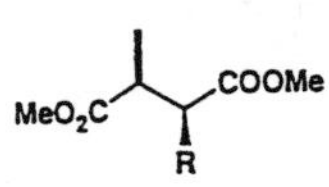

(4-139)

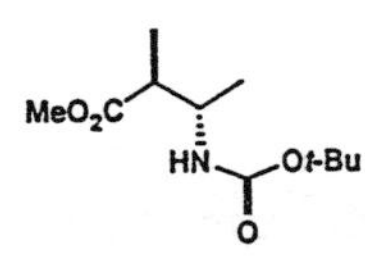

(4-140)

4.11.3. REDUCTION OF CARBON-OXYGEN AND CARBON-NITROGEN DOUBLE BONDS

The early development of enantioselective reduction of carbonyl groups showed strong parallels to the reactions studied for derivatised alkenes. Most of the common rhodium complexes available were tested, but successes were limited, and rather harsh conditions were generally required. Alkyl aryl ketones were reasonable substrates (Table *4.7*), but dialkyl ketones were invariably reduced with low optical efficiency.

TABLE 4.7
Asymmetric hydrogenation of acetophenone in the presence of rhodium complexes, $[Rh(diene)LL]^+$

Phosphine	Conditions	ee%	Configuration
R,R-DIOP[a]	70 atm, Et_3N	80	**S**
R,S-BPPFA	20°C, 50 atm, 65 h	15	**S**
R,S-BPPFOH, (**4-141**)	0°C, 50 atm, 8 h	43	**R**

[a] Catalyst prepared *in situ* from $[\{RhCl(diene)\}_2]$

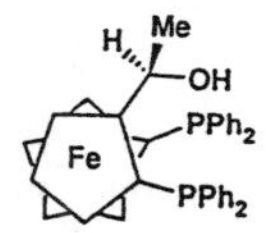

R,S-BPPFOH
(4-141)

α-Ketoesters have proved to be more popular and successful substrates (reaction (4.49)), in particular ketopantolactone, **4-142**, which may be converted into pantothenic acid (Table *4.8*). Other polar functional groups in the vicinity of the carbonyl can also usefully aid binding (reactions (4.50) [149], (4.51) [150] and (4.52) [151]). The products from the reduction of α-aminoketones may be used in the synthesis of important β-blocking agents.

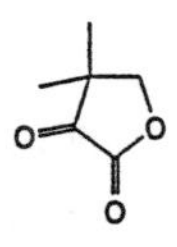

(4-142)

TABLE 4.8
Asymmetric hydrogenation of pantolacone, **4-142**, in the presence of $[\{RhCl(diene)\}_2]$/L

Phosphine		Conditions	ee%	Configuration
BPPM,	(**4-75**)	30°C, 50 atm, C_6H_6, 48 h	87	**R**
BCPM,	(**4-76**)	50°C, 50 atm, THF	92	**R**
S,S-DIOP,	(**4-73**)	30°C, 50 atm, C_6H_6, 48 h	39	**R**
DIOCP,	(**4-143**)	50°C, 50 atm, THF, 45 h	75	**R**
	(**4-144**)	50°C, 50 atm, THF, 45 h	45	**R**
	(**4-145**)	20°C, 1 atm, toluene	80	**S**

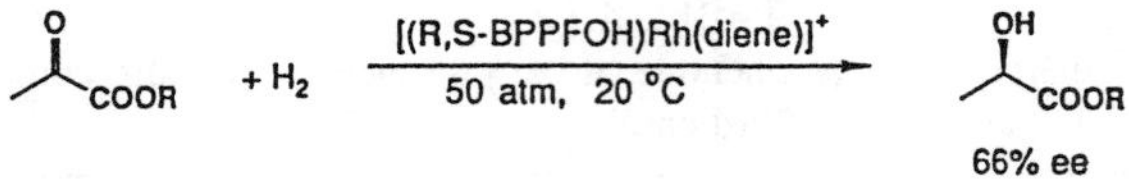

(4.49)

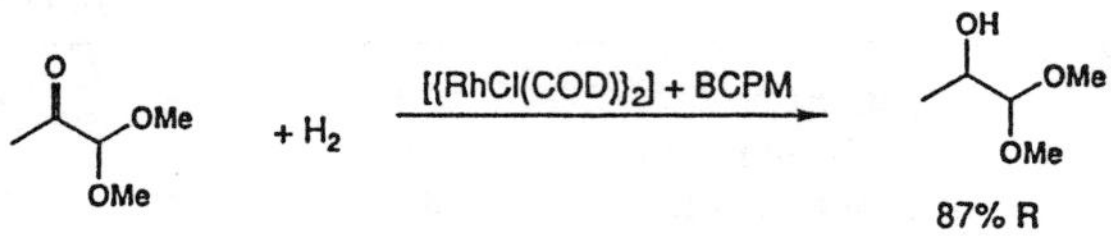

(4.50)

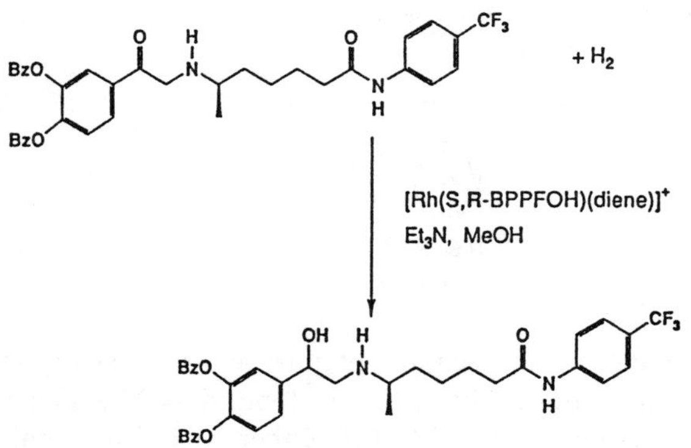

(4.51)

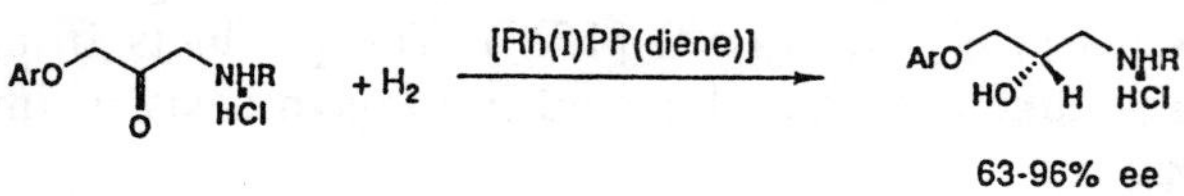

(4.52)

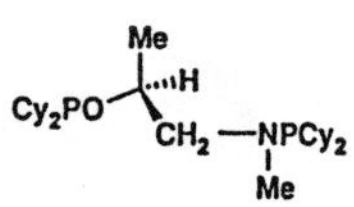

(4-145)

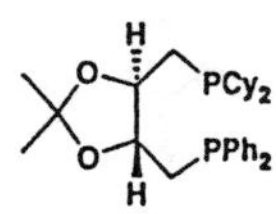

DIOCP
(4-143)

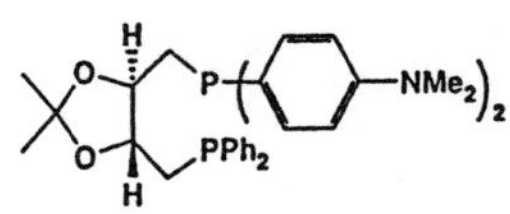

(4-144)

Much of this work has been superceded by the discovery that ruthenium BINAP derivatives are excellent catalysts for the enantioselective reduction of a wide range of functionalised ketones. In most cases halogen containing ruthenium complexes have proved to be superior to carboxylate derivatives. The range of functionalities which has been used to direct the reaction is large, and includes NR_2, OH, OR, COOR, and $CONR_2$. Compounds **4-146** to **4-151** are examples of chiral hydrogenation products obtained with **R**-BINAP ruthenium catalysts in alcoholic solvents. A wide range of β-keto esters have been used as substrates; most of the hydrogenations proceeded with 100% yield and in 100%

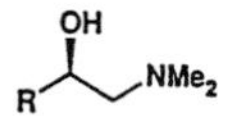

(4-146)
R= Me, CHMe$_2$, CMe$_3$, Ph
93-95% ee

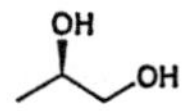

(4-147)
92% ee

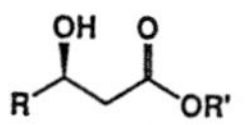

(4-148)
98-100% ee

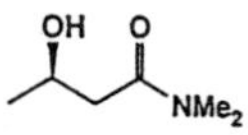

(4-149)
96% ee

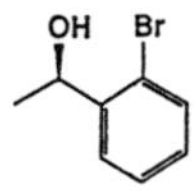

(4-150)
92% ee

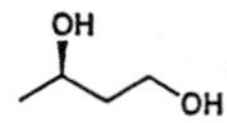

(4-151)
96% ee

optical purity. Since α,α-dialkylated products may be obtained in good optical yield it is evident that the reaction does not involve reduction of the enol. Tri- and tetrasubstituted carbon-carbon double bonds usually survive the reaction conditions. Reaction (4.53) was used in the synthesis of carnitine [152], (4.54) in the preparation of the spore germination inhibitor Gloeosporone [153] and (4.55) in the synthesis of FK506 [154]. Reduction of a racemic mixture of **4-152** and its enantiomer in the presence of a ruthenium derivative of S-BINAP gave a separable mixture of **4-153** and **4-154**, which were used in the synthesis of a novel chiral phosphine. Hydrogen was added to the *re*-face of the carbonyl group in both cases [155].

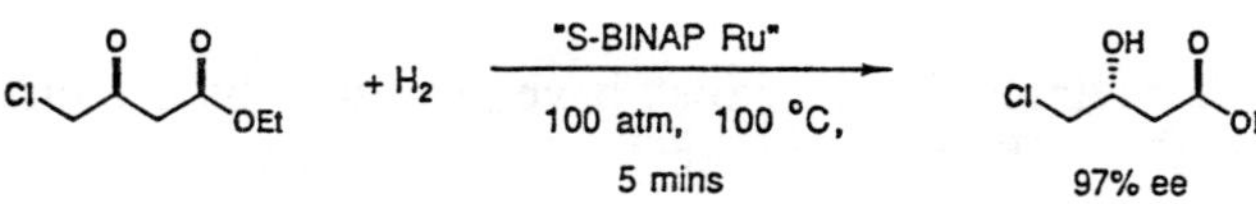

(4.53)

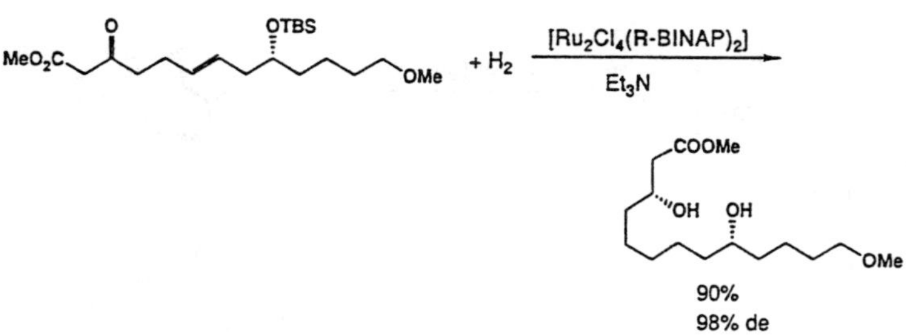

(4.54)

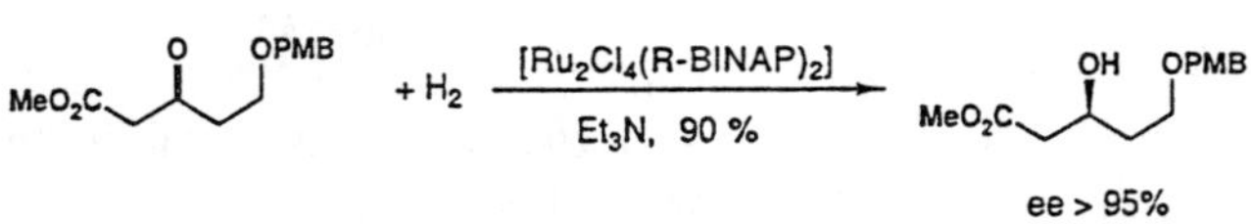

(4.55)

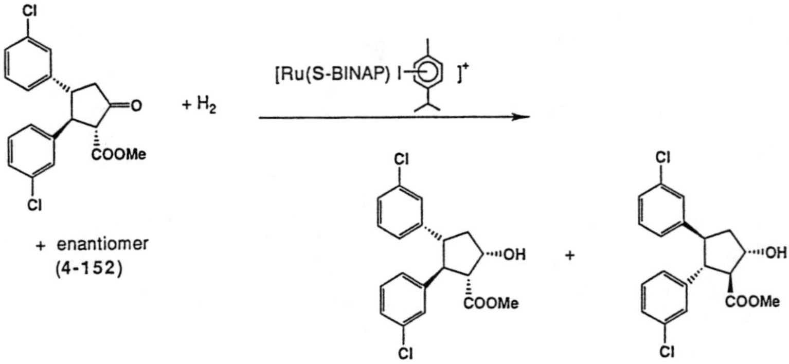

+ enantiomer
(4-152)

(4-153) (4-154)

The reduction of β-diketones is another useful reaction. When 2,4-pentanedione is reduced in the presence of an **R**-BINAP ruthenium catalyst, the almost optically pure **R,R**-diol, **4-155**, and the *meso*-compound, **4-156** were produced in a 99:1 ratio. The reaction proceeds *via* the **R**-hydroxyketone, **4-157**, formed in 98.5% enantiomer excess, but most of the **S**-isomer is removed by conversion to the *meso*-diol [156].

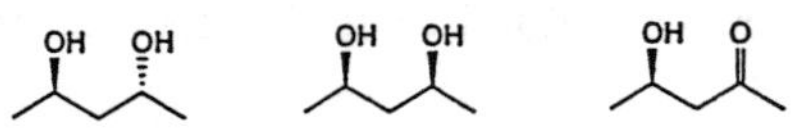

(4-155) (4-156) (4-157)

Some double asymmetric inductions have proved to be useful; thus, hydrogenation of **4-158** in the presence of an **R**-BINAP catalyst gave almost entirely **4-159**, a protected version of **S,S**-statine, **4-160** [157].

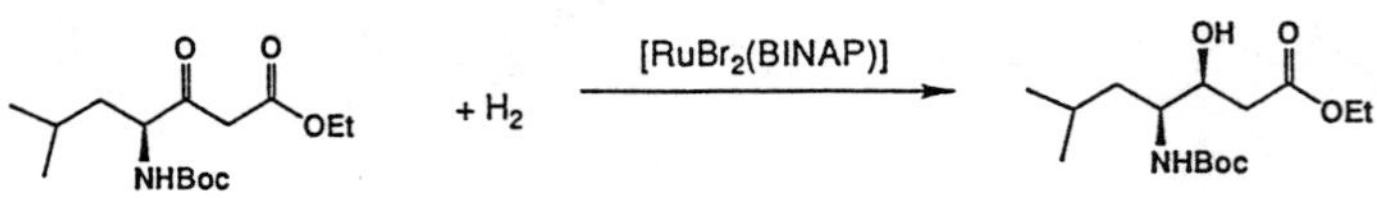

(4-158) (4-159)

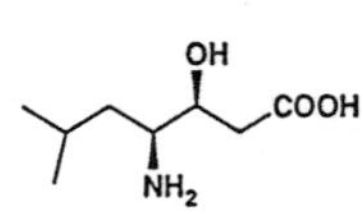

(4-160)

Since substituted β-ketoesters such as **4-161** are relatively labile, there exists the opportunity for a useful dynamic kinetic resolution, provided that equilibration is faster than hydrogenation. Thus, reaction of the equilibrating system **4-161a** and **4-161b** (R = NHCOMe) with hydrogen in the presence of a ruthenium BINAP complex gave mainly *syn*-product in 98% optical purity, leading to a useful synthesis of **L**-threonine. Substrates bearing alkyl substituents were reduced with good enantioselection, but relatively poor *syn/anti* selectivity [158,159]. A mathematical treatment of the reaction has been described, and applied to a range of case studies [160].

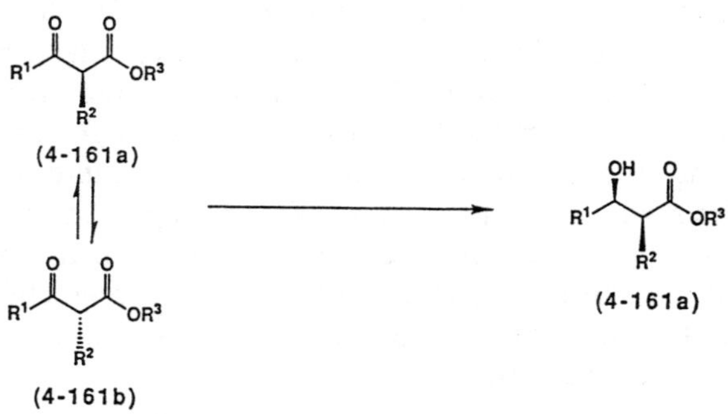

There have been rather few reports of successful enantioselective hydrogenation of imines. Early work using rhodium or ruthenium DIOP complexes as the catalysts gave rather modest enantiomer excesses, whilst a catalyst derived from [{RhCl(NBD)}$_2$] and **R**-PhCH$_2$CH(PPh$_2$)CH$_2$PPh$_2$ gave a rather unreproducible optical yield of up to 72% **S** for reduction of PhC(Me)=NCH$_2$Ph [161]. Some recent results are more interesting. The reduction of ArC(CH$_3$)=NCH$_2$Ph using a catalyst derived from [{RhCl(COD)}$_2$] and CyCH(PPh$_2$)CH$_2$PPh$_2$ resulted in the formation of ArCH(Me)NHCH$_2$Ph in low enantiomer excess. However, if iodide ion was added to the reaction mixture, the optical yield could be increased to 91% [162]. The structure of the true catalytic species is probably similar to that of [{IrHI$_2$(PP)}$_2$] which exists as a mixture of cisoid and transoid isomers; a range of phosphines proved to be useful in these systems [163]. Good results have also been obtained using [{RhCl(COD)}$_2$] and the sulphonated phosphine, **4-162** (Ar = 3-NaO$_3$SC$_6$H$_4$) [164]. In the reduction of PhC(CH$_3$)=NCH$_2$Ph, the best

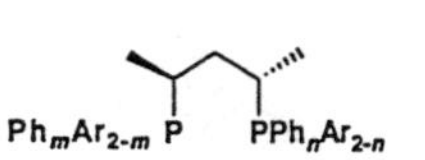

(4-162) m,n = 0,1,2

catalyst was derived from the bis monosulphonated ligand ($m = n = 1$, ee = 94%); with the complex derived from the disulphonated ($m = n = 2$) phosphine, the enantiomer excess obtained was very low [165].

4.12. Hydrogenolysis

Hydrogenolysis, the splitting of a bond AB to give AH and BH, has found most of its applications in synthesis in the removal of protecting groups, or other unwanted functionality, and the cleavage of small rings. Benzyl ethers, esters and amines are readily cleaved in the presence of heterogeneous catalysts, so that the benzyl group is useful for the protection of oxygen and nitrogen functionalities. Although there are homogeneous catalysts which fulfil the same function, applications have been few. Hydrogenolysis of H-AlaOBn was achieved in the presence of $K_3[Co(CN)_5]$ without racemisation [166] and the deprotection of **4-163** to give **4-164** in the presence of $PdCl_2$ was used in a synthesis of β-carboxyaspartic acid [167].

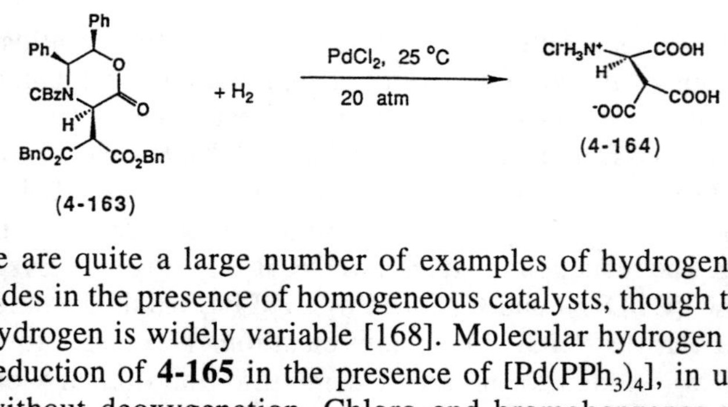

(4-163) (4-164)

There are quite a large number of examples of hydrogenolyses of aryl halides in the presence of homogeneous catalysts, though the source of the hydrogen is widely variable [168]. Molecular hydrogen was used in the reduction of **4-165** in the presence of $[Pd(PPh_3)_4]$, in up to 90% yield, without deoxygenation. Chloro and bromobenzenes were also reduced, and aldehyde and nitro groups tolerated if care was taken with the reaction conditions [169, 170]. $[RuCl_2(PPh_3)_3]$ proved to be a good catalyst for the site-specific deuteriolysis of 5-iodouracil [171], and $[RhHCl_2(PCy_3)_2]$ catalysed hydrogenolysis of chloroarenes under biphasic conditions [172].

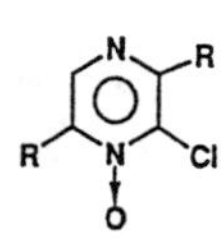

(4-165)

Transfer hydrogenolysis of **4-165** in the presence of $[Pd(PPh_3)_4]$ was accomplished using Na[HCOO] as the source of hydrogen; the products were used in sytheses of arglecin and argvalin [173]. Phase transfer condition proved to be most suitable for the related reduction of halobenzenes. Sources of hydrogen for transfer hydrogenolyses have included indoline, methanol and benzyl alcohol. Other hydrogenolyses of aryl halides involve hydride as the hydrogen source, either from $Na[BH_4]$, $Li[AlH_4]$, polymethylhydrosiloxane [174] or Grignard reagents.

The removal of hydroxyl groups from aryl rings had proved to be a rather intractable problem, but this may now be readily accomplished by hydrogenolysis of the triflates. Most of the reported reactions involve hydrogen transfer from methanoic acid, and are catalysed by palladium salts under phase transfer conditions. Halide, -OR, -CHO, -COOR and alkene functional groups are tolerated [175] Reaction (4.56) was used in a synthesis of the furanocoumarins [176], and (4.57) in the preparation of water soluble camptothecin analogues [177].

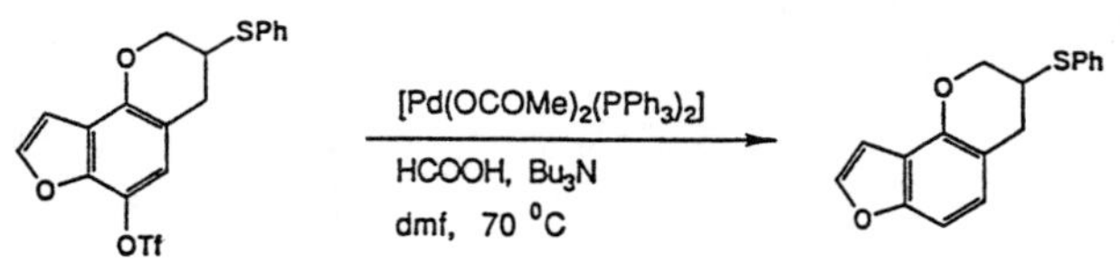

(4.56)

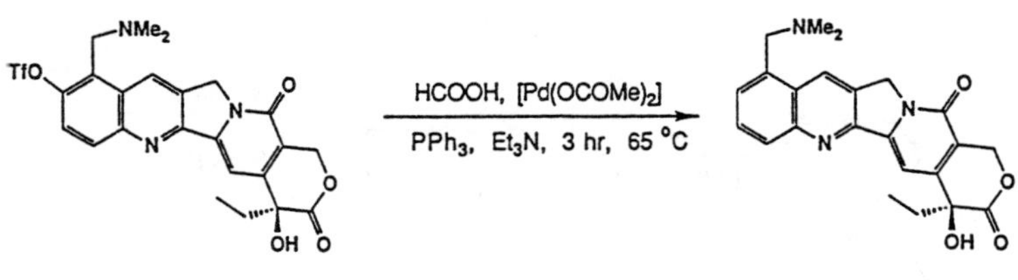

(4.57)

There have been rather fewer reports of hydrogenolyses of alkenyl derivatives, and the groups removed have been more varied. Triflates are good substrates, and their preparation from a carbonyl group and subsequent removal offer a uniquely mild alternative to the Clemensen or Wolff-Kishner reductions (reaction (4.58)) [178]. The souce of hydrogen may be methanoic acid, as in reactions (4.58) and (4.59), the latter being used in a regiocontrolled synthesis of the ergosterol B isomers [179], or Bu_3SnH, as in reaction (4.60) [180].

The hydrogenolysis of alkenyl sulphones may be accomplished using a Grignard reagent as the reductant and $[Ni(acac)_2]$ or $[Pd(acac)_2]$ as the catalyst, as for example in reaction (4.61). The reaction tolerates acetals and ethers, and has been widely used in pheromone synthesis

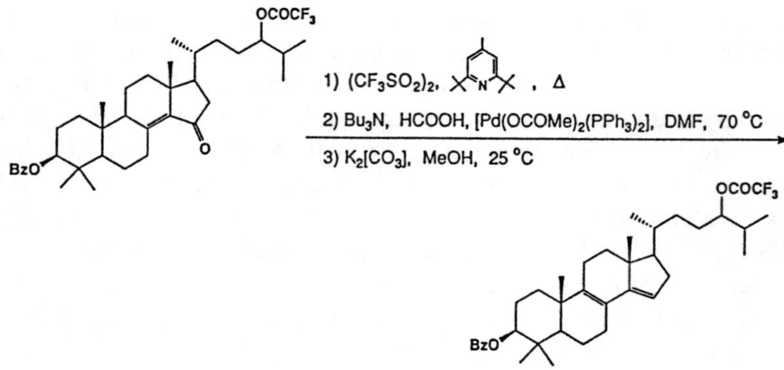

(4.58)

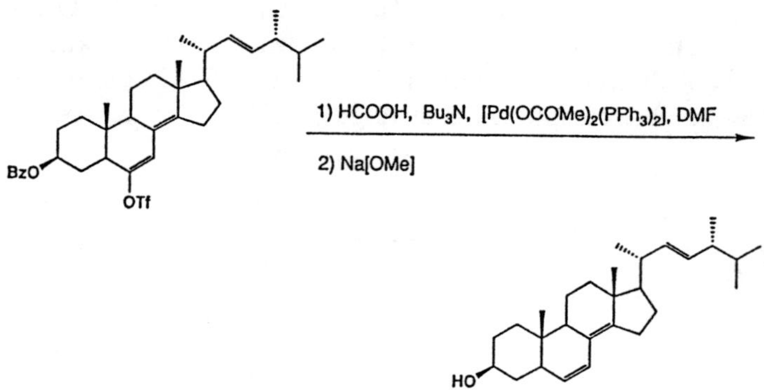

(4.59)

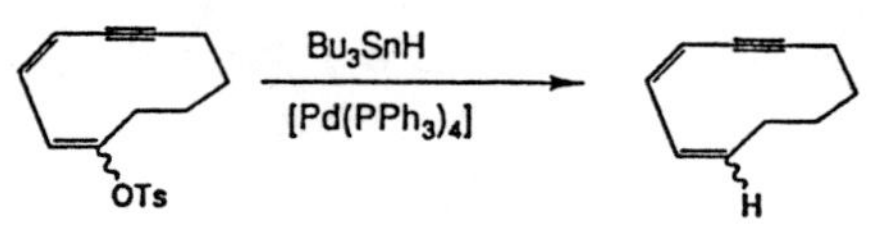

(4.60)

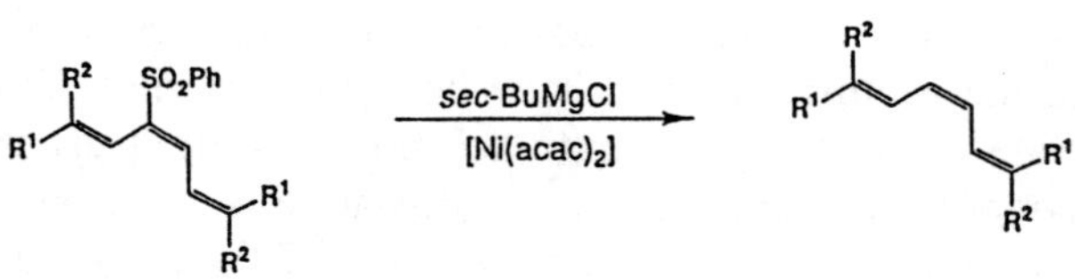

(4.61)

[181]. A related reaction of an alkenyl sulphide, (4.62), was used in a preparation of the Douglass Fir Tussock moth pheromone [182].

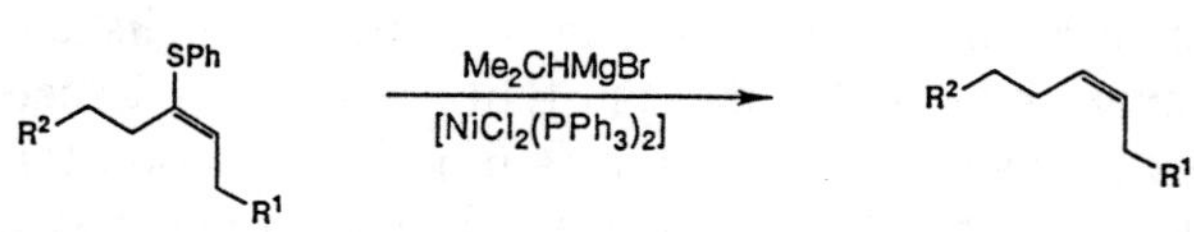

(4.62)

The hydrogenolysis of **4-166** using Et_3Al as the reductant and $[Pd(PPh_3)_4]$ as the catalyst gave **4-167**, though stereochemical control was limited [183].

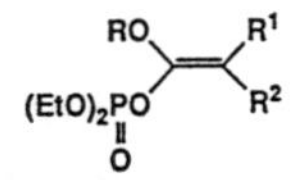

(4-166)

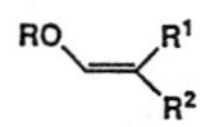

(4-167)

There have been a large number of reports of hydrogenolyses of allyl derivatives, most of them involving π-allyl complexes as the intermediates. In most cases the source of the hydrogen is a hydride donor. Examples of the common types are shown in reactions (4.63), (4.64) and (4.65); all proceed *via* palladium allyl intermediates [184–186]. Reaction (4.66) was used in a synthesis of the

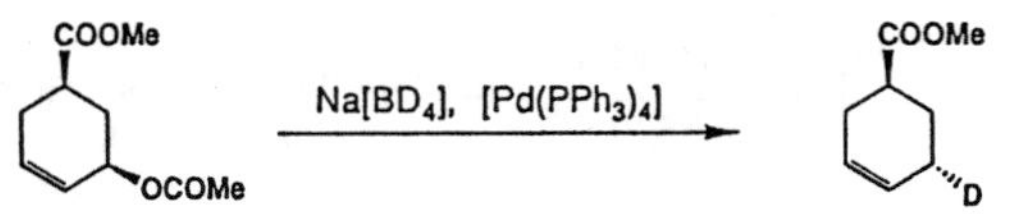

(4.63)

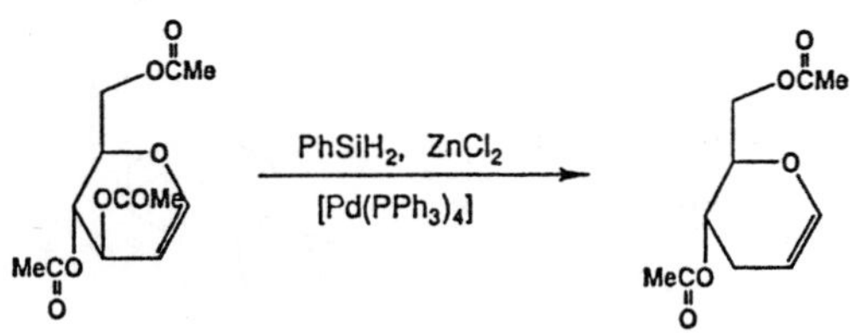

(4.64)

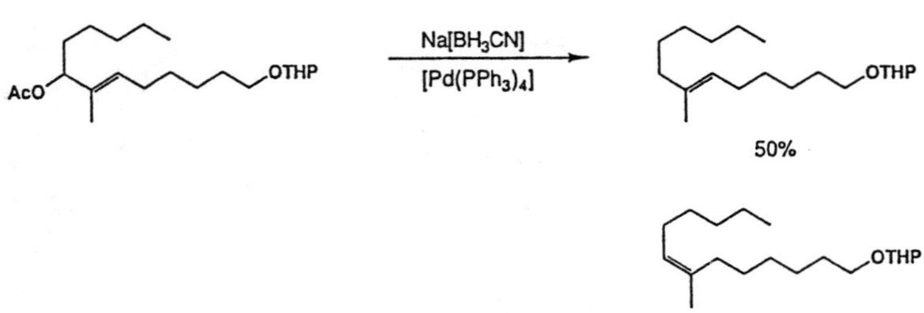

(4.65)

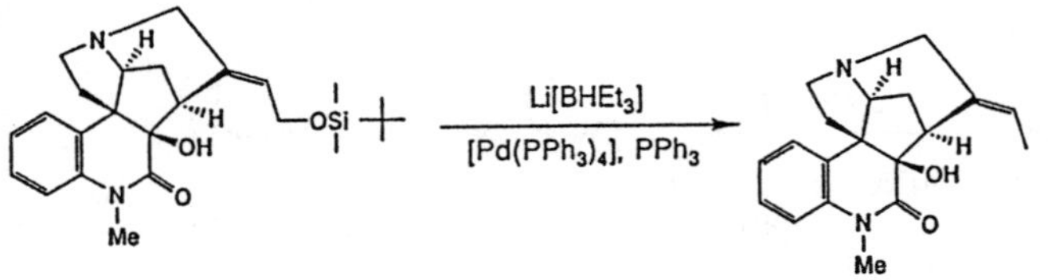

(4.66)

calebassinine skeleton [187], and (4.67) in the preparation of coenzyme Q [188]. 4α-Deuteriated and tritiated steroids were prepared by hydrogenolysis of an allylic carbonate (reaction (4.68)); the process showed a pronounced isotope effect [189, 190]. The regiochemistry of the reaction has been studied in some detail; alkyl phosphines have been said to be particularly useful in attaining selectivity towards 1-alkenes (reaction (4.69)) [191], and the source of the hydride donor is also important (reaction (4.70)) [192]. A number of these hydrogenolyses have been involved in the removal of allyl protecting groups (reactions (4.71) and (4.72)) [193, 194]. Methanoate has been used as a hydrogen donor in the hydrogenolysis of a dienyl triflate, which often proceeds with double bond migration [195] and complexes such as **4-168** have been isolated as intermediates [196].

There has been a few reports of related reactions of propargyl derivatives, exemplified by reaction (4.73) [197]. Good selectivity towards

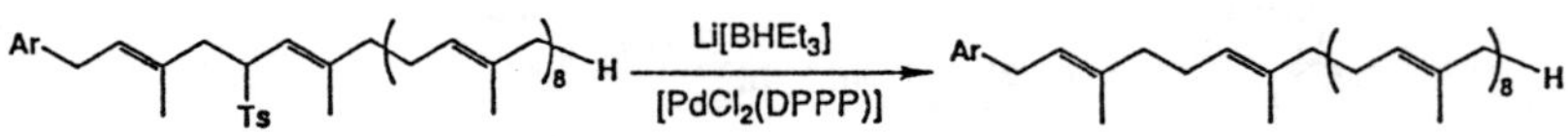

(4.67)

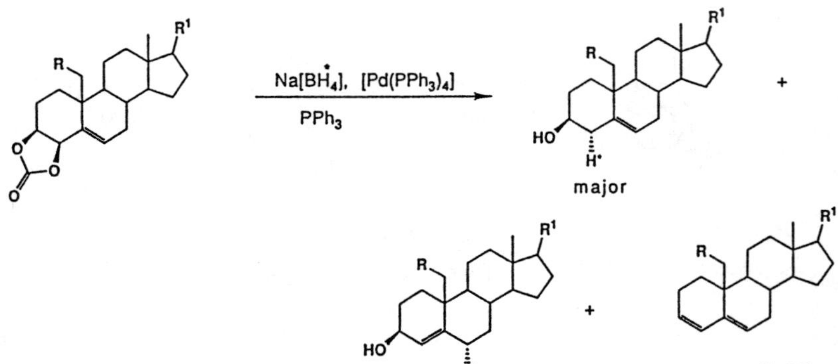

(4.68)

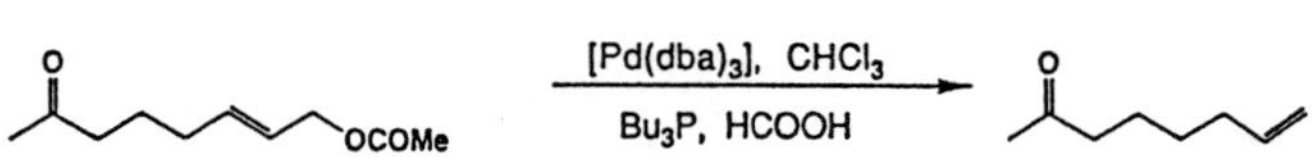

(4.69)

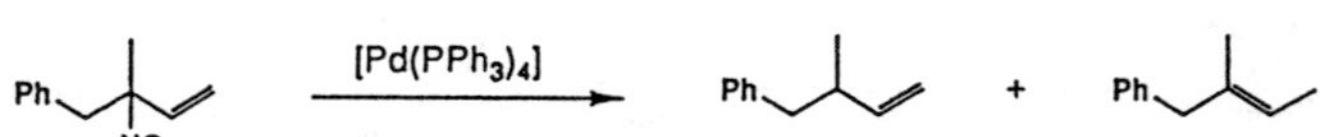

(4.70)

[H4N][HCOO]	92	:	8
Li[(*sec*-Bu)3BH]	1	:	99
BuZnCl	0	:	100

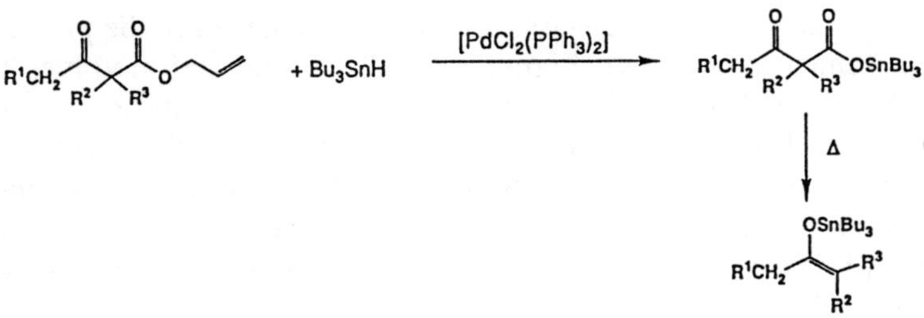

(4.71)

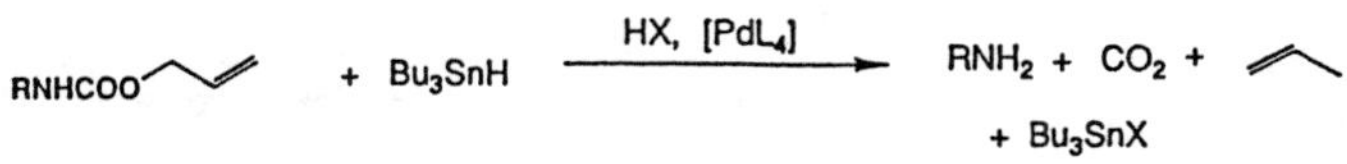

(4.72)

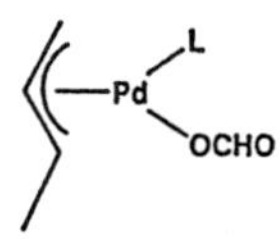

(4-168)

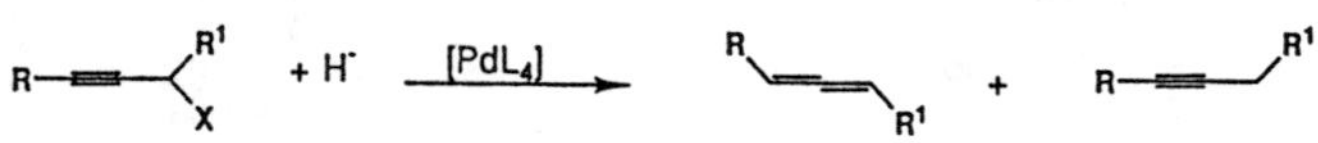

(4.73)

allenes seems to be achieved with hydrogen transfer agents such as methanoate as the hydrogen source, but some overreduction to the alkene does also occur (reaction (4.74)) [198].

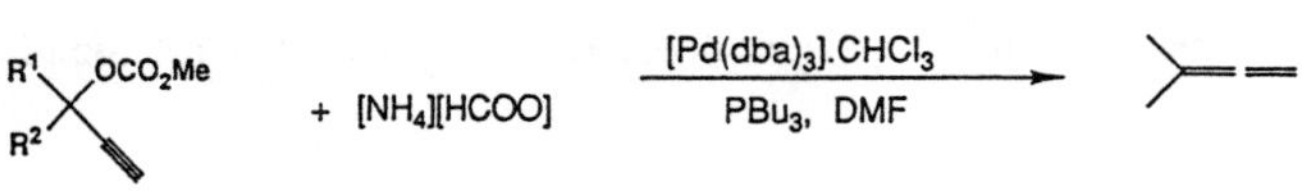

(4.74)

Reduction of acyl halides to aldehydes is normally achieved either *via* the heterogeneously catalysed Rosemund reduction, or by reduction with a hindered hydride reducing agent. The reaction may be realised in the presence of $Bu_3SnH/[Pd(PPh_3)_4]$ in good yield, but there are few reasons for preferring this route. An interesting hydrogenolysis of the imidoyl chloride, **4-169**, probably follows a similar mechanism [199].

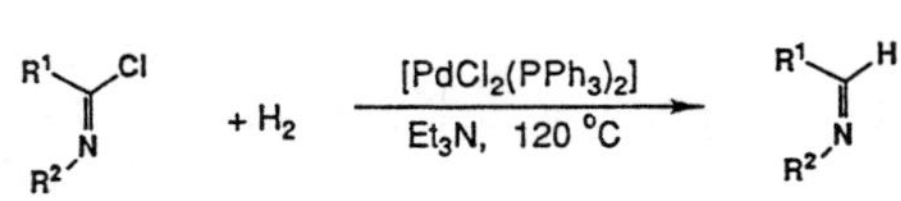

(4-169)

Hydrogenolysis of allylic epoxides has been achieved in the presence of palladium(0) complexes (Figure *4.6*). The initial step involves the formation of a palladium allyl complex by oxidative addition, and this is then reduced by hydrogen transfer from methanoate. Stereoselection for inversion depends on the phosphine and the solvent, but is generally good [200]. A similar process has been reported for cyclopropanes, (reaction (4.75), *via* **4-170**); again stereoselection was good for inversion of configuration [201].

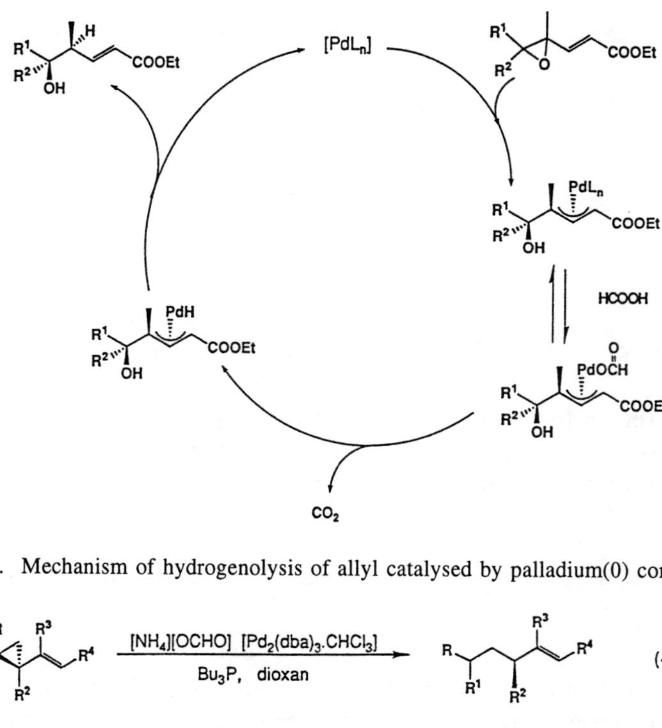

Fig. 4.6. Mechanism of hydrogenolysis of allyl catalysed by palladium(0) complexes.

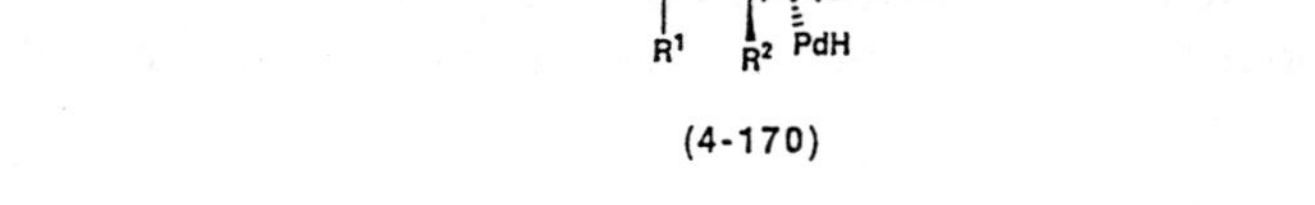

(4.75)

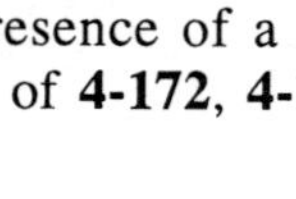

(4-170)

Hydrogenolysis of epoxides using molecular hydrogen has been reported to occur in the presence of a range of rhodium complexes. Thus **4-171** gave a mixture of **4-172**, **4-173** and **4-174** in the presence

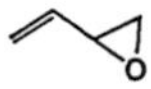

(4-171)

(4-172)

(4-173)

(4-174)

of $[Rn(NBD)(PR_3)_n]^+$, the proportions of the products varying with the nature of the phosphine [202]. An enantioselective version of the reaction has been reported, using **4-175** as the substrate, and a chiral cationic rhodium complex as the catalyst. **4-176** was formed as the main product, and deuterium incorporation studies indicated that the mechanism involved a direct ring opening, rather than reduction of a ketone formed by metal catalysed rearrangement. The best optical yield, 62%, was obtained using a complex of **4-177**. If the racemate of the *trans*-isomer is used as the substrate, an inactive product is obtained when the reaction is complete, but at 10% conversion, 50% optical yield was noted for the product [203].

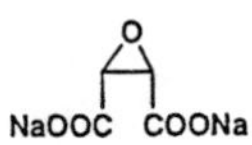

(4-175)

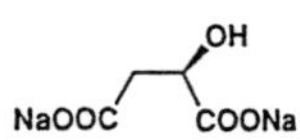

(4-176)

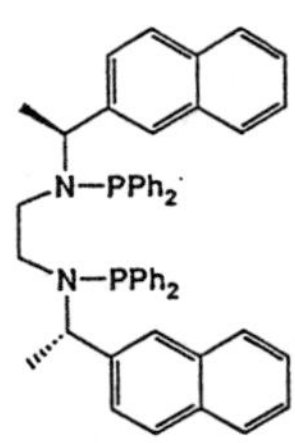

(4-177)

References

1. P.A. Chaloner, Handbook of Coordination Catalysis in Organic Chemistry, Butterworths, 1986
2. B.R. James, Homogeneous Hydrogenation, Wiley Interscience, New York, 1973
3. P. Rylander, Catalytic Hydrogenation in Organic Synthesis, Academic Press, New York, 1979
4. P.J. Brothers, Prog. Inorg. Chem., **28**, 1 (1981)
5. J. Halpern, Inorg. Chim. Acta, **50**, 11 (1981)
6. J. Halpern, Phosphorus Sulfur, **18**, 307 (1983)
7. M. Kato, M. Watanabe, B. Vogler, Y. Tooyama and A. Yoshikoshi, J. Chem. Soc., Chem. Commun., 1706 (1990)
8. D. E. Ward, Can. J. Chem., **65**, 2380 (1987)
9. H. Wagner and J. Römer, Z. Chem., **26**, 366 (1986)
10. W. Zhang and E.N. Jacobsen, Tetrahedron Lett., **32**, 1711 (1991)
11. R.H. Crabtree, P.C. Demou, D. Eden, J.M. Mihelcic, C.A. Parnell, J.M. Quirk and G.E. Morris, J. Am. Chem. Soc., **104**, 6994 (1982)

12. T. Kitamura, N. Sakamato and T. Joh, Chem. Lett., 379 (1973)
13. J.M. Grosselin, C. Mercier, G. Allmang and F. Grass, Organometallics, **10**, 2126 (1991)
14. L.I. Gvinter, L.N. Suvorova, S.S. Danišlova and L.Kh. Freidlin, J. Org. Chem. USSR, **22**, 67 (1986)
15. R.A. Sanchez-Delgado A. Andriollo and N. Valencia, J. Mol. Catal., **24**, 217 (1984)
16. M.N. Ricroch and A. Gaudemer, J. Organomet. Chem., **67**, 119 (1974)
17. E. Farnetti, J. Kašpar, R. Spogliarich and M. Graziani, J. Chem. Soc., Dalton Trans., 947 (1988)
18. F. Orsini, F. Pelizzoni and M. Forte, Gazz. Chim. Ital., **116**, 115 (1986)
19. J.A. Marco, M. Arno and M. Carda, Can. J. Chem., **65**, 630 (1987)
20. R.L. Danheiser, J.M. Morin and E.J. Salaski, J. Am. Chem. Soc., **107**, 8066 (1985)
21. C. Djerassi and J. Gutzwiller, J. Am. Chem. Soc., **88**, 4577 (1966)
22. J.W. Suggs, S.D. Cox, R.H. Crabtree and J.M. Quirk, Tetrahedron Lett., **22**, 303 (1981)
23. J.R. Mohrig, S.L. Dabora, T.F. Foster and S.C. Schultz, J. Org. Chem., **49**, 5179 (1984)
24. P.D. Clark, N.M. Irvine and P. Sarkar, Can. J. Chem., **69**, 1011 (1991)
25. M.M. Bhagwat and D. Devaprabhakara, Tetrahedon Lett., 1391 (1972)
26. A.J. Birch and K.A.M. Walker, J. Chem. Soc., C, 1894 (1966)
27. H. van Bekkum, F. van Rantwijk, G. van Minnen-Pathuis, J.D. Remijnse and A. van Veen, Rec. Trav. Chim. Pays-Bas, **88**, 911 (1969)
28. A.G. Hinze and D.J. Frost, J. Catal., **24**, 541 (1972); E.N. Frankel, E. Selke and C.A. Glass, J. Org. Chem., **34**, 3930, 3936 (1969)
29. P. Le Maux, J.Y. Saillard, D. Grandjean and G. Jaouen, J. Org. Chem., **45**, 4524 (1980)
30. J. Kwiakek, I.L. Mador and J.Y. Sayler, Adv. Chem., **37**, 201 (1963)
31. Y. Naruse, T. Esaki and H. Yamamoto, Tetrahedron, **44**, 4747 (1988)
32. M. Sodeoka, Y. Ogawa, Y. Kirio and M. Shibasaki, Chem. Pharm. Bull., **39**, 309 (1991)
33. T. Mase, M. Sodeoka and M. Shibasaki, Tetrahedron Lett., **25**, 5087 (1984)
34. T. Funabiki, M. Matsumoto and K. Tarama, Bull. Chem. Soc. Jpn., **45**, 2723 (1972)
35. P. Le Maux and G. Simonneaux, J. Mol. Catal., **26**, 195 (1984)
36. L. Crombie and S.J. Holloway, J. Chem. Soc., Perkin Trans., I, 2425 (1985)
37. R.R. Schrock and J.A. Osborn, J. Am. Chem. Soc., **98**, 2143 (1976)
38. H. Kosugi, M. Kitaoka, K. Tagami, A. Takahashi and H. Uda, J. Org. Chem., **52**, 1078 (1987)
39. J.S. Yu and I.P. Rothwell, J. Chem. Soc., Chem. Commun., 632 (1992)
40. S. Friedman, S. Metlin, A. Svedi and I. Wender, J. Org. Chem., **24**, 1287 (1959)
41. C.R. Landis and J. Halpern, Organometallics, **2**, 840 (1983)
42. R.A. Grey, G.P. Pez and A. Wallo, J. Am. Chem. Soc., **102**, 5948 (1980)
43. I. Amer H. Amer, R. Ascher, J. Blum, Y. Sasson and K.P.C. Vollhardt, J. Mol. Catal., **39**, 185 (1987)
44. R.H. Fish, E. Baralt and S.J. Smith, Organometallics, **10**, 54 (1991); E. Baralt, S.J. Smith, J. Hurwitz, I.T. Horváth and R.H. Fish, J. Am. Chem. Soc., **114**, 5187 (1992)
45. W.M. Kruse and L.W. Wright, Carbohydr. Res., **64**, 293 (1978)
46. T. Fuchikami, Y. Ubukata and Y. Tanaka, Tetrahedron Lett., **32**, 499 (1991)

47. E. Farnetti, M. Pesce, J. Kašpar, R. Spogliarich and M. Graziani, J. Chem. Soc., Chem. Commun., 746 (1986)
48. A. Bényer and F. Joó, J. Mol. Catal., **58**, 151 (1990)
49. D.J. Darensbouurg, F. Joó, M. Kannisto, A. Katho and J.H. Reibenspies, Organometallics, **11**, 1991 (1992)
50. S. Törös, L. Lollár, B. Heil and L. Markó, J. Organomet. Chem., **255**, 377 (1983)
51. F. Martinelli, G. Mestroni, A. Camus and G. Zassinovich, J. Organomet. Chem., **220**, 383 (1981)
52. P.A. Browne and D.N. Kirk, J. Chem. Soc., C, 1653 (1969)
53. J. Protiva, E. Klinotová, J. Klinot, M. Kušiak and A. Vystrčil, Coll. Czech. Chem. Commun., **51**, 2029 (1986)
54. E. Farnetti, G. Nardin and M. Graziani, J. Chem, Soc., Chem. Commun., 1264 (1989)
55. R.L. Chowdhury and J.-E. Bäckvall, J. Chem. Soc., Chem. Commun., 1063 (1991)
56. Y. Ishii, T. Ikariya, M. Saburi and S. Yoshikawa, Tetrahedron Lett., **27**, 365 (1986); Y. Hara and K. Wada, Chem. Lett., 553 (1991)
57. Y.N.C. Chan, D. Meyer and J.A. Osborn, J. Chem. Soc., Chem. Commun., 869 (1990)
58 A. Basu, S. Bhaduri, K. Sharma and P.G. Jones, J. Chem. Soc., Chem. Commun., 1126 (1987)
59. G.-Z. Wang and J.-E. Bäckvall, J. Chem. Soc., Chem. Commun., 980 (1992)
60. L. Markó and J. Bakos, J. Organomet. Chem., **81**, 411 (1974); Y. Watanabe, M. Yamamoto, T. Mitsudo and Y. Takegami, Tetrahedron Lett., 1289 (1978)
61. R.A. Grey, G.P. Pez and A. Wallo, J. Am. Chem. Soc., **103**, 7536 (1981)
62. J. Kwiatek, I.L. Mador and J.Y. Seyler, Adv. Chem., **37**, 201 (1963)
63. Y. Watanabe, T. Ohta, Y. Tsuji, T. Hiyoshi and Y. Tsuji, Bull. Chem. Soc. Jpn., **57**, 2440 (1984)
64. J.F. Knifton, J. Catal., **33**, 289 (1974)
65. E.I. Klabunovskii, D.D. Gogoladze, E.S. Levitina, E.I. Karpeiskaya, L.F. Godunova, L.N. Kaigorodova and G.O. Chivadze, Bull. Acad. Sci. USSR, Div. Chem. Sci., **36**, 1475 (1987)
66. M. Bartok, Stereochemistry of Heterogeneous Metal Catalysis, Wiley, New York, 1985
67. K. Haruda, Asymmetric Synthesis, Volume V, 346 (1985)
68. J.M. Brown, Angew. Chem. Int. Ed. Engl., **26**, 190 (1987)
69. J.M. Brown, I. Cutting and A.P. James, Bull. Soc. Chim. Fr., 211 (1988)
70. R.H. Crabtree and M.W. Davis, J. Org. Chem., **51**, 2655 (1986)
71. J.M. Brown and S.A. Hall, Tetrahedron Lett., **25**, 1393 (1984)
72. J.M. Brown and S. Hall, J. Organomet. Chem., **285**, 333 (1985)
73. A.S. Machado, A. Olesker, S. Castillon and G. Lukacs, J. Chem. Soc., Chem. Commun., 330 (1985)
74. G. Stork and D.D. Kahne, J. Am. Chem. Soc., **105**, 1072 (1983)
75. C.A. Hoeger, A.D. Johnston and W.H. Okamura, J. Am. Chem. Soc., **109**, 4690 (1987)
76. A. Kawai, O. Hara. Y. Hamada and T. Shioiri, Tetrahedron Lett., **29**, 6331 (1988)
77. A.J. Poss and M.S. Smyth. Tetrahedron Lett., **29**, 5723 (1988)
78. T.V. RajanBabu, W.A. Nugent, D.F. Taber and P.J. Fagan, J. Am. Chem. Soc., **100**, 7128 (1988)

79. A.B. Smith, Y. Yokoyama, D.M. Huryn and N.K. Dunlap, Tetrahedron Lett., **28**, 3659 (1987)
80. J.M. Brown and S.A. Hall, Tetrahedron, **41**, 4639 (1985)
81. A.J. Schultz and N.J. Green, J. Am. Chem. Soc., **113**, 4931 (1991)
82. A.G. Schultz and P.J. McCloskey, J. Org. Chem., **50**, 5905 (1985)
83. P.J. McCloskey and A.G. Schultz, Heterocycles, **25**, 437 (1987)
84. J. Lejeune, J.T. Lallemand, T. Prangé and L. Ricard, Tetrahedron Lett., **32**, 1063 (1991)
85. J.M. Brown, P.L. Evans and A.P. James, Org. Synth., **68**, 64 (1988)
86. D.A. Evans and M.M. Morrisey, J. Am. Chem. Soc., **106**, 3866 (1984)
87. D.A. Evans, R.L. Dow, T.L. Shih, J.M. Takacs and R. Zahler, J. Am. Chem., Soc., **112**, 5290 (1990)
88. D.A. Evans, S.L. Bender and J. Morris, J. Am. Chem. Soc., **110**, 2506 (1988)
89. D.A. Evans and S.L. Bender, Tetrahedron Lett., **27**, 799 (1986)
90. J.M. Brown, A.P. James and L.M. Prior, Tetrahedron Lett., **28**, 2179 (1987)
91. N.W. Alcock, J.M. Brown and A.P. James, Acta Crystallogr., Sect. C, **45C**, 734 (1989)
92. K. Yamamoto, M. Takagi and J. Tsuji, Bull. Chem. Soc. Jpn., **61**, 319 (1988)
93. D.H. Birtwistle, J.M. Brown, R.H. Herbert, A.P. Jones, K.-F. Lee and R.J. Taylor, J. Chem. Soc., Chem. Commun., 194 (1989)
94. W.D. Lubell and H. Rapoport, J. Am. Chem. Soc., **110**, 7447 (1988)
95. D. Ando, C. Bevan, J.M. Brown and D.W. Price, J. Chem. Soc., Chem. Commun., 592 (1992)
96. L. Markó and J. Bakos, Aspects Homogeneous Catalysis, **4**, 145 (1981)
97. B. Bosnich, Chem. Scripta, **25**, NS45 (1985)
98. H. Brunner, J. Organomet. Chem., **300**, 39 (1986)
99. J.D. Morrison, Asymmetric Synthesis, Volume 5, Academic Press, New York, 1985
100. H. Buschmann, H.-D. Scharf, N. Hoffmann and P. Esser, Angew. Chem., Int. Ed. Engl., **30**, 477 (1991)
101. M.J. Burk, J.E. Feaster and R.L. Harlow, Organometallics, 9 (1990) 2653
102. M.J. Burk, J. Am. Chem. Soc., **113**, 8518 (1991)
103. H. Takahashi amd K. Achiwa, Chem. Lett., (1987) 1921; Chem. Pharm. Bull., **36**, 3230 (1988)
104. C. Döbler, H.-J. Kreuzfeld and H. Pracejus, J. Organomet. Chem., **344**, 89 (1988)
105. B.M. McCulloch, J. Halpern, M.R. Thompson and C.R. Landis, Organometallics, **9**, 1392 (1990)
106. J.M. Brown, P.A. Chaloner and G.A. Morris, J. Chem. Soc., Perkin Trans. II, 1583 (1987)
107. D.G. Allen, S.B. Wild and D.L. Wood, Organometallics, **5**, 1009 (1986)
108. N.W. Alcock, J.M. Brown, A.E. Derome and A.R. Lucy, J. Chem. Soc., Chem. Commun., 575 (1985)
109. J.M. Brown and P.J. Maddox, J. Chem. Soc., Chem. Commun., 1278 (1987)
110. J.M. Brown and P.L. Evans, Tetrahedron, **44**, 4905 (1988)
111. P.L. Bogdan, J.J. Irwin and B. Bosnich, Organometallics, **8**, 1450 (1980)
112. T. Ikariya, Y. Ishii, H. Kawano, T. Arai, M. Saburi, S. Yoshikawa and J. Akutagawa, J. Chem. Soc., Chem. Commun., 922 (1985)
113. J.A. Cabeza, C. Cativiela, M.D. Diaz de Villegas and L.A. Oro, J. Chem. Soc., Perkin Trans., I, 1881 (1988)

114. N.W. Alcock, J.M. Brown and P.J. Maddox, J. Chem. Soc., Chem. Commun., 1532 (1986)
115. L.F. Godunova, E.S. Levitina, E.I. Karpeiskaya and E.I. Klabunovskii, Bull. Acad. Sci, USSR, Div. Chem. Sci., **30**, 595 (1981)
116. A.T. Baryshnikov, V.M. Belikov, V.K. Latov and M. B. Saporovskaya, Bull. Acad. Sci. USSR, Div. Chem. Sci., **30**, 1206 (1981)
117. L.O. Nindakova, F.K. Schmidt, E.I. Klabunovskii and V.A. Pavlov, Bull. Acad. Sci. USSR, Div. Chem. Sci., **33**, 671 (1984)
118. S. Takeuchi and Y. Ohgo, Bull. Chem. Soc. Jpn., **57**, 1920 (1984)
119. I. Ojima, T. Kogure, N. Yoda, T. Suzuki, M. Yatabe and T. Tanaka, J. Org. Chem., **47**, 1329 (1982)
120. I. Ojima and N. Shimizu, J. Am. Chem. Soc., **108**, 3100 (1986)
121. I. Ojima, N. Yoda, M. Yatabe, T. Tanaka and T. Kogure, Tetrahedron, **40**, 1255 (1984)
122. H. Levine-Pinto, J.L. Morgat, P. Fromageot, D. Meyer, J.C. Poulin and H.B. Kagan, Tetrahedon, **38**, 119 (1982)
123. U. Schmidt, D. Weller, A. Holder, and A. Lieberknecht, Tetrahedron Lett., **29**, 3227 (1988)
124. U. Schmidt, A. Lieberknecht, H. Griesser, R. Utz, T. Buettler and F. Bartkowiak, Synthesis 361 (1986)
125. U. Schmidt, R. Meyer, V. Leitenberger, A. Lieberknecht and H. Griesser, J. Chem. Soc., Chem. Commun., 275 (1991); A.-S. Carlström and T. Frejd, J. Chem. Soc., Chem. Commun., 1216 (1991)
126. H. Brunner, E. Graf, W. Leitner and K. Wutz, Synthesis 743 (1989)
127. T. Chiba, A. Miyashita, H. Nohira and H. Takaya, Tetrahedron Lett., **32**, 4745 (1991)
128. J.M. Brown and D. Parker, J. Org. Chem., **47**, 2722 (1982)
129. H. Kawano, Y. Ishii, T. Ikariya, M. Saburi, S. Yoshikawa, Y. Uchida amd H. Kumobayashi, Tetrahedron Lett., **28**, 1905 (1987)
130. M.D. Fryzuk and B. Bosnich, J. Am. Chem. Soc., **101**, 3043 (1979)
131. T. Hayashi, N. Kawamura and Y. Ito, J. Am. Chem. Soc., **109**, 7876 (1987); Tetrahedron Lett., **29**, 5969 (1988)
132. R. Noyori and H. Takaya, Acc. Chem. Res., **23**, 345 (1990)
133. H. Muramatsu, H. Kawano, Y. Ishii, M. Saburi and Y. Uchida, J. Chem. Soc., Chem. Commun., 769 (1989)
134. K. Mashima, K. Kusano, T. Ohta, R. Noyori and H. Takaya, J. Chem. Soc., Chem. Commun., 1208 (1989)
135. T. Ohta, H. Takaya and R. Noyori, Tetrahedron Lett., **31**, 7189 (1990)
136. M.T. Ashby and J. Halpern, J. Am. Chem. Soc., **113**, 589 (1991); M.T. Ashby, M.A. Khan and J. Halpern, Organometallics, **10**, 2011 (1991)
137. T. Tsukahara, H. Kawanao, Y. Ishii, T. Takahashi, M. Saburi, Y. Uchida and S. Akutagawa, Chem. Lett., 2055 (1988)
138. J. Solodar, J. Org. Chem., **43**, 1787 (1978)
139. P. Le Maux, V. Massonneau and G. Simonneaux, Tetrahedron, **44**, 1409 (1988)
140. P. Le Maux, V. Massonneau and G. Simonneaux, J. Organomet. Chem., **284**, 101 (1985)
141. V. Massonneau, P. Le Maux and G. Simonneaux, J. Organomet. Chem., **327**, 269 (1987)
142. D.A. Evans, M.M. Morrissey and R.L. Dow, Tetrahedron Lett., **26**, 6005 (1985.

143. J. Bourson and L. Oliveros, J. Organomet. Chem., **229**, 77 (1982)
144. J. Bakos. I. Tóth, B. Heil and L. Markó, J. Organomet. Chem., **279**, 23 (1985)
145. R.L. Halterman, K.P.C. Vollhardt, M.E. Welker, D. Bläser and R. Boese, J. Am. Chem. Soc., **109**, 8105 (1987)
146. M. Kitamura, I. Kasahara, K. Manabe, R. Noyori and H. Takaya, J. Org. Chem., **53**, 708 (1988)
147. M. Kitamura, I. Kasahara, K. Manabe, R. Noyori and H. Takaya, J. Org. Chem., **53**, 708 (1988)
148. J.M. Brown and A.P. James, J. Chem. Soc., Chem. Commun., 181 (1987)
149. H. Takahashi, T. Morimoto and K. Achiwa, Chem. Lett., 855 (1987)
150. H.P. Märki, Y. Crameri, R. Eigenmann, A. Krasso, H. Ramuz, K. Bernauer, M. Goodman and K.L. Melmon, Helv. Chim. Acta, **71**, 320 (1988)
151. H. Takahashi, S. Sakurabara, H. Takeda and K. Achiwa, J. Am. Chem. Soc., **112**, 5876 (1990)
152. M. Kitamura, T. Ohkuma, H. Takaya and R. Noyori, Tetrahedron Lett., **29**, 1555 (1988)
153. S.L. Schreiber, S.E. Kelly, J.A. Porco, T. Sammakia and E.M. Suh, J. Am. Chem. Soc., **110**, 6210 (1988)
154. M. Nakatsuka, J.A. Ragan, T. Sammakia, D.B. Smith, D.E. Uehling and S.L. Schreiber, J. AM. Chem. Soc., **112**, 5583 (1990)
155. N. Fukuda, K. Mashima, Y. Matsumura and H. Takaya, Tetrahedron Lett., **31**, 7185 (1990)
156. M. Kitamura, T. Ohkuma, S. Inoue, N. Sayo, H. Kumobayashi, S. Akutagawa, T. Ohta, H. Takaya and R. Noyori, J. Am. Chem. Soc., **110**, 629. (1988)
157. T. Nishi, M. Kitamura, T. Ohkuma and R. Noyori, Tetrahedron Lett., **29**, 6327 (1988)
158. J.P. Genet, C. Pinel, S. Mallart, S. Juge, S. Thorimbert and J.A. Laffitte, Tetrahedron Asymmetry, **2**, 555 (1991)
159. K. Mashima, Y. Matsumura, K. Kusano, H. Kumobayashi, N. Sayo, Y. Hori, T. Ishizaki, S. Akutagawa and H. Takaya, J. Chem. Soc., Chem. Commun., 609 (1991)
160. M. Kitamura, M. Tokunaga and R. Noyori, J. Am. Chem. Soc., **115**, 144 (1993)
161. S. Vastag, J. Bakos, S. Törös, N.E. Takach, R.B. King, B. Heil and L. Markó J. Mol. Catal., **22**, 283 (1984)
162. G. Kang, W.R. Cullen, M.D. Fryzuk, B.R. James and J.P. Kutney, J. Chem. Soc., Chem. Commun., 1466 (1988)
163. Y.N.C. Chan and J.A. Osborn, J. Am. Chem. Soc., **112**, 9400 (1990)
164. J. Bakos, A. Orosz, B. Heil, M. Laghmari, P. Lhoste and D. Sinou, J. Chem. Soc., Chem. Commun., 1684 (1991)
165. C. Lensink and J.G. de Vries, Tetrahedrom Asymmetry, **3**, 235 (1992)
166. G. Losse and H.-U. Stiehl, Z. Chem., **21**, 188 (1981)
167. R.M. Williams, P.J. Sinclair and W. Zhai, J. Am. Chem. Soc., **110**, 482 (1988)
168. A.R. Pinder, Synthesis 425 (1980)
169. Y. Akita, A. Inoue, K. Ishida, K. Terui and A. Ohta, Synth. Commun., **16**, 1067 (1986)
170. Y. Akita, A. Inoue, Y. Mori and A. Ohta, Heterocycles, **24**, 2093 (1986)
171. J. Roček, V. Sváta and L. Lešetický, Coll. Czech. Chem. Commun., **50**, 1244 (1985)
172. V.V. Grushin and H. Alper, Organometallics, **10**, 1620 (1991)
173. A. Ohta, Y. Aoyagi, Y. Kurihara, A. Kojima, K. Yuasa and M. Shimizaki, Heterocycles, **27**, 437 (1988)

174. I. Pri-Bar and O. Buchman, J. Org. Chem., **51** (1986
175. Q. Chen and Y. He, Synthesis 896 (1988).
176. W.D. Wulff, J.S. McCallum and F. Kunig, J. Am. Chem. Soc., **110**, 7419 (1988)
177. W.D. Kingsbury, J.C. Boehm, D.R. Jakas, K.G. Holden, S.M Hecht, G. Gallagher, M.J. Caranfa, F.L. McCabe, L.F. Faucette, R.K. Johnson and R.P. Hertzberg, J. Med. Chem., **34**, 98 (1991)
178. R.E. Dolle, S.J. Schmidt, K.F. Erhard and L.I. Kruse, J. Am. Chem. Soc., **111**, 278 (1989)
179. R.E. Dolle, S.J. Schmidt, and L.I. Kruse, Tetrahedron Lett., **29**, 1581 (1988)
180. M. Hanack and R. Rieth, Chem. Ber., **120**, 1659 (1987)
181. M. Capet, T. Cuvigny, C. Hervé du Penhoat, M. Julia and G. Loomis, Tetrahedron Lett., **28**, 6273 (1987)
182. B.M. Trost and P.L. Ornstein, Tetrahedron Lett., **22**, 3463 (1981)
183. F. Charbonnier, A. Moyano and A.E. Greene, J. Org. Chem., **52**, 2303 (1987)
184. E. Keinan and N. Greenspoon, Tetrahedron Lett., **23**, 241 (1982)
185. N. Greenspoon and E. Keinan, J. Org. Chem., **53**, 3723 (1988)
186. A. Bernardi, W. Cabri, G. Poli and L. Prati, J. Chem. Res., S, 52 (1986)
187. G. Palmisano, B. Danieli, G. Lesma and M. Mauro, J. Chem. Soc., Chem. Commun., 1564 (1986)
188. M. Mohri, H. Kinoshita, K. Inomata, H. Kotake, H. Takagaki and K. Yamazaki, Chem. Lett., 1177 (1986)
189. K. Mori and S. Harashima, Tetrahedron Lett., **32**, 5995 (1991)
190. M.H. Rabinowitz and C. Djerassi, J. Am. Chem. Soc., **114**, 304 (1992)
191. J. Tsuji, I. Shimizu and I. Minami, Chem. Lett., 1017 (1984)
192. N. Ono, I. Hamamoto, A Kaminura and A. Kaji, J. Org. Chem., **51**, 3734 (1986)
193. F. Guibé, Y.T. Xian, A.M. Zigna and G. Balavoine, Tetrahedron Lett., **26**, 3559 (1985)
194. O. Dangles, F. Guibé, G. Balavoine, S. Lavielle and A. Marquet, J. Org. Chem., **52**, 4984 (1987)
195. D.A. Holt, M.A. Levy, H.-K. Yen, H.-J. Oh, B.W. Metcalf and P.J. Wier, BioMed. Chem. Lett., **1**, 27 (1991)
196. M. Oshima, I. Shimizu, A. Yamamoto and F. Ozawa, Organometallics, **10**, 1221 (1991)
197. Y. Colas, B. Cazes and J. Gore, Bull. Soc. Chim. Fr., 165 (1987)
198. J. Tsuji, T. Sugiura and I. Minami, Synthesis 603 (1987)
199. M. Tanaka and T. Kobayashi, Synthesis 967 (1985)
200. M. Oshima, H. Yamazaki, I. Shimizu, M. Nisar and J. Tsuji, J. Am. Chem. Soc., **111**, 6280 (1989); K. Mori and S. Harashima, Tetrahedron Lett., **32**, 5995 (1991)
201. I. Shimizu and F. Aida, Chem. Lett., 601 (1988)
202. H. Fujitsu, E. Matsumura, E. Shirahama, K. Takeshita and I. Mochida, J. Chem. Soc., Perkin Trans., I, 755 (1982)
203. A.S.C. Chan and J.P. Coleman, J. Chem. Soc., Chem. Commun., 535 (1991)

CHAPTER 5

HYDROGENATION IN AQUEOUS SYSTEMS

5.1. Introduction

5.1.1. WHY AQUEOUS? THE SCOPE OF HYDROGENATIONS IN AQUEOUS SYSTEMS

Throughout the history of organometallic catalysis, water has been generally regarded as a medium which should be strictly avoided. Ironically, the history of organometallic chemistry started with the discovery of a water soluble complex $[PtCl_3(C_2H_4)]^{3-}$ (Zeise's salt, 1827) and there have been also other exceptions to the rule that organometallic compounds are moisture sensitive. The ability of $[HCo(CO)_4]$ and $[HCo(CN)_5]^{3-}$ to catalyse hydrogenation was first reported fifty years ago; both are rather soluble in water, which is in fact the only suitable solvent for hydrogenations catalysed by the latter complex. So whilst it is still true that most organometallic compounds are best handled in a moisture-free environment, there is no *à priori* reason to exclude water as solvent for hydrogenation reactions.

All of this has been well known for some time. Nevertheless the $[HCo(CN)_5]^{3-}$-catalysed hydrogenations, and the early examples of the successful use of water soluble phosphine complexes for reduction of unsaturated substrates were mostly regarded as curiosities, and exceptions to the general rule. A very sound basis for this neglect was that most of the important substrates to be hydrogenated were not water soluble and there seemed to be no practical need for catalysts soluble in aqueous solvents. This situation has changed dramatically in the last fifteen, and especially fast in the last five years.

The main impetus for this change is that there is an ever-increasing demand for complete recovery of precious metal catalysts in industrial processes. With the currently very high price of rhodium, losses of catalyst simply cannot be tolerated. The attachment of soluble metal

P.A. Chaloner, M.A. Esteruelas, F. Joó and L.A. Oro, Homogeneous Hydrogenation 183–239.

complexes to insoluble supports has so far failed to give catalysts suitable for large scale use, although there is still much research effort in this area. On the other hand, the success of the biphasic hydroformylation of 1-propene on an industrial scale [1] triggered tremendous interest in organometallic catalysis in water, including hydrogenations.

Another reason for the current interest, although perhaps less well known, is the combination of homogeneous organometallic catalysis and phase transfer catalysis [2]. Hydrogenation reactions are not amongst the most interesting examples, but their investigation did play an important role in the development of this relatively new field of catalysis.

Research into aqueous hydrogenations has also been stimulated by a recently recurred interest in bio-mimetic processes, and in other reactions of biochemical or biological importance. The coupling of a metal complex catalysed reaction with a simple enzymatic transformation may model more complicated enzyme-catalysed reactions, and may yield a chiral end product [3]. Additionally, the hydrogenation of biological membranes has become a powerful tool in studies of membrane biochemistry and cell physiology [4]. The use of aqueous systems for these studies is very straightforward and often essential (as in case of *in vivo* cell hydrogenations).

Last but not least water is a protic, very polar solvent, with an ability to form strong hydrogen bonds both to itself and to solutes (including catalysts and substrates) having appropriate atoms for H-bonding. In many ways it is a unique solvent, deserving investigation of its use as a solvent for organometallic catalysis in its own right. As we shall see later, in several cases these unique properties exert pronounced effects on the final outcome of catalytic hydrogenations.

With regard to the *scope* of hydrogenations in aqueous systems, virtually all kinds of unsaturated functionalities can be reduced. The selectivity is also impressive; as a result of recent investigations, very high regio-, stereo-, and enantioselectivity has been observed for reactions in true aqueous solutions. In comparison with non-aqueous systems the scope of the reaction is increased in that water soluble substrates may now be reduced. Even more important, in hydrogen transfer reactions water soluble donors (ascorbate, methanoate, EDTA, etc) can be used, as can $CO + H_2O$ as the hydrogen source.

There have been a number of reviews of the use of transition metal complexes as catalysts under phase transfer conditions [2, 7], of water soluble transition metal phosphine complexes in catalysis [5–8], on hydrogenation of biological membranes [4], and on hydrogenations with $CO + H_2O$ as hydrogen source [9].

5.1.2. TERMS AND DEFINITIONS

In general, we speak of an *aqueous system* when the liquid phase in which the catalytic hydrogenation takes place is comprised either of a homogeneous aqueous solution or of a mixture of an aqueous and a water-immiscible organic solution.

In homogeneous systems water may be the sole or major solvent. In the latter case water-miscible co-solvents may be used to facilitate dissolution of reactants and/or the catalyst. Homogeneous reactions are obviously the most amenable to mechanistic studies, but in several instance co-solvents are used simply to speed up the reaction by increasing the concentration of reactants. On the other hand, it has been shown that small amounts of water may have a substantial effect on rates or selectivity. It is noted here, that the concentration of H_2O in organic solvents containing 1% v/v water is 0.55 M. This value falls into the usual concentration range of the substrates and far exceeds catalyst concentrations. These systems are generally not considered here, since in these cases water is behaving more a reactant than as a co-solvent. It should be realized, however, that water may also act as a reactant in purely aqueous solutions, and it is not always easy to distinguish between its molecular and bulk effects.

When two immiscible phases are used to dissolve all the reactants and the catalyst, the reaction may take place in either (or in both) phase(s) or at the liquid/liquid interface, as illustrated on Scheme *5-1*. Generally, these systems are designed in such a way that some of the reactants and/or the catalyst dissolve well in one of the phases but have negligible solubility in the other.

In *phase transfer catalysed* reactions the transfer of reactants across the interface is facilitated by appropriate phase transfer agents (quaternary ammonium or phosphonium salts, crown ethers, etc.). When no such agent is used the reaction is termed as *biphasic*. The importance of both of these types of catalytic processes is in that they can lead to clean product separation and/or catalyst recycling. The use of two immiscible solvents also offers other advantages, such as protection of sensitive substances, controlled addition of substrates, removal of inhibitory (by)products, etc.

In many cases, catalytic hydrogenations proceed in *microheterogeneous* systems. Either the compounds participating in the chemical transformation, or appropriate additives may form micelles, liposomes or lyotropic liquid crystalline phases, in which hydrogenation takes place under the direct influence of the immediate environment. Together with

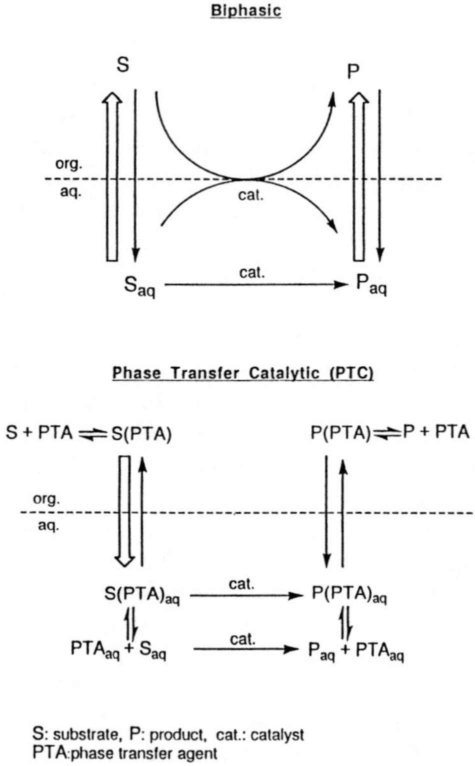

Scheme 5.1.

the use of colloidal metals as catalysts for hydrogenation in aqueous solutions, these microheterogeneous systems are not treated here in detail. We emphasize, however, that with certain substrates, such as polar lipids in aqueous systems, microheterogeneity cannot be avoided.

5.2. Water soluble hydrogenation catalysts

5.2.1. LIGANDS FOR WATER SOLUBLE CATALYSTS

The solubility of a metal complex in water is the consequence of its ionic or highly polar nature. In many cases this solubility is provided by appropriate ligands, having ionic or polar substituents. By far the most important ligands are the various sulfonated phosphines and protonated or quaternary ammonium phosphines, but other phosphines and non-phosphine ligands (having for example carboxylate groups) have also been used in hydrogenations in aqueous systems.

5.2.1.1. *Sulfonated mono- and diphosphines*

Examples of sulfonated phosphines are shown in Table *5.1* together with references to their preparation or use in hydrogenation.

TABLE 5.1
Sulfonated phosphines for aqueous catalysis

$Ph_{3-n}PAr_n$ n=1: TPPMS, **1**
n=2: TPPDS, **2**
n=3: TPPTS, **3**

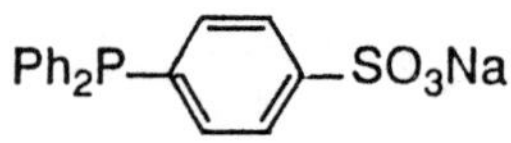

4

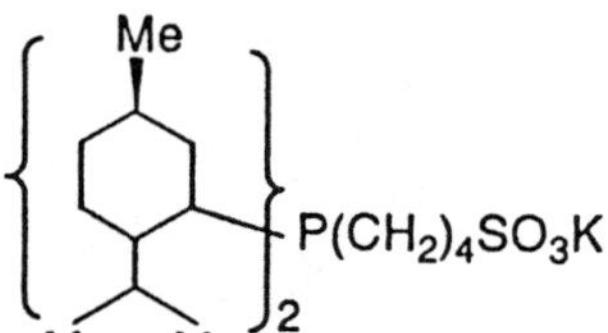

5

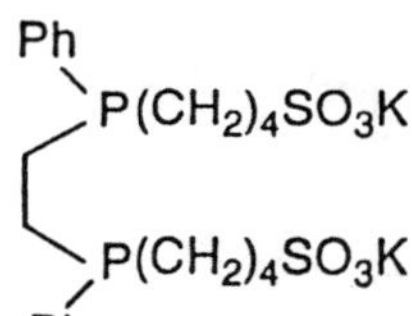

6

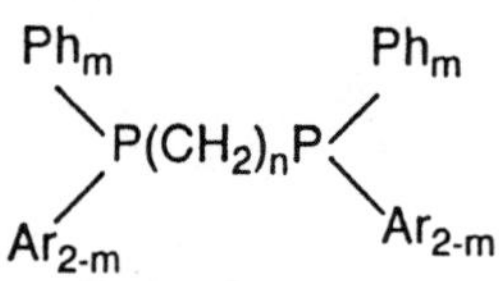

7; n= 2,3 or 4; m = 0 or 1

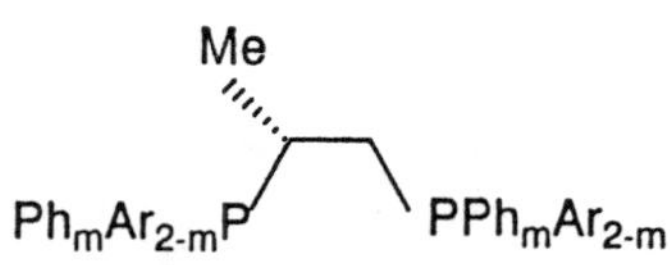

8; m= 0 or 1

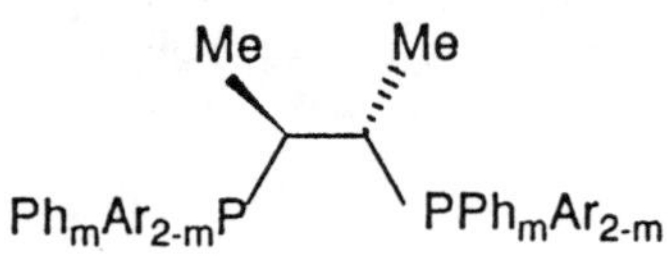

9; m= 0 or 1

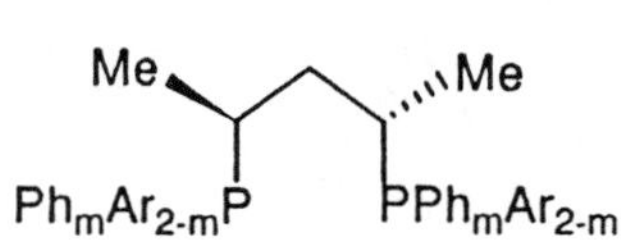

10; m= 0 or 1

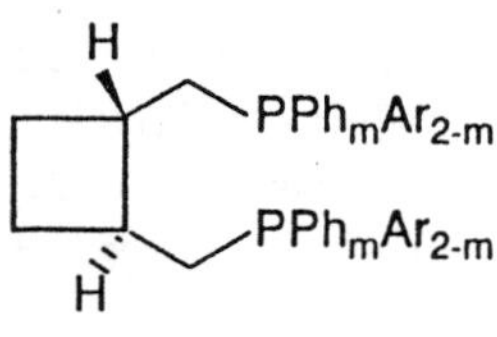

11; m= 0 or 1

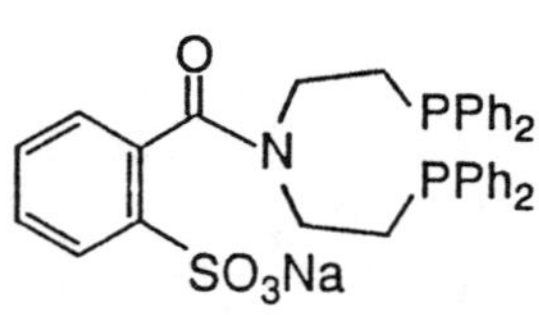

12

TABLE 5.1 *(continued)*

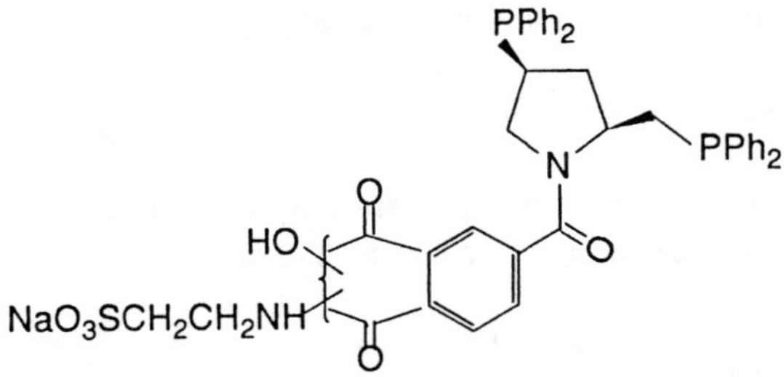

13

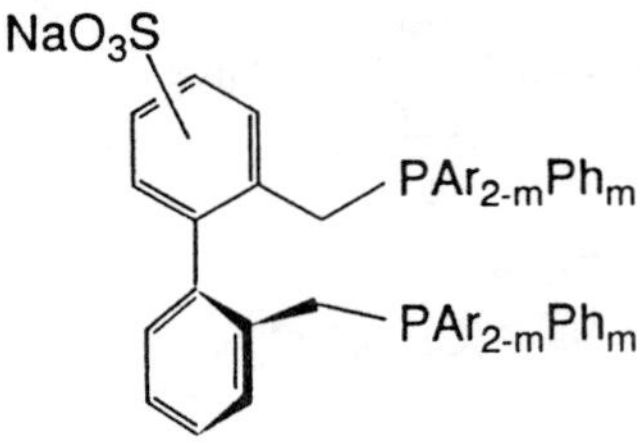

14; m=0 or 1

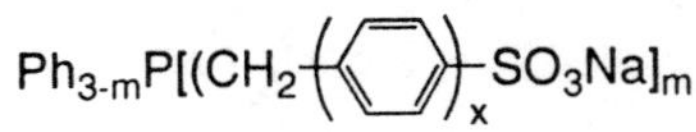

15; n=1,2,3 or 6
m=1,2 or 3
x= 0 or 1

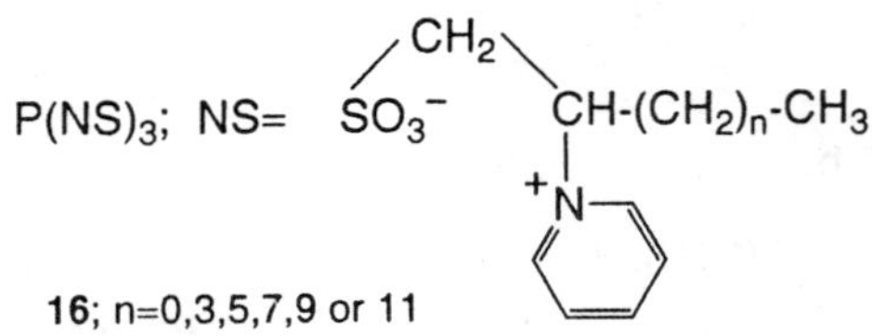

16; n=0,3,5,7,9 or 11

References: **1** [13-44, 60, 108, 109, 110, 159, 160, 164]; **2** [13, 30, 31, 46]; **3** [30, 31, 35, 39, 46-59, 63, 64, 92, 176, 178, 186]; **4** [12]; **5** [11]; **6** [11]; **7** [10, 35]; **8** [10, 68]; **9** [10, 65-68, 192]; **10** [10, 66, 68, 103, 192]; **11** [10, 35, 65-68, 176, 192]; **12** [70, 71]; **13** [100]; **14** [187]; **15** [45, 185]; **16** [189]

Sulfonated phosphines are rather soluble in aqueous systems at any pH. Their salts with lipophilic cations (*e.g.* quaternary ammonium salts) can be extracted into non-polar organic solvents. The ligands themselves tend to form micelles, but since complex formation uses the phosphine which is necessary as an acceptor for H-bond formation, complexes of these ligands give true solutions in water, and generally their solubility is greater than that of the ligands.

Direct sulfonation of aromatic phosphines uses 20–60% fuming sulfuric acid and yields 3-sulfonic acids. There is only one example of use of 4-diphenylphosphino benzene sulfonic acid in catalysis; the ligand can be prepared [12] by reaction of $K[PPh_2]$ with $K[4\text{-}ClC_6H_4SO_3]$.

The solubility of the sodium salt of 3-diphenylphosphino benzene sulfonic acid (TPPMS) in water, at room temperature, is ca. 0.03 M (12 g/l) [108], that of the potassium salt is 0.001 M (0.4 g/l) [108]. The Na-salt is poorly soluble in cold ethanol, but dissolves well in tetrahydrofuran or warm ethanol. It is virtually insoluble in non-polar organic solvents.

The solubilities of the sodium salts of higher sulfonated derivatives of triphenylphosphine (TPPDS and TPPTS) in water are so high that they can be isolated only after several successive precipitations by methanol, or by solvent extraction of the free acids followed by neutralization. The extent of sulfonation can be followed by ^{31}P-NMR spectroscopy [13, 68, 69] and reversed phase liquid chromatography [10], and the phosphines (and their complexes) can be purified by gel-chromatography [51–53].

It should be noted here, that direct sulfonation in fuming sulfuric acid of aryl phosphines containing acid sensitive groups (*e.g.* DIOP) is impossible, so water soluble analogues of these phosphines are prepared by other means.

Another general method [70, 71] to obtain phosphines bearing SO_3^--groups is shown in Scheme *5.2*. *Bis*(2-diphenylphosphinoethyl)amine

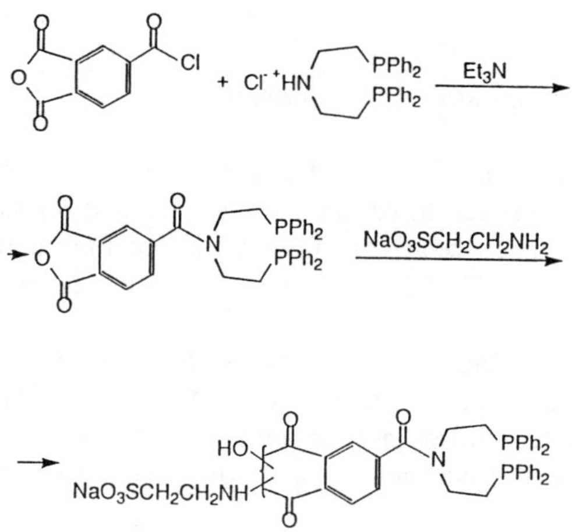

18

Scheme 5.2.

is reacted with a double acylating agent, then further with a nucleophile bearing polar or ionic substituents. Some of the chelating bisphosphines prepared this way are included in Tables 1, 2 and 4.

The reaction of alkali metal phosphides with sultons yields various sulfoalkyl- and sulfoarylphosphines [11], as shown in Scheme *5.3*. Phosphides prepared from mono- and diphosphines react equally well, making this method a valuable general procedure. Sulfoalkylation of tris(2-pyridylphosphine) with various sultons results in the formation of highly water soluble betains, **16** [189].

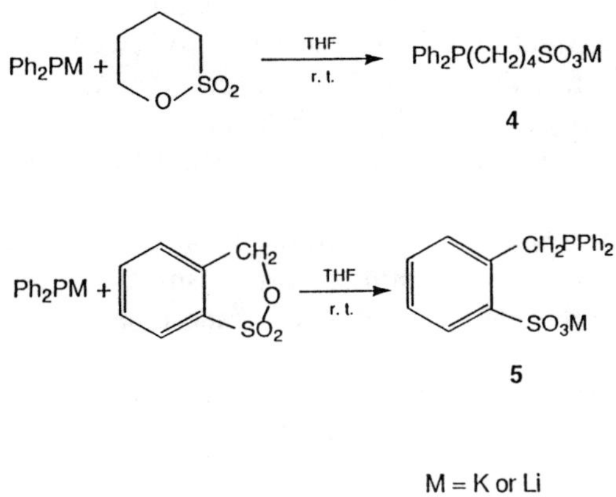

Scheme 5.3.

5.2.1.2. *Other water soluble phosphines*

Despite the fact that the complex formation equilibria of *carboxyl* substituted phosphines have been investigated in considerable detail [74] there have been relatively few studies of the catalytic properties of these complexes [75, 77, 78]. Some of the ligands are shown in Table *5.2*.

Diphenylphosphinobenzoic acids can be relatively easily prepared either from chlorodiphenylphosphine [104] or from $K[PPh_2]$ [105]. As with the sulfonated phosphines, carboxylated phosphines can be obtained from phosphinoalkylamines by acylation with an appropriate agent [70–73].

Phosphinocarboxylic acids are poorly soluble in water, but their alkali metal salts dissolve well in aqueous solvents.

The water soluble *ammonium* substituted phosphines can be prepared

TABLE 5.2
Carboxylated phosphines for aqueous catalysis

$Ph_nP(CH_2COOH)_{3-n}$

17

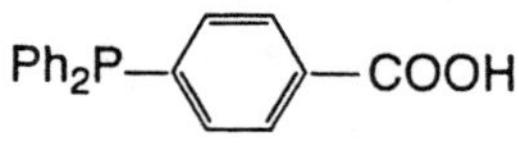

18

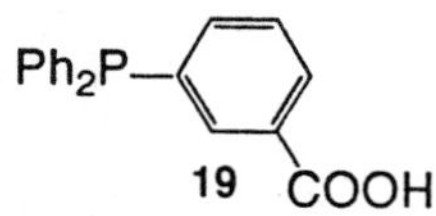

19

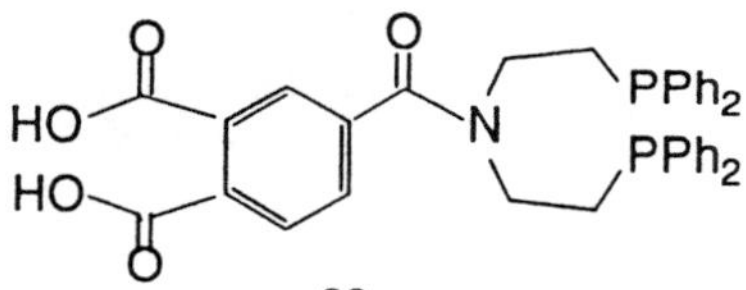

20

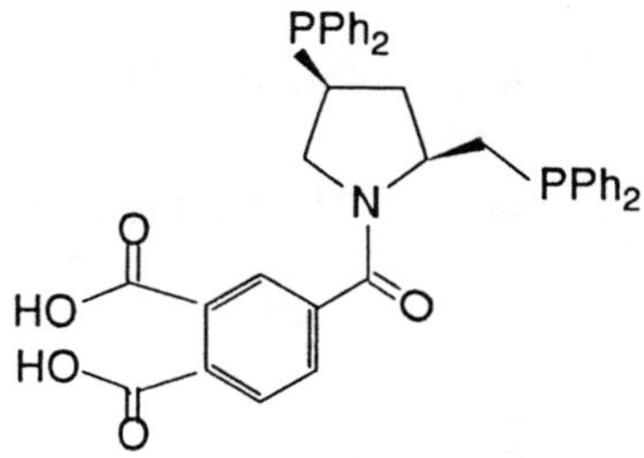

22

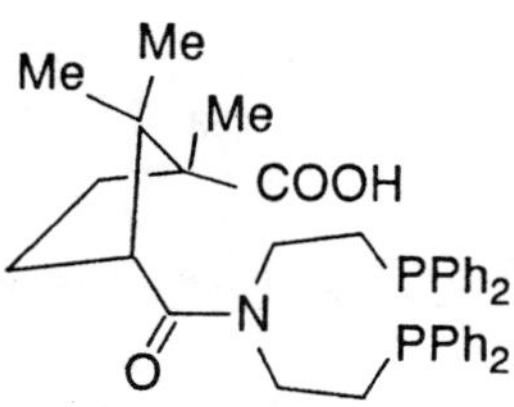

21

References: **17** [75, 76, 141a]; **18** [77, 78]; **19** [35]; **20** [3, 70, 71]; **21** [71, 73]; **22** [100]

by *protonation* or *alkylation* of the parent amino substituted phosphine. There have been various investigations of the uses of complexes of such ligands in homogeneous catalysis in aqueous solvents [79–82]. Table *5.3* shows some of the ligands available.

Protonation or alkylation of amino substituted phosphines very often yields quaternary phosphonium rather than ammonium salts as the products [106, 107]. Thus multistep procedures, such as quaternization of an intermediate phosphine oxide, followed by reduction to the ammonium-phosphine have been devised, as shown on the Scheme *5.4* [79–80].

TABLE 5.3
Amino and ammonium phosphines for aqueous catalysis

$[Ph_2PCH_2CH_2NMe_3]^+$

23

$P(C_6H_4\text{-}4\text{-}NMe_2)_3$

24

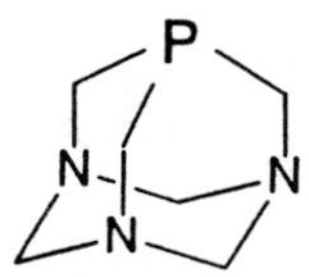

25

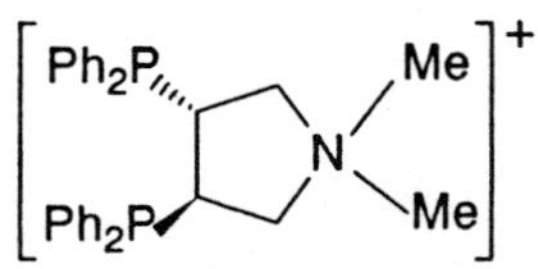

26

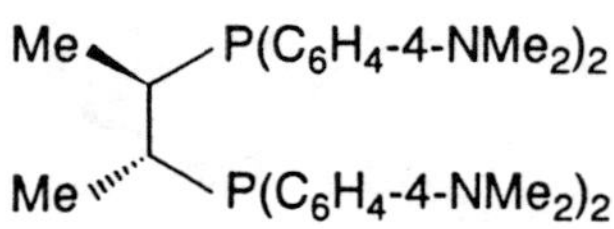

27

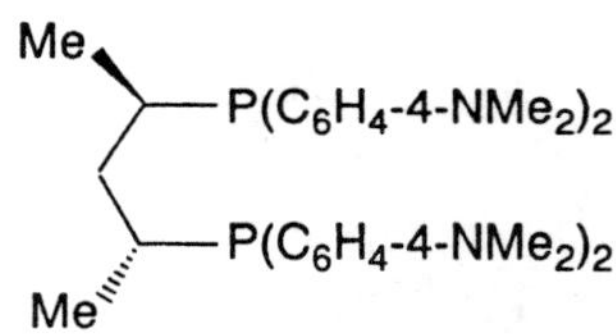

28

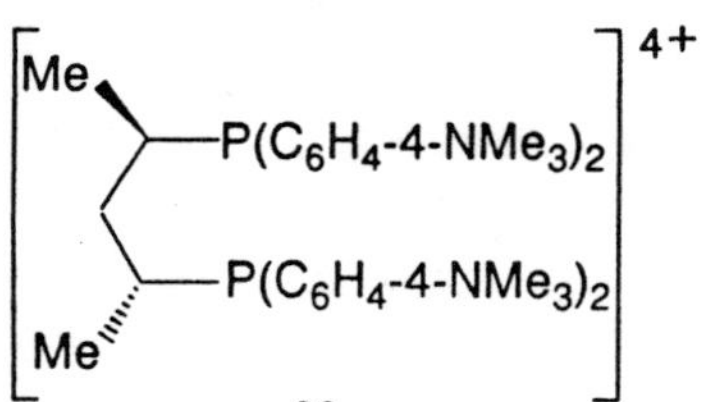

29

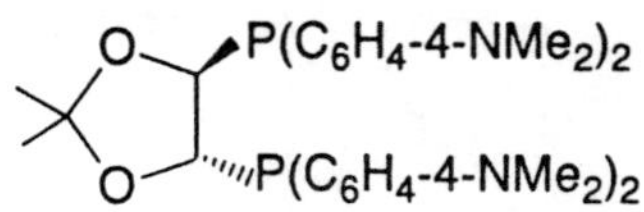

30

References: **23** [79, 81]; **24** [83]; **25** [84-87]; **26** [91]; **27** [88,90]; **28** [88, 90]; **29** [88-91]; **30** [89, 90]

A non-chelating amino phosphine ligand can be N-methylated [91] with either CH_3I or $[(CH_3)_3O][BF_4]$ while coordinated through phosphorus to a transition metal centre (Scheme 5.5), leading to water soluble complexes [88–91].

Protonation by a strong acid with a non-coordinating anion (*e.g.* $H[BF_4]$ has been used both with the free ligands and with their complexes [88a]. In order to immobilize the ligand, strongly acidic anion exchangers (*e.g.* Nafion-H) can also be used [88b].

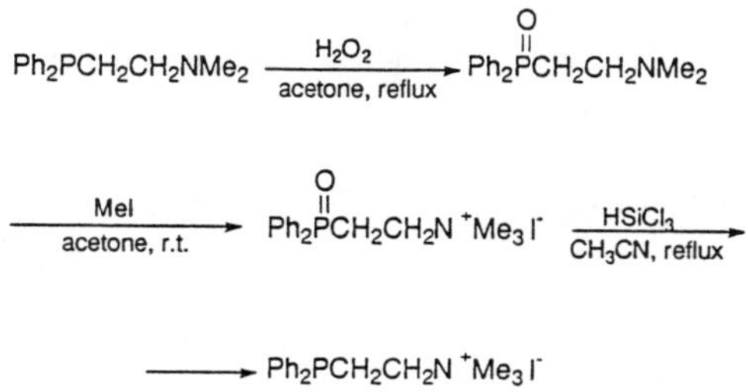

Scheme 5.4.

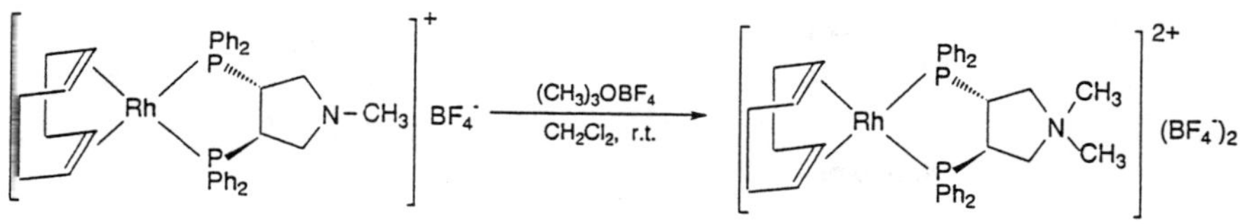

Scheme 5.5.

The water solubility of *hydroxyphosphines*, and those with *open chain and cyclic polyether* substituents is variable ranging from slight [98] to unlimited [33, 95–97]. Members of this group are listed in Table 5.4. Such ligands can be prepared by addition of aldehydes or unsaturated alcohols or their ethanoates across the P-H bond [95], by acylation of phosphinoalkylamines [71–73], or by Grignard reactions [97].

5.2.1.3. *Non-phosphine ligands*

In many cases the solubility of a metal complex is a result of its ionic nature [194], and any ligand which can stabilize a lower oxidation state metal ion in an aqueous environment retaining a net charge on the complex ion may be useful. Common examples include halide, cyanide, carboxylate, cyclopentadienyl, pentamethylcyclopentadienyl, and carbonyl ligands and combinations of these. Certain dyes and indicators, such as indigosulfonic acid, and Alizarin Red have also been investigated.

In general however, the scope for a systematic study of the effects of ligand modification on the catalytic reaction is rather limited.

TABLE 5.4
Miscellaneous ligands for aqueous catalysis

$P(CH_2OH)_3$
31

$Ph_2PCH_2CH_2OH$
32

$P(CH_2CH_2OCH_2CH_2OCH_2CH_2OCH_3)_3$
33

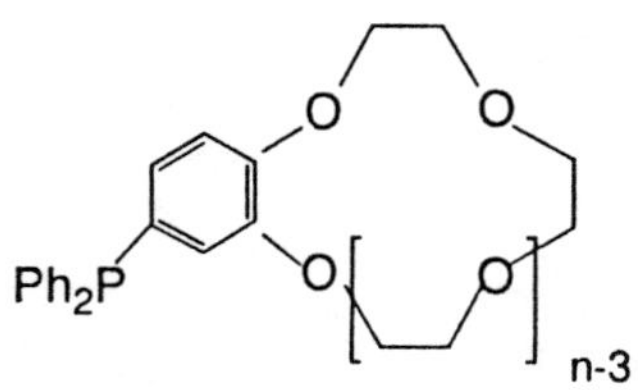

35; n= 4, 5, 6 or 7

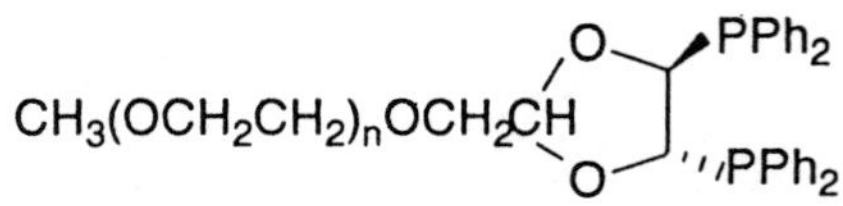

36

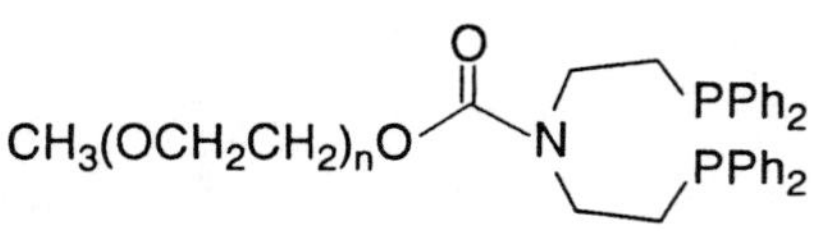

37

$Ph_2P(CH_2)_{11}PO_3H_2$

39

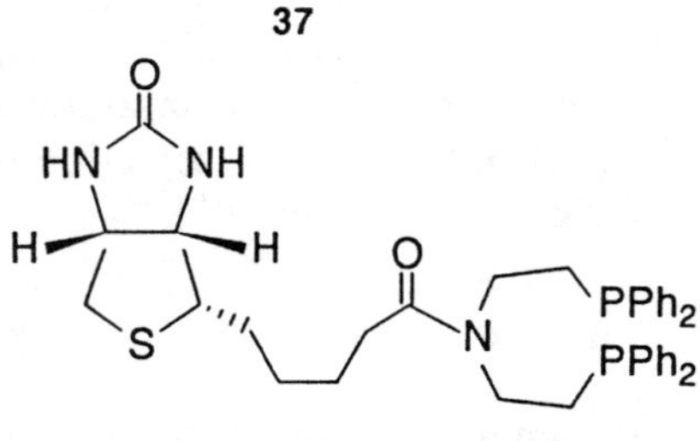

38

$[Ph_2P(CH_2)_nPMe_3]X$

40; n=2,3,6 or 10
$X=Cl^-, [NO_3]^-, [PF_6]^-$

References: **31** [94-96]; **32** [93]; **33** [70, 71, 190]; **34** [33]; **35** [98, 99]; **36** [102]; **37** [70, 71, 73]; **38** [71-73]; **39** [184, 185]; **40** [81]

5.2.2. EFFECT OF THE AQUEOUS ENVIRONMENT ON THE FORMATION OF CATALYTICALLY ACTIVE COMPLEXES

Systematic studies on the coordination chemistry of TPPMS [5, 13, 14b, 18, 28, 103, 109] and TPPTS [50–53, 63] have shown that sulfonation makes little difference to the complexation properties of these ligands. The infrared and NMR spectroscopic properties of the resultant complexes are very similar when using sulfonated and non-sulfonated forms of otherwise identical phosphines. There are, however, interesting changes in the number of coordinated phosphine ligands. In several cases, this number is *smaller* than in analogous complexes with non-sulfonated ligands. Examples include [$RuCl_2$(TPPMS)$_2$] [18], [$RuCl_2$(TPPTS)$_2$] [50, 53], [Pd(TPPMS)$_3$] [60], and [Ni(TPPTS)$_3$] [53]. Steric requirements for the bulky SO_3^- substituent, and the electronic repulsion of the charged ligands (especially in case of TPPTS) may both contribute to the decrease in the number of coordinated phosphines. Tolman cone angles for the TPPTS ligand were calculated from crystallographic data for [Fe(CO)$_4$TPPTS] and [Na-kryptofix-221]$_3$[W(CO)$_3$TPPTS], and were found 166–178°C, depending on the nature of the counterion. This is compared to 131.6° for PPh_3 in [Mo(CO)$_5$(PPh_3)] determined in the same study [47b].

It is very important for catalysis, that in case of the corresponding [RhH(CO)P_3] complexes not only is the number of coordinated phosphine ligands *not* decreased by sulfonation of PPh_3, but that the binding of TPPTS appears to be *stronger* than the binding of PPh_3. Variable temperature ^{31}P NMR spectroscopic studies [63] established the activation energy for phosphine dissociation in aqueous solution from [RhH(CO)(TPPMS)$_3$] as 126±4 kJ mol^{-1}, in contrast to 80±4 kJ mol^{-1} for [RhH(CO)(PPh_3)$_3$]. Considering the almost identical IR and NMR spectroscopic parameters, this reduced dissociation was explained by hydrogen bonding between adjacent sulfonated ligands in the complex; direct evidence is, however, lacking.

Detailed kinetic studies of dissociation processes in aqueous solutions of [M(CO)$_5$TPPTS] and *cis*-[M(CO)$_4$(TPPTS)$_2$] (M = Mo, W) in miscible aqueous/organic media were carried out by *in situ* infrared spectroscopy employing a cylindrical internal reflectance reactor [47.b]. A stabilizing effect against dissociation was observed in *cis*-[Mo(CO)$_4$(P)$_2$] when PPh_3 was replaced by TPPTS where the Na^+ ions were free to interact with the sulfonate groups. By way of contrast, when the sodium ions were encapsulated by a cryptand, kryptofix-221, an approximately 10-fold acceleration of TPPTS dissociation was noted relative to the PPh_3 analog as could be expected taking the larger steric bulk of TPPTS.

In general, the coordination chemistry and the stoichiometric reactions of ammonium-alkyl- or ammonium-arylphosphines is similar to that of the corresponding neutral phosphines. There are, however, some interesting exceptions. With the monophosphine, AMPHOS (**23**) no clean reaction with [{RhCl(C_2H_4)$_2$}$_2$] could be observed, so a Wilkinson-type [RhClP_3] complex could not be obtained [81]. Similarly,

$[Rh(NBD)(AMPHOS)_3]^{4+}$ could not be prepared from $[Rh(NBD)(AMPHOS)_2]^{3+}$ and an excess of AMPHOS [81], in contrast to the behavior of other phosphines [111]. There seems to be no particular problem with preparations involving the tetrasubstituted ammonium diphosphines **27–30** [89] despite the fact that they yield complex ions with high positive charges (up to 5+).

Reductive elimination of dihydrogen from the AMPHOS complex (equation *5.6*) takes place even under 1 bar of hydrogen, both in methanol and in water [81]:

$$[RhH_2(AMPHOS)_2(solvent)_2]^{3+} \longrightarrow [Rh(AMPHOS)_2(solvent)_2]^{3+} \quad (5.6)$$

In water the reaction is very fast, and no dihydride complex may be detected by NMR spectroscopy a few minutes after dissolution of the complex. This is in striking contrast to the case of analogous complexes of PPh_2Me, where no reductive elimination of H_2 occurs [112]. No detailed study of this reaction has appeared, so it is difficult to assess the contributions of the ammonium substitution of the ligand, the low solubility of dihydrogen in water and the presence of H^+ in the (presumably) slightly acidic aqueous solutions.

In case of certain metal ions there is a pronounced tendency to undergo hydrolysis even in non-alkaline solutions. Prolonged reaction of $RhCl_3.3H_2O$, $[RhCl(PPh_3)_3]$ or $[\{RhCl(COD)\}_2]$ with TPPTS in water or in an aqueous/organic solvent mixture invariably leads to formation of high proportions of $[Rh(OH)(TPPTS)_3]$ [51, 52] in addition to $[RhCl(TPPTS)_3]$. Similarly, $[Rh(acac)(CO)_2]$ yields $[Rh(OH)(CO)(TPPTS)_2]$ on stirring with TPPTS in aqueous solution at room temperature [52]. There is no systematic study of such hydrolysis equilibria reported, but they do effect the catalytic performance of a given complex. Therefore, the use of appropriate(!) buffers is an essential precaution.

5.3. Hydrogenation of organic substrates in aqueous systems

As mentioned in the Introduction, aqueous solutions were first used during the early period of investigations on homogeneous hydrogenation. This was straightforward, since the catalysts of that time were mainly simple transition metal salts or water soluble complexes. For mechanistic studies of H_2-activation and hydrogenation, homogeneous solutions were preferred, so most of the substrates used in these investigations were also soluble in water. In addition to high valent metal ions, such as Fe(III), reducible substrates included unsaturated organic compounds, such as unsaturated acids (*e.g.* E- and Z-2-butene dioic and E-3-phenylpropenoic acid) or alcohols (*e.g.* 2-propene-1-ol). The same substrates, together with oxo-acids (*e.g.* 2-oxo-propanoic acid) are still used for mechanistic investigations.

It is not the purpose of this book to review early work on hydrogenation in much detail; this was superbly done by James [113, 114]. Here we shall only consider representative examples, some of which have found new applications in a modified form. The main part of this Chapter is devoted to catalysts containing more complex ligands (mainly phosphines) which allow fine tuning of the performance of the catalyst.

5.3.1. CATALYSTS WITH SIMPLE IONS AS LIGANDS

5.3.1.1. *Ruthenium salts*

Solutions of $RuCl_3$ in 3 M aqueous HCl catalyse the hydrogenation of Fe(III) to Fe(II) at 80°C temperature and 0.6 bar H_2 pressure [115a]. Ru(IV) is hydrogenated autocatalytically to Ru(III), but the latter is not reduced further by H_2. Variation of the [HCl] concentration revealed, that hydrolysis (equation *5.7*) results in a complex of lower catalytic activity.

$$[RuCl_{3-n}]^{3-n} + H_2O \rightleftharpoons [Ru(OH)Cl_{n-1}]^{3-n} + H^+ + Cl^- \qquad (5.7)$$

The mechanism of hydrogen activation involves a heterolytic splitting of H_2, and all the experimental findings could be explained by assuming the reactions depicted in Scheme *5.8*, provided that k_2, $k_3 > k_{-1} > k_1$, k_4.

Since there is practically no reduction of Ru(III) to Ru(II), it can be expected that in the absence of a reducible substrate, such as Fe(III), isotopic exchange with the solvent H_2O takes place when D_2 is used instead of H_2. Indeed, under the same conditions the combined rate of HD and D_2 formation approached the corresponding rate of oxidation of D_2 by Fe(III) [115b]. This supports the view that the two reactions proceed through related mechanisms (Scheme *5.9*).

An important feature of the isotopic exchange reaction should be noted, with special reference to the hydrogen activation by the enzyme hydrogenase (section 7.5). It was observed that the resulting gas mixture always contained approximately 15–30% H_2 relative to HD, even when

$$\mathrm{Ru(III)} + \mathrm{H_2} \underset{k_{-1}}{\overset{k_1}{\rightleftharpoons}} \{\mathrm{Ru(III).H^-}\} + \mathrm{H^+} \qquad \text{5-8.1}$$

$$\{\mathrm{Ru(III).H^-}\} + 2\,\mathrm{Ru(IV)} \xrightarrow{k_2} 3\,\mathrm{Ru(III)} + \mathrm{H^+} \qquad \text{5-8.2}$$

$$\{\mathrm{Ru(III).H^-}\} + 2\,\mathrm{Fe(III)} \xrightarrow{k_3} \mathrm{Ru(III)} + 2\,\mathrm{Fe(II)} + \mathrm{H^+} \qquad \text{5-8.3}$$

$$\{\mathrm{Ru(III).H^-}\} + \mathrm{Ru(III)} \xrightarrow{k_4} 2\,\mathrm{Ru(II)} + \mathrm{H^+} \qquad \text{5-8.4}$$

Scheme 5.8.

$$\{\mathrm{Ru(III)Cl^-}\} + \mathrm{D_2} \xrightarrow{k_1} \{\mathrm{Ru(III)D^-}\} + \mathrm{D^+} + \mathrm{Cl^-} \qquad \text{5-9.1}$$

$$\{\mathrm{Ru(III)D^-}\} \xrightarrow[\mathrm{H^+ + Cl^-}]{k_{-1}} \{\mathrm{Ru(III)Cl^-}\} + \mathrm{HD} \qquad \text{5-9.2}$$

$$\{\mathrm{Ru(III)D^-}\} \xrightarrow[\mathrm{H_2O}]{k_2} \{\mathrm{Ru(III)H^-}\} + \mathrm{HDO} \qquad \text{5-9.3}$$

$$\{\mathrm{Ru(III)H^-}\} \xrightarrow[\mathrm{H^+ + Cl^-}]{k_{-1}} \{\mathrm{Ru(III)Cl^-}\} + \mathrm{H_2} \qquad \text{5-9.4}$$

$$\{\mathrm{Ru(III)D^-}\} + 2\,\mathrm{Fe(III)} + \mathrm{Cl^-} \longrightarrow \{\mathrm{Ru(III)Cl^-}\} + 2\,\mathrm{Fe(II)} + \mathrm{D^+} \qquad \text{5-9.5}$$

$$\{\mathrm{Ru(III)D^-}\} + \mathrm{H_2O} + \mathrm{Cl^-} \longrightarrow \{\mathrm{Ru(III)Cl^-}\} + \mathrm{HD} + \mathrm{OH^-} \qquad \text{5-9.6}$$

Scheme 5.9.

the concentration of Ru(III) or HCl approached zero. These observations suggest, that H_2 is a primary product of the reaction and does not originate from a second exchange of HD with H_2O. Also, there must be a H^+-independent way of exchange, presumably that in equation 5-9.6.

E- and Z-2-butene dioic acids were efficiently hydrogenated to butane dioic acid in aqueous hydrochloric acid solutions, using ruthenium(II) chloride as the catalyst [116]. The mechanism involves formation of a ruthenium(II)-olefin complex, followed by the heterolytic activation of H_2 on the same complex. In the temperature range of 60–90°C the rate could be described by the equation $d(H_2)/dt = k[H_2][Ru(II)(alkene)]$. E-2-butene dioic acid was reduced *faster* ($k = 3.6 \pm 0.6\ M^{-1}s^{-1}$) than the Z-isomer ($k = 2.3 \pm 0.1\ M^{-1}s^{-1}$) and the activation energies were found to be 71 $kJ.mol^{-1}$ and 59 $kJ.mol^{-1}$, respectively. Very interestingly, with both of the alkenes there was no deuterium incorporation to the product when the reduction was carried out using D_2; however, in D_2O solutions, but with H_2 as the reductant, exclusive formation of *dideuterated* product was observed. This implies, that the monohydrido intermediate formed in the hydrogen activation step exchanges its hydrogen with the solvent before undergoing rearrangement.

5.3.1.2. *Hydridopentacyanocobaltate(III)*

As was mentioned in Chapter 2, the cobalt-hydride, readily formed in the reduction of $[Co(CN)_5]^{3-}$ (equation *5.10*) is a very active and selective catalyst [113, 117] for the hydrogenation of conjugated dienes to monoenes, mainly with 1,4-addition of hydrogen (Scheme *5.11*).

$$2[Co(CN)_5]^{3-} + H_2 = 2[HCo(CN)_5]^{3-} \quad (5.10)$$

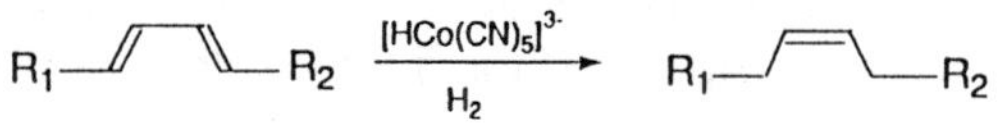

Scheme 5.11.

As a result of very detailed investigations, it became clear that use of this catalyst for preparative purposes has limitations:

a) An excess of the substrate inhibits hydrogenation
b) Solutions of the catalyst "age" rapidly with concomitant loss of activity
c) Solutions of the catalyst are highly basic, and this cannot be always tolerated by the substrate.

In the PTC version [118, 119] of hydrogenations by $[HCo(CN)_5]^{3-}$, the substrate was added in a separate organic phase (usually toluene). Tetramethylammonium bromide or triethylbenzylammonium bromide were used as phase transfer agents. Very importantly, in addition to their PT effect, the quaternary ammonium salts *stabilized* the catalyst, which could then be stored in solution for several days. Dienes were hydrogenated on a preparative scale with the usual 1,4-addition of hydrogen, and α,β-unsaturated ketones could be selectively reduced at the olefinic double bond. In case of aldehydes, such as E-3-phenylpropenal, the yield of the saturated aldehyde is only 10–20%, owing to side reactions such as condensation and polymerization.

Both neutral (Brij 35) or ionic (SDS, CTAB) surfactants substantially increased the rate of hydrogenation of 2-phenylpropenoic acid and its esters in a $ClCH_2CH_2Cl/H_2O$ solvent system [120]. The rate increase could be attributed to the incorporation of the substrate to the micelle, leading to increased local substrate concentration.

Cyclodextrins are water soluble oligomers of D-glucose which have recently received much attention. This interest arises from the fact that cyclodextrins are versatile host molecules for many compounds, and through such complexation they alter the solubility and reactivity of the guest compounds. There are several examples of their use as "reverse" phase transfer agents [121, 144], carrying water-insoluble substrates from an organic to an aqueous phase.

It was found [122] that β-cyclodextrin was an effective phase transfer agent in the hydrogenation of conjugated dienes by $[HCo(CN)_5]^{3-}$ in a benzene/water solvent system. A variety of dienes could be hydrogenated to monoenes with good to excellent selectivity for the product of *1,2-addition* of hydrogen. With a proper choice of KOH concentration and with the addition of hydrated lanthanide salts (of which La and Ce were superior to Yb) 2,3-dimethyl-1,3-butadiene was hydrogenated to 2,3-dimethyl-1-butene with 100% yield and 97% selectivity (Scheme *5.12*). The β-cyclodextrin phase transfer agent could be replaced by PEG 400 with equally good results.

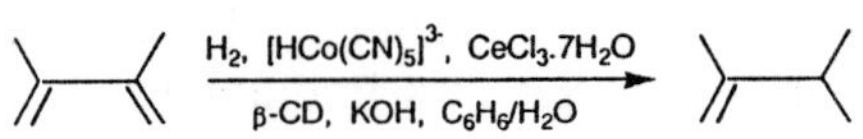

Scheme 5.12.

Hydrogenation of α,β-unsaturated acids [123] and their derivatives was also studied in the presence of β-cyclodextrin. In several cases the cyclodextrin certainly exerted a phase transfer function, but more importantly, the complexation of the substrates by CD also influenced the selectivity of hydrogenations. As an example, propenoic acid gave mixtures of monomeric and dimeric products when reacted in homogeneous solutions, resulting in a yield of only 16% of propanoic acid. The yield of the reduced acid was increased to 81% in the presence of β-cyclodextrin.

The use of highly basic solutions can be circumvented by adsorbing $[HCo(CN)_5]^{3-}$ onto strongly basic anion exchangers [124]. In a neutral aqueous slurry of the anchored catalyst propenal was hydrogenated to propanal.

$[HCo(CN)_5]^{3-}$ catalyses the hydrogenation of nitro compounds [113, 125]. Aliphatic substrates are reduced to amines, whilst nitroarenes give the products of reductive dimerization, *i.e.* azo and hydrazo compounds. Ketoximes and oximes of 2-oxo-acids are hydrogenated to amines [125]. An important related reaction is the reductive amination [126] of

2-oxo-acids (Scheme *5.13*). This reaction is carried out in an aqueous NH_3 solution at a temperature of 40–50°C, 70 bar H_2 pressure, with approx. 90% yield of 2-amino-acids.

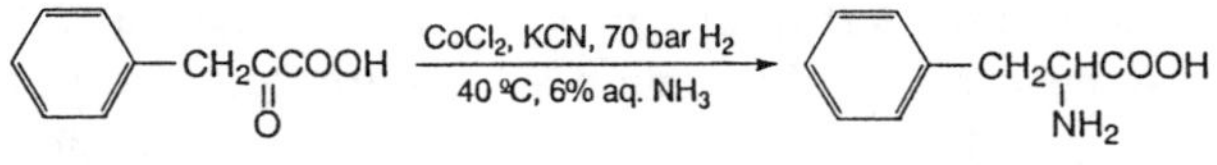

Scheme 5.13.

5.3.1.3. $[Q^+]_p[MX_n]^{p-}$ *Ion-pairs*

The ion-pairs formed in solutions of Group VIII metal halides and quaternary ammonium salts with long chain substituents, can be extracted into non-polar organic solvents where they catalyse a range of reactions (isomerization, hydrogenation, dehydrogenation, etc). The catalytically active species are not always well characterized and the formation of stabilized metal colloids and halide and/or hydroxide bridged polynuclear complexes cannot be excluded.

One of the best studied of these catalysts is the $RhCl_3.3H_2O$/Aliquat-336 system, containing the $[(C_8H_{17})_3N(CH_3)]^+[RhCl_4(H_2O)_2]^-$ ion-pair, which actively hydrogenates arenes in water/$ClCH_2CH_2Cl$ at 30°C and 1 bar total pressure [127, 128]. The yields are usually greater than 95% and, except for a few examples, the selectivities are 100%. A unique observation is that the reduction of nitrobenzene affords, albeit in only 5% yield, nitrocyclohexane, in addition to the major product, aniline (90%) [127]. This is the only example in the literature on preferential hydrogenation of an aromatic bond to a nitro function by a homogeneous catalyst.

Addition of a tertiary amine (*e.g.* Et_3N) to the above system and replacement of 1,2-dichloroethane with diethyl ether results in a catalyst mixture [129] which is very active for the hydrogenation of alkenes (including sterically hindered ones), for the hydrogenation of arenes (benzene, toluene, phenol, and methyl benzoate) to the corresponding cyclohexane derivatives, and also for the hydrogenation of nitriles, aldehydes, and nitro compounds. In some cases Rh could be replaced by Ru, but Co and Fe were much less effective.

The treatment of a mixture of an aqueous solution of $[NH_4]_2[PdCl_4]$ and a water soluble chelating macromolecule, such as polyvinylpyrrolidinone (PVP) or polyethyleneimine (PEI) with H_2 or $Na[BH_4]$ results in a highly active catalyst for hydrogenation of alkenes and alkynes. It is not clear whether the catalytically active solution contains an ion-pair of the protonated macromolecule and $[PdCl_4]^{2-}$, another anionic palladium derivative, or whether it is a stabilized metal colloid. Very useful regio- and stereoselectivities were observed [130]. In the presence of added alanine, Pd-PVP hydrogenated the linoleate and linolenate content of soy-bean oil to oleate with 97% stereoselectivity to the *cis*-product. With no additive, Pd-PEI reduces internal alkynes, such as 2-pentyne to give exclusively *cis*-alkenes, whilst in the presence of benzonitrile the sole product of the same reaction is the *trans*-alkene. A very important feature of this reaction system is in that the macromolecular catalyst can be retained in the reactor while all other ingredients can be separated by ultrafiltration.

5.3.2. NON-ENANTIOSELECTIVE HYDROGENATIONS CATALYSED BY COMPLEXES OF WATER SOLUBLE PHOSPHINES AND OTHER COMPLEX LIGANDS

5.3.2.1. *Catalysts containing sulfonated phosphine ligands*

Detailed kinetic studies of hydrogenations catalysed by Ru [16, 18] and Rh [17, 19] complexes served to establish the relationship between the catalytic activities of analogous complexes with PPh_3 and with TPPMS.

Catalysis by ruthenium complexes of sulfonated phosphines [16, 18] was investigated in 0.1 M aqueous HCl solutions at 60°C. $[RuCl_2(TPPMS)_2]$, $[RuHCl(TPPMS)_3]$, and $[RuH(O_2CCH_3)(TPPMS)_3]$ catalysed the hydrogenation of unsaturated and 2-oxo-acids, such as E- and Z-2-butene dioic acid, 2-oxo-propanoic and 2-oxo-butane dioic acid. Kinetic studies revealed that the catalytically active species for reduction of alkenes was $[RuHX(TPPMS)_2]$ whilst for 2-oxo-acids it was $[RuHX(TPPMS)_3]$. The mechanism of hydrogenation of unsaturated acids (Scheme *5.14*) was found to be virtually identical to that of the hydrogenation of alkenes by $[RuHCl(PPh_3)_3]$ in non-aqueous solvents (Chapter 2).

Unlike the Ru-complexes, the hydride, $[RhH_2Cl(TPPMS)_3]$ formed by the reaction of $[RhCl(TPPMS)_3]$ with H_2, was stable in aqueous solution only at room temperature. At 50°C the solution became darker in colour and after longer heating metal precipitation ocurred. However, if the substrate was added before the admission of H_2, rapid hydrogenation of unsaturated and 2-oxo-acids was observed [17, 19] (Table *5.5*).

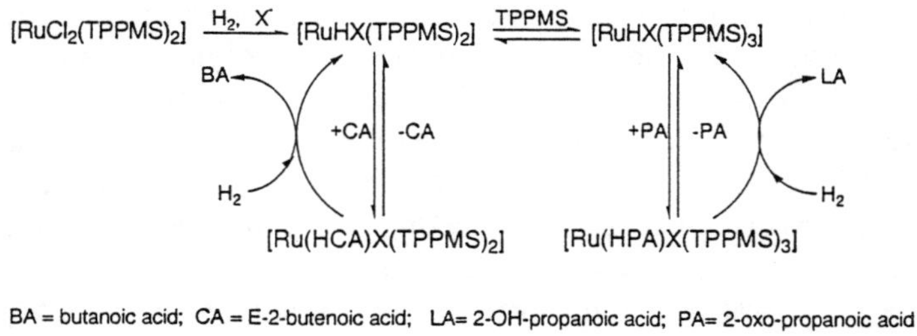

Scheme 5.14.

TABLE 5.5
Initial turnover frequency (r_0) and half-time ($t_{1/2}$) of hydrogenations A) catalysed by $[RhCl(TPPMS)_3]$ [19]

Substrate (S)	102[S] (M)	r_0 (h^{-1})	$t_{1/2}$ (min)
E-2-Butene dioic acid[b]	3	1270	6.5
Z-2-Butene dioic acid	3	53	13.5
E-2-butenoic acid	3	180	7.0
E-3-Phenylpropenoic acid	0.7	46	8.5
2-Propen-1-ol	5	111	82.0
Na-2-oxo-propanoate	3	35	16.5
2-Oxo-pentanoic acid	3	18	42.0

[a] Conditions: 0.01 mmol $[RhCl(TPPMS)_3]$, 0.8 bar H_2, 60 °C, 10 ml aqueous solution
[b] Same as a) but 0.001 mmol $[RhCl(TPPMS)_3]$

A very interesting feature of the hydrogenation of Z-2-butene dioic acid catalysed by $[RhCl(TPPMS)_3]$ is in that the rate is independent of the concentration of added phosphine. This is in sharp contrast to the findings with $[RhCl(PPh_3)_3]$ in alkene hydrogenations (Chapter 2). The rate of hydrogenation of E-2-butene dioic acid is also very slightly effected by the presence of an excess of TPPMS, whilst reduction of E-2-butenoic acid is strongly inhibited by an excess of ligand [19a]. Later it was shown that the reaction of the substrate olefin with excess phosphine scavenges the inhibition by the latter [196]. Such reactions, yielding phosphonium salts ([36] see also section 5.8) may play a more important role in catalysis than it is generally held.

In the first studies which used a *biphasic* reaction mixture for alkene hydrogenation, $[RhCl(TPPMS)_3]$ was used as catalyst [29]. It was established that in order to achieve useful reaction rates co-solvents were necessary in the case of water-immiscible alkenes. Both [RuHCl(TPPMS)$_3$] and [RhHCl(TPPMS)$_3$] were found to be active catalysts for the hydrogenation of hexenes and cyclohexene at 25–80°C, and 3 bar H_2 pressure [28]. Since the activity of the Ru-complex was lower, hydrogenation was accompanied by more pronounced isomerization than in case of the Rh catalyst.

When $RhCl_3$ was heated under reflux in methanol:water = 1:1 together with 3 equivalents of TPPTS, followed by evaporation of the reaction mixture to dryness, an air-stable brown solid was obtained, which catalysed the hydrogenation of unsaturated acids in aqueous solutions [54]. Based mainly on NMR spectroscopic data it was established, that the catalyst was a mixture of various Rh(I)-phosphine complexes, phosphine oxide and colloidal rhodium [55]. Using this catalyst a preparative scale method was developed for the biphasic hydrogenation of alkenes [54]. In the 25 substrates studied, C=C double bonds could be selectively reduced in the presence of keto, carboxyl, ester, hydroxyl and amine functionalities (Table *5.6*), and it was strongly influenced by steric effects. The reaction of [{RhCl(1,5-hexadiene)}$_2$] with H_2 in a phase transfer catalytic system, comprising of benzene or hexane as the organic phase, an aqueous phase buffered to pH 7.4–7.6 and a quaternary ammonium salt gave a catalyst mixture which was highly active and selective for the hydrogenation of arenes under exceedingly mild conditions (room temperature, 1 bar total pressure) [143]. There was no reaction in the absence of a phase transfer catalyst, and $RhCl_3.3H_2O$, [$Rh_2(O_2CCH_3)_4$] and [{RhCl(CO)$_2$}$_2$] were found inert under these conditions. The process is applicable to a variety of functionalized arenes (ketones, esters, amides) and heteroarenes. Phenol was hydrogenated to a mixture of cyclohexanol and cyclohexanone. Similar to the process discussed above [54] the latter reactions may also involve colloidal rhodium as the catalyst.

When [$RhHCl_2(PCy_3)_2$] was used as catalyst precursor in a biphasic system [182] α,β-aldehydes were selectively reduced at the carbon-carbon double bond under very mild conditions. The catalytically active species is formed by HCl elimination [141b] from the intermediate product of oxidative addition of dihydrogen, and the equilibrium (equation *5.15*) is shifted towards the formation of the dihydride in the presence of a base (water).

$$[RhHCl_2(PCy_3)_2] + H_2 \rightleftharpoons [RhH_2Cl(PCy_3)_2] + HCl \qquad (5.15)$$

TABLE 5.6
Hydrogenations catalyzed by $RhCl_3$/TPPTS [54]

Substrate	Product	Reaction time (h) [a]
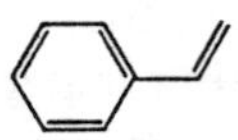	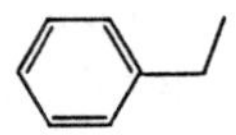	40
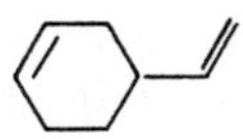	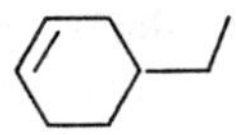	20
	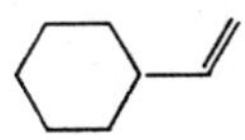	40
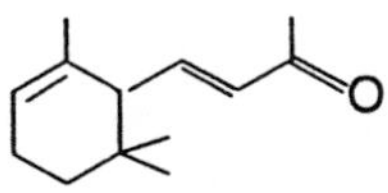	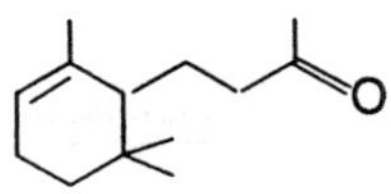	60
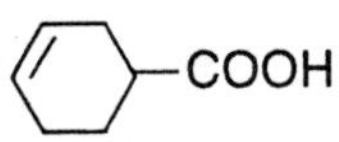	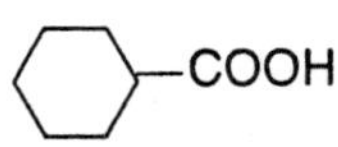	18
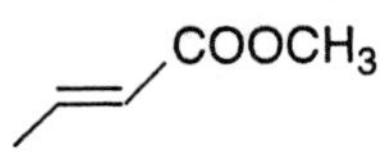	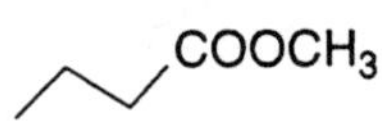	16
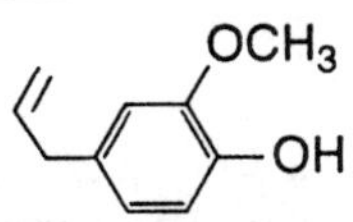	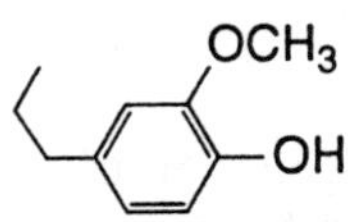	24

a) Reaction time for 100% conversion

In contrast to $[RhCl(PPh_3)_3]$ only slight decarbonylation of aldehydes occurs even under more forcing conditions. An additional advantage of using $[RhHCl_2(PCy_3)_2]$ as the catalyst precursor is that this monohydride is air stable and easy to prepare.

Alkynes and dienes were reduced to monoenes under catalysis by $[Pd(OH)_2(TPPMS)_2]$ or $[Pd(O_2CCH_3)(OH)(TPPMS)_2]$ at 20°C, and 1 bar H_2, in aqueous solutions or in emulsions [38]. However, under the reaction conditions the Pd-complexes were degraded to give phosphine-bridged clusters, and finally $[Pd(OH)_2]$.

Much work has been devoted to investigating the hydrogenation of α,β-unsaturated aldehydes to unsaturated alcohols. Although there are examples in the literature of such transformation, the selectivity and ease of reduction using Ru(II)-TPPTS derivatives [48, 92] is unprecedented. 3-Methyl-2-buten-1-al was reduced to 3-methyl-2-buten-1-ol, 2E- and 2Z-3,7-dimethyl-2,6-octadienal to 2E- and 2Z-3,7-dimethyl-2,6-octadienol, E-3-phenylpropenal to E-3-phenyl-2-propen-1-ol, and propenal to 2-propen-1-ol (Scheme *5.16*). The reaction took place in neutral or slightly basic solutions, with both the starting material and the product in the organic phase, at 35–50°C, and 20–50 bar H_2 pressure. No phase transfer agent was used. In sharp contrast to the Ru-based catalysts, $[RhCl(TPPTS)_3]$ selectively gave saturated aldehydes [49]. No detailed comparative study of Ru- and Rh-complexes in aldehyde hydrogenation has been published, and therefore this difference still awaits explanation.

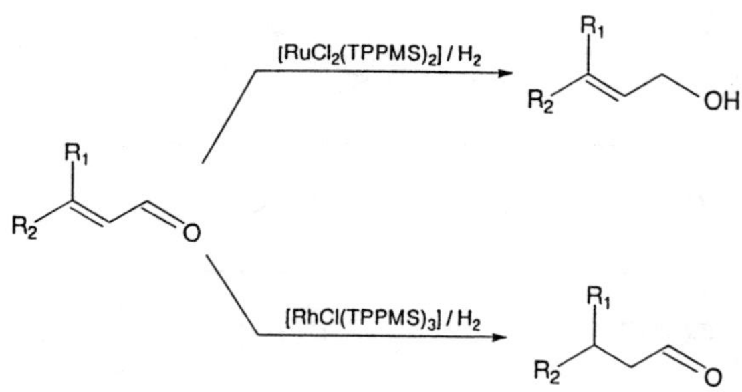

Scheme 5.16.

The kinetics of hydrogenation of propanal by various well characterized Ru(II)-TPPTS complexes was investigated in detail and compared to the findings with the organosoluble (PPh_3) analogs [92]. Very interesting salt effects have been revealed. Addition of NaI to the system not only increased the rate of reduction, but changed the reaction kinetics completely. With $[RuCl_2(TPPTS)_2]_2$ as catalyst precursor the following activation parameters were found a) without NaI: E_a = 79.5 kJ.mol^{-1}, $\Delta H^{\neq}_{375}$ = 76.4 kJ.mol^{-1}, $\Delta S^{\neq}_{375}$ = 52.7 kJ.mol^{-1}; with NaI E_a = 50.1 kJ.mol^{-1}, $\Delta H^{\neq}_{308}$ = 47.6 kJ.mol^{-1}, $\Delta S^{\neq}_{308}$ = 98.2 kJ.mol^{-1}. Obviously, the presence of salt caused changes in important step(s) of the catalytic cycle. Formation of the intermediate $[RuHI(TPPTS)_3]$ was postulated, and indeed, the isolated complex gave the same results as the *in situ* prepared $[RuHI(TPPTS)_3]$. However, when the Na^+ was sequestred by complexation with cryptands, the effect of added NaI was largely diminished, showing a very important effect of the cation, too. The proposed mechanism considers C-coordination of the aldehyde facilitated by the interaction of the aldehyde oxygen and the cation of the added salt.

5.3.2.2. *Catalysts containing amino phosphine ligands*

Hydrogenation of alkenes catalysed by $[Rh(NBD)(AMPHOS)_2]^{3+}$ in water, methanol or in aqueous/organic biphasic systems has been studied in detail [81]. In methanol the rate of hydrogenation of Z-2-butene dioic acid was almost the same as when $[Rh(NBD)(PPh_3)_2]^+$ was used. In water, however, the rate was decreased by a factor of 6 relative to methanol, presumably because of the lower solubility of H_2 in water. As with the PPh_3 complex, addition of an excess of AMPHOS decreased the rate considerably. Another similarity was that in biphasic systems 1-hexene reacted faster than styrene. Additionally, there was some isomerization to internal hexenes (10–20%, depending on reaction conditions). AMPHOS and its Rh-complex are only slowly oxidized, so that phase separation, recovery and reuse of the catalyst solution could be undertaken in air, without any substantial loss of catalytic activity.

$[RuCl_2(PTA)_4]$ catalysed the slow hydrogenation of aldehydes under 28 bar H_2 at 80°C temperature [87]. E-3-Phenylpropenal was selectively reduced to E-3-phenyl-2-propen-1-ol. On the other hand, $[RhCl(PTA)_3]$ did not catalyse the reduction of the aldehyde function, but was very active in hydrogenation of carbon-carbon double bonds. Thus the selectivity pattern is the same as that with Rh- and Ru- complexes of sulfonated phosphines. Hydrogenation of 1-phenyl-2-propene with the $[RhCl(PTA)_3]$ catalyst proceeded with a 30 mol H_2/mol Rh.h initial turnover frequency at 50°C temperature and 1 bar total pressure and was accompanied by extensive isomerization to E- and Z-1-phenyl-1-propene [87].

One of the very first attempts at catalyst recovery was using extraction into an aqueous phase [82]. Aminoalkylphosphine complexes of Co, Rh and Ru were used for hydrogenation of alkenes (propene, 1-hexene) and aldehydes. The reaction proceeded in an organic solution but in the presence of an aqueous phase. On acidification of the latter the protonated catalyst accumulated in the aqueous phase and could be separated from the product. The free catalyst was liberated by base and reextracted into the organic phase to catalyse further substrate reduction. With sufficiently basic phosphines, such as that in $[\{Co(CO)_3(P\{CH_2CH_2CH_2NMe_2\}_3)\}_2]$, extraction could be achieved with aqueous solution saturated with CO_2 at 1 bar pressure. In order to reextract the complex into the organic phase it was necessary simply to remove the dissolved CO_2 by warming. Thus there was no overall consumption of acid or base.

5.3.2.3. *Hydrogenations with miscellaneous catalysts*

$[(NBD)Rh(phosphos)_2][NO_3]$ complexes (phosphos = $[Ph_2P(CH_2)nPMe_3][NO_3]$, n = 2,3,6 and 10) were studied as catalysts in the hydrogenation of Z-2-butene dioic acid [81a]. All four catalysts exhibited high activity for the hydrogenation of 1-hexene. The ligand chain length influenced the activity, giving rates in the order of n $6 > 10 \geq 3 > 2$. It is interesting, though was not explained, that the extent of substrate isomerization was also at minimum with n = 6. The catalytic activity of these complexes varied in the same order with ligand chain length when tethered to a cation exchange resin (Amberlyst 15) [81b].

Several Rh(I) complexes of the various diphosphines prepared by acylation of *bis*(2-diphenylphosphinoethyl)amine were tested [3, 70–73] for hydrogenation of unsaturated acids (including acylated dehydroaminoacids), 2-oxo-propanoic acid, 2-propen-1-ol and flavin mononucleotide. In 0.1 M phosphate buffer at pH = 7.0, at 25°C temperature and 2.5 bar H_2 pressure, initial rates were in the range of 1.6–200 mol H_2/mol Rh.h, and the highest absolute turnover was 4070. Thiols inhibited catalytic hydrogenation by these complexes.

Although complexes of hydroxyalkylphosphines have been investigated in several catalytic reactions, there is no successful hydrogenation in aqueous solution described in the literature. Attempts were made to achieve catalysis using a water soluble phosphine complex with $P(CH_2OH)_3$ as ligand [94]; however, its Rh(I) complex proved inactive in hydrogenation of Z-2-butene dioic acid in water. Among the complexes of phosphines with polyether chains, $[Rh(NBD)L]^+[O_2CCF_3]^-$ with L = **33** (n = 16), was a moderately active catalyst for hydrogenation of 2-propen-1-ol [70].

A very interesting phenomenon was observed in case of complexes of phosphines having crown ether substituents [98, 99]. These phosphines served as ligands in Rh(I) complexes, but at the same time, also functioned as phase transfer agents for the E-3-phenylpropenoate salts to be reduced. There was no direct relation found between the rate of hydrogenation of Li- Na-, K-, and Cs-salts of Z-3-phenylpropenoic acid and the ability of the crown-ether phosphines to extract the cations (determined from extraction equilibria of 1,3,5-trinitrophenolates). However, the rate of hydrogenation, catalysed by the $[RhClL_2]$ catalysts prepared *in situ* was 50-times higher with L = crown-phosphine, than with PPh_3 in the presence of benzo-18-crown-6 added separately. This effect is certainly due to the fact that the pendant crown ether moiety in the complex secures the substrate in close proximity to the metal ion.

A biphasic reaction can also be carried out in this way, so that the catalyst is dissolved in the organic phase, and the substrate is found in the aqueous phase. A model reaction was the hydrogenation of aqueous 2-butene-1,4-diol [29] catalysed by $[RhCl(PPh_3)_3]$ dissolved in benzene. In a similar system an unusual effect of water was observed. In dry benzene, with $RhCl[P(p\text{-tolyl})_3]_3$ as catalyst, hydrogenation of 3,8-nonadienoic acid afforded mostly 3-nonenoic acid, while in a biphasic system 8-noneoic acid was obtained selectively [191].

Another interesting example is the hydrogenation of acrylonitrile/butadiene/styrene (ABS) copolymers [136] to achieve better thermal stability. The polymer was reduced in form of an aqueous emulsion using several Rh-complexes, including $[RhHCl(PPh_3)_3]$, $[RhH(CO)(PPh_3)_3]$, and $[Rh(COD)(PPh_3)_2]^+[BF_4]^-$. The slightly water soluble *cationic* complex partitions more efficiently through the bulk aqueous phase into the micelles of anionic detergents surrounding the polymer than do the neutral catalyst molecules. This is reflected in the conversions: for example, 70% conversion was achieved with $[Rh(COD)(PPh_3)_2]^+[BF_4]^-$ by comparison with 20% with $[RhCl(PPh_3)_3]$.

Some hydrogenation catalysts contain ligands with extended conjugated π-systems, usually bound to Pt(II), Pd(II) or Rh(III) central ions. Such ligands include 1-phenyl-azo-2-naphtol [131], indigosulfonic acid [132] and the sodium salt of 1,2-dioxy-9,10-anthraquinone-3-sulfonic acid (Alizarin Red) [133].

Based on the results of EPR and NMR spectroscopic investigations [134–135] of the Pd(II) complex of Alizarin Red, $[Pd(QS)_2]$, it was proposed that activation of dihydrogen takes place at the metal ion, but that the substrate is reduced by electron transfer from the metal to the unsaturated moiety *via* the ligand. During the induction period for the hydrogenation, the dihydroxyanthraquinone ligand in $[Pd(QS)_2]$ is reduced to the semiquinone radical. Such radicals are rapidly oxidized by nitro compounds, and indeed, these complexes are active catalysts for hydrogenation of nitro groups under very mild conditions (room temperature, 1 bar H_2). In contrast with the better known phosphine complexes, where oxidative addition of dihydrogen and reductive elimination of the product involves two-electron changes, $[Pd(QS)_2]$ and similar complexes are able to transfer electrons in one-electron steps, which complements the preferred one-electron reduction steps for nitro compounds. Other characteristics of reductions involving $[Pd(QS)_2]$ *i.e.* the very slight temperature dependence of the rate, hydrogenolysis of carbon-halogen bonds, and sensitivity to radical scavangers, are also in accord with the formation of radicals during the hydrogenation process.

5.3.3. ENANTIOSELECTIVE HYDROGENATIONS IN AQUEOUS SYSTEMS

The usual targets of enantioselective hydrogenation (amides of dehydroaminoacids, prochiral ketones, etc.) are themselves water insoluble, although salts of the prochiral unsaturated acids dissolve in aqueous systems. Therefore hydrogenations are usually run in aqueous slurries or emulsions. Frequently, solvent mixtures (mainly alcohols in water) are used to obtain homogeneous solutions, but excellent enantioselectivities may also be observed in biphasic systems. All these techniques can be used on a synthetic scale, and indeed, as a result of work mostly over the last 5 years products of almost complete enantiomeric purity are often obtained from hydrogenations in aqueous media. On the other hand, the relative scarcity of kinetic studies in truly homogeneous aqueous solutions does not allow us to draw general conclusions as to the role of water in enantioselective hydrogenations.

The *effect of water* on enantioselectivity may vary widely from system to system, as exemplified by the following observations. In hydrogenation of 2-ethanamido-propenoic acid catalysed by complexes of aminoalkyl-ferrocenylphosphines the optical yield *increased* with increasing concentration of water in methanol [137]. For example, the *e.e.* was 8% in pure methanol, but 87% in a 1:3 water:methanol mixture. This increase is attributed to salt formation between the carboxylate group of the substrate and the protonated amine of the ligand, resulting in a rigid transition state structure. However, protonation may also lead to a *decrease* in the *e.e.*, as was found with chiral rhodium complexes of pyrrolidinephosphines [138]. Finally, there was *no change* in the enantioselectivity of the hydrogenation of Z-2-ethanamido-3-phenylpropenoic acid [139] catalysed by $[Rh(COD)(NORPHOS)]^+[ClO_4]^-$ on changing the solvent from pure ethanol (95% *e.e.*) to a 1:1 ethanol:water mixture (94% *e.e.*).

Rhodium complexes of the various sulfonated diphosphines **8–11**, prepared *in situ* from $[\{RhCl(COD)\}_2]$ and the ligand, were successfully used for enantioselective hydrogenation of enamides (max. 88% *e.e.*), of phenylethanone (28% *e.e.*) and of 1-phenylbuten-3-one (58%

e.e.). Dehydropeptides were reduced in a two-phase system with [{Rh(COD)Cl}$_2$] associated with **10** and **11**. Diastereoselectivities of up to 87% were obtained with tetrasulfonated BDPP [192].

In a systematic study of the influence of various solvents on the enantioselectivity in the reduction of dehydroaminoacids [67], a good linear relationship of log(%S/%R) *vs.* S_p, the solvofobicity parameter [62] of the solvent was found. However, the effect is not well understood, since there are examples, when by contrast, enantioselectivity was the same both in pure alcohols and in water alcohol mixtures.

The hydrogenation of propenoic acid derivatives in ethyl ethanoate/D_2O mixtures, catalysed by rhodium(I) complexes of the sulfonated phosphines **1**,**3**,**7**, and **11** resulted in selective *monodeuteration* at the position α to the ethanamido and the ester groups (Scheme *5.17*) [35]. By comparison, the product obtained by reduction of the same substrate in C_2D_5OD using [RhCl(PPh_3)$_3$] or [Rh(COD){$Ph_2P(CH_2)_4PPh_2$}]$^+ClO_4^-$ as the catalyst showed no deuterium incorporation.

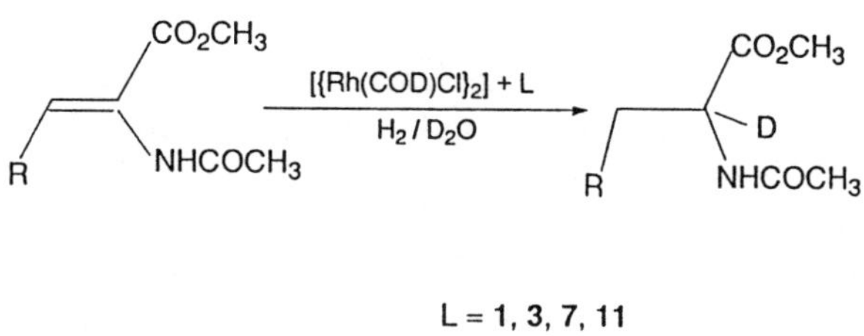

Scheme 5.17.

As is usual in asymmetric hydrogenation, the highest optical yields were achieved with complexes of *chelating diphosphines*. Until recently only rhodium complexes had been studied, but in light of recent results using ruthenium BINAP derivatives this latter metal is worthy of further attention.

On quaternization of the ligand (3**R**,4**R**)-3,4-*bis*(diphenylphosphino)-1-methylpyrrolidine [91], bound in the complex [Rh(COD)L]$^+$ (Scheme *5.5*), a very active catalyst for the hydrogenation of Z-2-ethanamido-3-phenylpropenoic acid (and its Na-salt) was obtained. The reactions were run at 22°C, a pressure of 50 bar H_2, in either aqueous or methanolic solutions, and in aqueous slurries. In aqueous systems (**S**)-N-ethanoyl-phenylalanine was obtained in 88–96% *e.e.* this being very close to the 100% enantioselectivity achieved in methanol. Together with the uniformly high rate (1000 mol H_2/mol Rh.h) these data show, that modification of the ligand and the presence of water had no significant influence on the mechanism of hydrogenation.

Similarly, enantioselectivities of as high as 97% were obtained [88–90] for the reduction of propenoic acid derivatives in either aqueous slurries of the substrate or in biphasic systems. The complexes $[Rh(NBD)L]^+$ were used as catalysts (L = **28–30**). The reaction was slower in water than in methanol by a factor of about 100, due to the lower concentrations of both the substrate and dihydrogen.

When the parent diphosphines (BDPP, CHIRAPHOS, PROPHOS) are substituted in a stepwise manner (*e.g.* by sulfonation), or when their 4-dimethylamino derivatives **27**, **28** and **30** are quaternized by protonation or methylation, new chiral centers at the phosphorus atoms are formed, leading to the appearance of new optical isomers of the ligands. However, both with the sulfonated [68], and with the ammonium substituted [89] phosphines it was found that the enantioselectivities obtained using the cationic rhodium complexes $[Rh(NBD)L]^+$ were not affected by the degree of substitution or quaternization. It is noteworthy, that the phosphines were sulfonated *prior* to complexation, whereas the aminophosphines were quaternized when already bonded to rhodium. These results show, that the important stereochemistry at rhodium is determined by the δ or λ configuration of chelate ring and, is unaffected by distant substituents.

Prochiral imines, such as ArC(Me) = NCH_2Ph (Ar = Ph, 2-MeO-C_6H_4, 3-MeO-C_6H_4 and 4-MeO-C_6H_4) were hydrogenated to the corresponding amines with extremely high enantioselectivities (up to 96% *e.e.*) in H_2O/ethyl ethanoate biphasic systems, using Rh^I complexes associated with sulfonated BDPP [103a]. Interestingly, the highest optical yield was achieved using partially sulfonated ligand (degree of sulfonation 1.4–1.65) while the use of the almost completely sulfonated BDPP (degree of sulfonation 3.75) led to a rather low *e.e.* (19%). Further detailed studies [103b] revealed that it was the catalyst with *mono-sulfonated* (2S,4S)-bis(diphenylphosphino)pentane ligand which was highly discriminative (94% *e.e.*) while the one with di-sulfonated BDPP gave only 2% enantiomeric excess.

In the early attempts to achieve asymmetric hydrogenation in aqueous systems [101, 102] asymmetric polyoxa-1,2- and polyoxa-1,4-diphosphines such as **36** served as ligands in the complexes $[Rh(COD)(L\text{-}L)]^+$. For hydrogenation of 2-ethanamido-propenoic, Z-2-ethanamido-3-phenyl-propenoic and 1-propene-2,3-dicarboxylic acids, enantioselectivities up to 75% were obtained in ethanol, but only up to 41% in water. The same substrates were also selectively hydrogenated using the same type of catalyst with the acylated 2-(diphenylphosphino-methyl)-4-diphenylphosphinopyrrolidine [100] derivatives, **13** and **22** as ligands. Interestingly, in this case the enantioselectivity was independent of whether the solvent was ethanol or water.

Acylated *bis*(2-diphenylphosphinoethyl)amine derivatives gave rise to moderately enantioselective catalysts, $[Rh(NBD)L]^+$, for hydrogenation of 2-ethanamido-propenoic acid. With L = **21**, 30% *e.e.* was obtained [71, 73]. In a very remarkable approach, the same secondary amine was reacted with biotin to give the diphosphine **38**. The cationic rhodium complex $[Rh(NBD)(L\text{-}L)]^+$, where L-L = **38**, was an active catalyst for hydrogenation of 2-ethanamido-propenoic acid in aqueous solution, and, in the presence of avidin, a globular protein, the reaction proceeded with 44% *e.e.* It is known that avidin forms an extremely stable molecular complex with biotin [140], and in this way, provides a chiral local environment for the pendant Rh-complex. In the presence of free biotin, enantioselectivity dropped below 5%, since the catalyst was forced out of the protein environment.

Cationic complexes of the formula $[L_2Rh(COD)][BF_4]$ with *water insoluble* phosphines [*e.g.*(-)-DIOP] were studied as catalysts for hydrogenation of Z-methyl 2-ethanamido-3-phenylpropanoate in water/methanol mixtures and in water itself in the presence of various surfactants. In water, both anionic (SDS) and neutral (Triton X-100) amphiphiles could be used to achieve (or even surpass) the high rates and enantioselectivities (93% *e.e.*) than what was observed in true methanolic solutions.

Reports are scarce on the use of chiral *monophosphines* in aqueous systems. It was mentioned [11], that the catalyst, prepared *in situ* from $[\{RhCl(COD)\}_2]$ and **5** gave a maximum of 68% *e.e.* in hydrogenation of Z-2-ethanamido-3-phenylpropenoic acid, both in aqueous solutions and in biphasic systems.

Esters of aminoacids were acylated [141a] with diphenylphosphinoethanoic acid, **17** (n = 2). Complexes prepared *in situ* from these chiral phosphines and $[\{RhCl(COD)\}_2]$ were moderately active and selective for the hydrogenation of phenylethanone in biphasic systems (max. 22% *e.e.*).

5.4. Hydrogenation of biological membranes

Biological membranes are major constituents of the cells, and many functions of the living state are directly or indirectly influenced by their composition and organization. Cell membranes are built up mainly of lipids and proteins, but for several purposes they can be modelled by fine aqueous dispersions of polar lipids only, termed liposomes.

The study of the relationship between the composition and the actual physical state of the membranes, and the various membrane associated properties requires investigation of the biochemical and physiological consequences of membrane modifications. There are several methods for such modifications, such as thermal acclimation, dietary manipulations, etc. Hydrogenation of the unsaturated fatty acids esterified in polar lipids is an outstandingly selective procedure for this purpose, which is fast enough to allow the detection of immediate response of the cells. This technique emerged in the last 15 years [4, 147] and since the natural environment for cells and liposomes is an aqueous suspension, most of the catalysts applied are water soluble.

It is very important to note here, that the "answers" of a living cell to environmental effects (including hydrogenation) are of compensative nature, *i.e.* their outcome diminishes the damage, caused in the homeostatic state by the original intrusion. Under unchanged metabolic conditions, following a non-lethal hydrogenation, the cells restore the original distribution of fatty acids in their membranes, and it is the "time window" allowed by this compensatory process, which can be used for study of membranes with altered properties.

In principle, the chemistry of membrane hydrogenation involves the simple reduction of isolated double bonds in relatively complex molecules (Scheme *5.18*) under very mild conditions. What makes this problem difficult is the presence of a large number of other compounds, and the high degree of spatial and functional organization of the living cell. The ample possibility of side reactions combined with the profound sensitivity of living cells put strict requirements on the activity and selectivity of the catalyst.

For studies of membrane hydrogenations only a few catalysts have ever been applied, such as $[RhCl(PPh_3)_3]$ [153], $[RhCl(TPPMS)_3]$ [25], $[RuCl_2(TPPMS)_2]$ [26], and most of all, $[Pd(QS)_2]$ [135, 147–152].

In many respects, under hydrogenation conditions membrane lipids behave like simple esters of unsaturated acids. For example, it was several times observed that polyunsaturated lipids reacted faster [26] than monoenoic ones. Generally, double bonds with Z-geometry are reduced

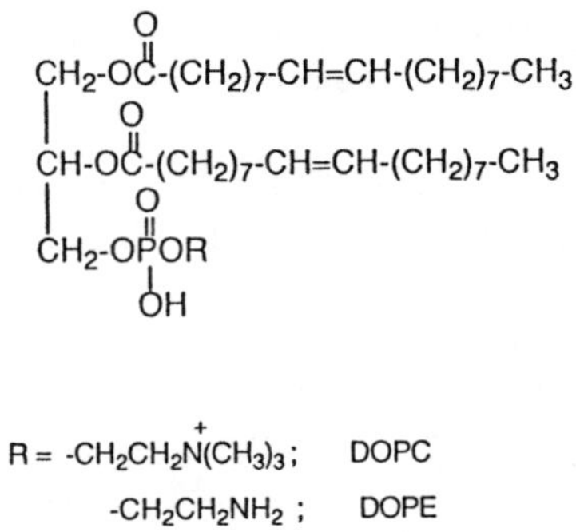

Scheme 5.18.

faster than their E-counterparts [148]. Consequently, unnatural lipids, containing isomeric E-fatty acids may accumulate. In addition to the formation of geometric isomers, positional isomers may also be produced.

Cells and liposomes are microheterogeneous systems, in which mass transport, as opposed to inherent reactivity, may be rate determining in hydrogenation reactions. In certain cases *site-selectivity*, *i.e.* sequential hydrogenation [150] of different organelles of the cell, was achieved by making use of the relatively slow penetration of the catalyst across the boundary membranes. Accessibility may impose selectivity even on the reaction of very closely related substrates. For example, under identical conditions, hydrogenation of dioleoylphosphatidylethanolamine (DOPE) was consistently slower, than that of dioleoylphosphatidylcholine (DOPC) [148]. At the temperature of the reaction DOPE forms closely packed hexagonal type II lyotropic liquid crystalline phases in contrast with the more accessible bilayer structure of DOPC, and this is reflected in the rate of hydrogenation.

A very important problem is the removal of the catalyst from the membranes after hydrogenation. Though anionic catalysts bind very strongly to proteins, and amphiphilic catalysts partition into the non-aqueous phase of lipids, and therefore a *complete* removal is very unlikely, the residual amount of catalyst could, in many cases, be decreased as much as not to interfere with the functions of, or associated with the membrane. However, this must be thoroughly checked in every single case of *in vivo* hydrogenation.

5.5. Transfer hydrogenation and hydrogenolysis in aqueous systems

Transfer hydrogenation is discussed in detail in Chapter 2, and examples are also given in Chapter 3, so there is no need to repeat the general features. However, there is an increasing number of papers dealing with transfer hydrogenation in aqueous systems, and these are summarized here. Not surprisingly, many of the catalysts capable of breaking an E-H bond in a hydrogen donor molecule (E = C, O, N, etc.), are also able to catalyse hydrogenolytic reactions of substrates having carbon-halogen or carbon-oxygen bonds. Therefore hydrogenolysis with either molecular hydrogen, or by hydrogen transfer from donor, is also treated in this section.

Several of the usual hydrogen donors, such as ascorbic acid or methanoic acid, their salts, EDTA, or 2-propanol are rather or at least sufficiently soluble in water. In addition, H_2O itself can also serve as a source of hydrogen. In the last two decades extensive research has been conducted on the photochemical splitting of water. These investigations are well discussed elsewhere [155, 195], and we will consider only the examples, for which hydrogenation of unsaturated substrates has been achieved in these photochemical systems. Hydrogenations with CO/H_2O mixtures is discussed in the next section (5.6).

Methanoic acid and methanoates were amongst the most effective donors used for reduction of alkenes, with $[RuCl_2(PPh_3)_3]$ or $[RhCl(PPh_3)_3]$ as catalysts in non-aqueous systems [156–158]. Several papers describe [159, 160] the reduction of alkenes and aldehydes with HCOOH and methanoates in aqueous solutions, using Rh, Ru, Pt, and Pd complexes of TPPMS, the most active catalyst being $[RuCl_2(TPPMS)_2]$. Ketones and esters were not reduced. In solutions of methanoic acid, Na-methanoate and 1:1 HCOOH:Na[HCOO] mixtures the yields of heptane from 1-heptene over 2 h at room temperature were 37%, 34% and 97% (!), respectively. Since the reaction mixtures were heated in *closed* ampoules, it seems likely, that the actual reductant was H_2, supplied by the decomposition of methanoic acid. Ultraviolet irradiation accelerated the reduction of alkenes, using $[RhCl(TPPMS)_3]$ as catalyst, at room temperature. In other experiments [22] no reduction of the carbon-carbon double bond with Na[HCOO] was found in *open* systems with the same catalysts.

Prochiral derivatives of propenoic acid were reduced by hydrogen transfer from aqueous solutions of M[HCOO] ($M = K^+$, Na^+ and $[NH4]^+$) catalysed by Rh(I) complexes containing TPPTS or the tetrasulfonated cyclobutanediop (**11**) as ligand [176]. The reactions were run at 50°C for 15–67 hr, during which conversions of 48–100% were achieved. Use of the chiral sulphonated phosphine afforded the products with an enantiomeric excess of up to 43%, this being similar to the values obtained in the biphasic hydrogenations catalysed by the same rhodium complex [68].

Aldehydes were reduced in a phase transfer catalytic system [161], having $[RuCl_2(PPh_3)_3]$ as the catalyst in the organic phase (usually chlorobenzene), and the hydrogen donor (Na-methanoate) in the aqueous phase. $[R_4N]^+$-methanoates were used as phase transfer agents. The rate showed a maximum as a function of the aldehyde concentration in the organic phase (Fig. *5.1*) and using neat aldehyde as the organic phase resulted in an inconveniently slow reaction.

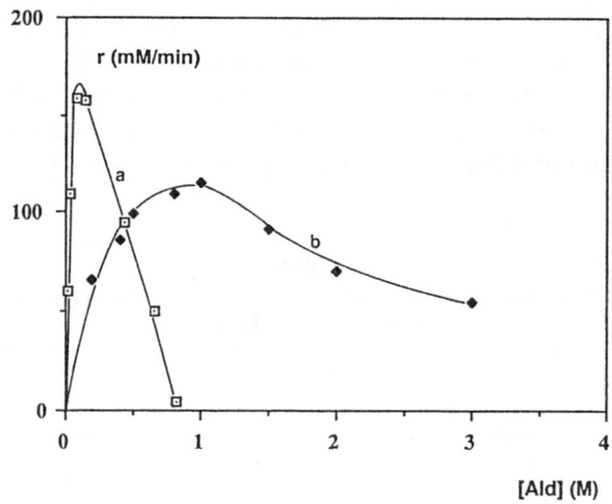

Fig. 5.1. Reduction of **a**) 4-Me-benzaldehyde and **b**) benzaldehyde by hydrogen transfer from $Na[HCOO]/H_2O$ catalyzed by **a**) $[RuCl_2(PPh_3)_3]$ and **b**) $[RuCl_2(TPPMS)_2]$/TPPMS. Conditions:

a) 2.4 mM Ru, 14.4 mM PT catalyst, in 5 ml organic phase (solvent: chlorobenzene), 50°C

b) 2.0 mM Ru, 40 mM TPPMS, in 5 ml organic phase (solvent: chlorobenzene), 80°C.

The aqueous phase in both cases: 3 ml 5 M Na[HCOO].

The same reaction was studied using $[RuCl_2(TPPMS)_2]$ as the catalyst [22]. In this genuine biphasic system no phase transfer agent was needed. The high excess of methanoate protected the catalyst against oxidation, and inhibition by the substrate was reduced by the limited solubility of the aldehydes in the aqueous phase. (This is another example of chemical protection by phase separation [142].) Consequently, no organic solvent was necessary. Both the execution and workup of this process proved to be particularly simple.

The reaction tolerates the presence of substituents on the aromatic ring other than *ortho*-hydroxy groups (Table *5.7*), but the most important feature is that ketones and carbon-carbon double bonds are completely inert to reduction. Thus unsaturated aldehydes are reduced to unsaturated alcohols.

In a similar biphasic system $[RuCl_2(PTA)_4]$ was found to be a rather active catalyst for reduction of aldehydes to alcohols using $Na[HCOO]/H_2O$ as the hydrogen source [87]. For example, the initial turnover frequency for benzaldehyde to benzyl alcohol was 22 mol/mol Ru.h at 80°C. Unsaturated aldehydes were reduced exclusively to unsaturated alcohols. Thus the catalyst showed the same selectivity as did $[RuCl_2(TPPMS)_2]$. $[RhCl(PTA)_3]$ was inert in reduction of the aldehyde function, however, it catalysed hydrogen transfer from methanoate to carbon-carbon double bonds (Scheme *5.19*). At 80°C, reduction of E-3-phenylpropenal afforded a product mixture containing 94.4% 3-phenylpropanal, 5.0% 3-phenylpropenol and 0.6% 3-phenyl-propanol.

TABLE 5.7
Reduction of aldehydes by hydrogen transfer[a] from $Na[HCOO]/H_2O$ catalysed by $[RuCl_2(TPPMS)_2]$ [22]

Substrate	Reaction time (h)	Yield[b] (%)
C_6H_5CHO	1.5	94
$4\text{-}CH_3C_6H_4CHO$	1.5	99
$4\text{-}OCH_3C_6H_4CHO$	1.5	90
$4\text{-}BrC_6H_4CHO$	1.5	94
$4\text{-}N(CH_3)_2C_6H_4CHO$	1.5	98
$2\text{-}NO_2C_6H_4CHO$	2	90
$2\text{-}HOC_6H_4CHO$[c]	3	–
$C_6H_5CH{=}CHCHO$[d]	2	92
$(CH_3)_2C{=}CH(CH_2)_2C(CH_3){=}CHCHO$[d,e]	7	95

[a] Conditions: 0.01 mmol $[RuCl_2(TPPMS)_2$, 0.1 mmol TPPMS, 1 mmol aldehyde (neat), 3 ml 5 M Na[HCOO] in water, 80 °C.
[b] Yields of isolated compouns. Conversions as established by gas chromatography exceeded 98%.
[c] 63% of the starting material was recovered.
[d] Exclusive reduction of the aldehyde group.
[e] Mixture of E- and Z-isomers (2:1). No isomerization was observed.

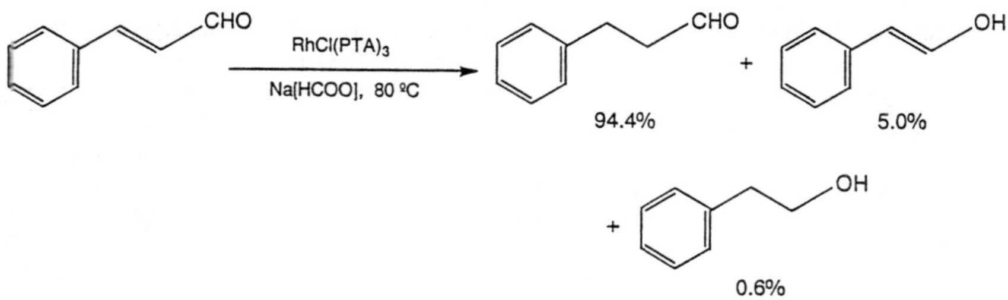

Scheme 5.19.

In contrast to the reduction of unsaturated aldehydes, α,β-unsaturated ketones were selectively reduced at the carbon-carbon double bond in a PTC system [162] comprised of 1,2-dichlorobenzene as the organic phase and a solution of Na[HCOO], as the aqueous phase, using $[RuCl_2(PPh_3)_3]$ as the catalyst, and $[R_4N]^+X^-$ as the PT agent (Scheme *5.20*).

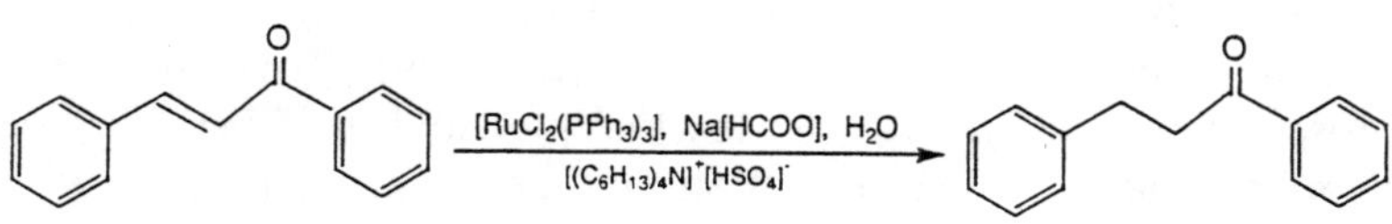

Scheme 5.20.

Hydrogen transfer from polymethylhydrosiloxane [163] to substituted alkenes, alkynes, α,β-unsaturated aldehydes, ketones and esters was catalysed by the $[Q]^+[RhCl_4]^-$ ion-pair. Reduction of the unsaturated carbonyl compounds took place exclusively at the –C = C- bond (Table *5.8*).

The same system catalysed the Rosenmund reduction of acid chlorides, as shown in Scheme *5.21*. This is one of the rare examples of homogeneous catalysis of the Rosenmund reduction. Instead of the usual 100–180°C, the PTC hydrogen transfer from PMHS proceeded at room temperature. The product aldehydes were not reduced further, but the reaction of unsaturated acid chlorides afforded saturated aldehydes.

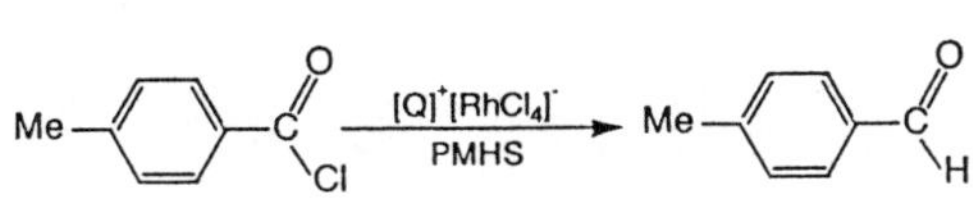

Scheme 5.21.

TABLE 5.8
Reduction of various compounds by hydrogen transfer from PMHS catalysed by the $RhCl_3$-Aliquat-336 catalyst[a] [163]

Substrate	Products	Yield[b] (%)	Reaction time (min)
$C_6H_5C{\equiv}CC_2H_5$	Z-$C_6H_5CH{=}CHC_2H_5$	16.1	30
	E-$C_6H_5CH{=}CHC_2H_5$	1.8	
	$C_6H_5CH_2CH_2C_2H_5$	3.0	
$C_6H_5CH{=}CH_2$	$C_6H_5C_2H_5$	24.8	150
$C_6H_5CH{=}CHCHO$	$C_6H_5(CH_2)_2CHO$	26.6	60
2-Cyclohexen-1-one	Cyclohexanone	7.2	30
$C_6H_5CH{=}CHCOOC_2H_5$	$C_6H_5(CH_2)_2COOC_2H_5$	3.9	30
4-$CH_3C_6H_4COCl$	4-$CH_3C_6H_4CHO$	15.0	30

[a] Conditions: 1-mmol substrate, 10–15 mg $RhCl_3.3H_2O$, 3 mg Aliquat-336, 1 ml 1,2-dichloroethane, 100 ml PMHS, 20 °C.
[b] Determined by gas chromatography.

An aqueous solution of ascorbic acid buffered to pH = 5–6 produced H_2 on illumination with visible light if $[Ru(bipy)_3]^{2+}$ was used as the photosensitizer and $[RhCl(TPPMS)_3]$ as the catalyst [164]. This same system is suitable for hydrogenation of simple alkenes [43], as well as of phospholipid dispersions [27]. Other H-donors (L-sorbose, D-glucose, D-lactose, 2,3,5,6-di-O-isopropylidene-mannofuranose, EDTA) could not replace ascorbic acid. Hydrogenation of biological membranes (see also 5.4) by hydrogen transfer from biocompatible donors may have some potential for such reactions in higher organisms.

Hydrogenolysis of allylic chlorides and ethanoates was studied in biphasic systems [33] with the water soluble catalysts $[PdCl_2L_2]$, L = **1**, **19** or **34**, as shown in Scheme *5.22*. Hydrogenolysis of E-3-phenylpropenoic acid chloride and 2-propenyl-E-3-phenylpropenoate with aqueous Na-methanoate at 100°C proceeded with practically complete conversion with all three catalysts. Among the products 52–99% of isomeric phenylpropenes were detected. In the reduction of E-3-phenylpropenoic acid chloride the methanoate ester was also detected as by-product. In reactions 1–7 h long, when the catalyst was

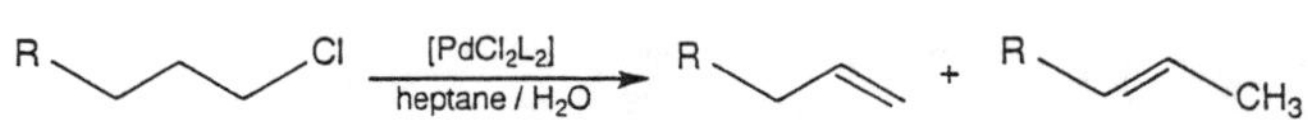

Scheme 5.22.

$[PdCl_2\{P(nBu)_3\}_2]$ a conversion of only 16–65% was attained, due to the insolubility of the catalyst in the aqueous phase.

The latter observations suggest that the reaction is taking place in the aqueous phase. Indeed, with ingeniously devised experiments it was demonstrated that in the hydrogenolysis of allyl acetate propene was evolved only from the aqueous phase. In addition to its chemical effect, the Pd-catalyst facilitated the transfer of the substrate *from* the organic to the aqueous phase, also playing the role of a phase transfer agent.

3-Chloro-1-phenylpropene was reductively dehalogenated in a *n*-heptane/H_2O solvent mixture using Pd(II) complexes with the sulfonated phosphines **15** and **16** as catalysts [145]. Addition of polyether detergents influenced both the rate and the selectivity of the reaction; the best selectivity for 1-phenylpropene (90%) was achieved by the use of triethylene glycol.

The complexes $[PdCl_2L_2]$, where L = **35**, having a crown ether moiety in the phosphine ligand, catalysed the hydrogenolysis [99] of 1-chloromethyl-naphtalene with K[HCOO] or Na[HCOO] (Scheme *5.23*). Both the solid methanoates, and their aqueous solutions could be employed. When K[HCOO] was used as the reductant and the phosphine contained a benzo-18-crown-6 moiety (n = 6 in **35**), the conversion was double than that with the complex for which n = 5, under identical conditions. Thus the catalyst acted bifunctionally, bringing the reactants into close proximity, which resulted in the increased rates. By comparison, $[PdCl_2(PPh_3)_2]$ was catalytically inactive with solid K[HOOC], and gave only low conversions with an aqueous methanoate solution.

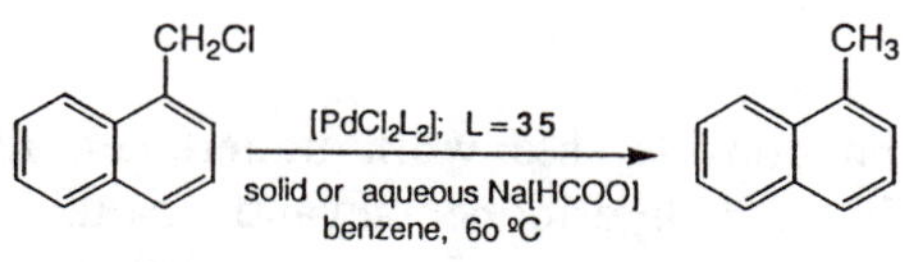

Scheme 5.23.

In agreement with the latter observations it was found that addition of $[R_4N]^+X^-$ phase transfer agents significantly accelerated the hydrogenolysis of aryl halides with methanoates [165]. Benzyl alcohol could also be used as H-source. When a mixture of Na[HCOO], $[PdCl_2(PPh_3)_2]/PPh_3$ and $[(C_6H_{13})_4N]^+[HSO_4]^-$ was heated in D_2O at 100°C for 15 h, the resulting catalyst was markedly selective for the reduction of bromoarenes, yielding d_1-arenes with 95% isotopic purity (Scheme *5.24*).

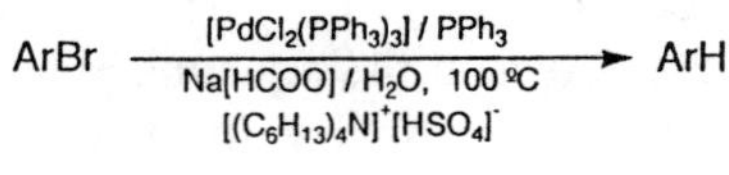

Scheme 5.24.

The rhodium(III) complexes $[RhHCl_2L_2]$ (L = PCy_3 or P*i*Pr_3) were found to be excellent catalyst precursors for the biphasic hydrogenolysis of chloroarenes [146]. The reaction occurs under mild conditions with high yields. The catalyst tolerates the presence of a variety of functional groups (*e.g.* R, OR, CF_3, COAr, COOH, NH_2). Some chloro heterocycles (*e.g.* 5-chloro-1-ethyl-2-methylimidazole) are readily dehalogenated, but 2-chlorothiophene does not react at all.

Prochiral epoxides have been hydrogenolysed [61a] using rhodium complexes of the chiral sulfonated phosphine **10**, as catalysts (Scheme *5.25*). The highest optical yield was achieved with a mixture of mono-, di- and trisulfonated BDPP in aqueous/organic biphasic system. In water/methanol mixtures hydrogenation of sodium epoxysuccinate with cationic rhodium complexes $[(NBD)Rh(P_2)]X$ (P_2 = various water insoluble chelating chiral phosphines, X = Cl^-, and $[BF_4]^-$) afforded chiral hydroxybutene-2-dioate with up to 62% enantiomeric purity [61b].

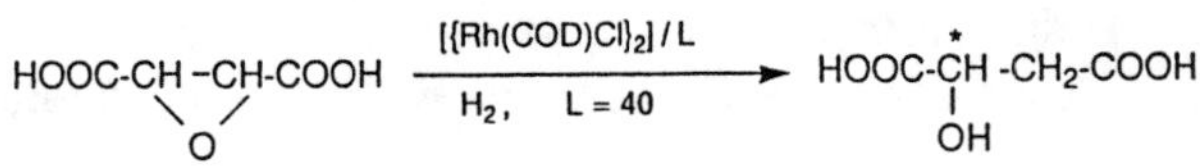

Scheme 5.25.

There has been considerable work undertaken in exploring the hydrogenolytic ability of hydridopentacyanocobaltate and hexacyanodinickelate [113]. It was recently found [166] that, in a PTC system with β-cyclodextrin as the phase transfer agent, allylic alcohols were deoxygenated by $[HCo(CN)_5]^{3-}$, giving selectively *trans*-alkenes in 51–91% yield (Scheme *5.26*, Table *5.9*). This is an important difference in the results of the same reaction without cyclodextrin, in which

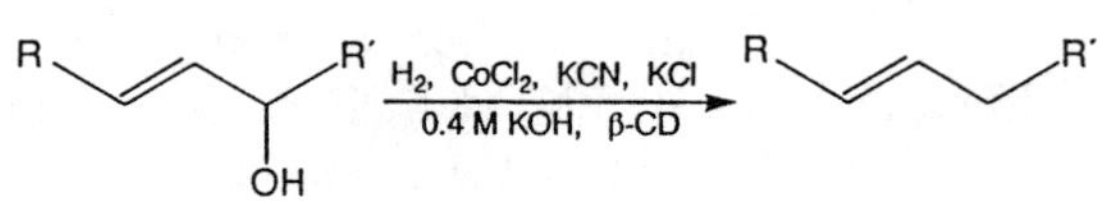

Scheme 5.26.

TABLE 5.9
Hydrogenolysis of allylic alcohols catalysed by $[HCo(CN)_5]^{3-}$ and β-cyclodextrin[a]

Substrate	Products	Yield[b] (%)
2-hexen-1-ol[c]	E-2-hexene	91
	1-hexene	4
2-hexen-1-ol[d]	E-2-hexene	66
	1-hexene	4
3-phenylpropenol	E-1-phenylpropene	87 (80)
	3-phenyl-1-propanol	12 (11)
3-phenylpropenol[e]	E-1-phenylpropene	(6)
	3-phenyl-1-propanol	(3)

[a] Conditions: 5 mmol substrate, 10 ml 0.4 M KOH, 5.5% β-cyclodextrin, 1.65 mmol $CoCl_2$, 8.55 mmol KCN, 3.60 mmol KCl, room temperature.
[b] Yields determined by gas chromatography (yields of isolated products are in parenthesis)
[c] Use of 6.0/1.0 or 7.0/1.0 ratio of $[CN^-]/[Co]$ gave only recovered 2-hexen-1-ol.
[d] No β-cyclodextrin.
[e] Using H_2O instead of 0.4 M KOH.

case a *cis/trans* isomer mixture is formed, and the isomer distribution depends on the $[CN]^-/Co$ ratio. It was postulated, that β-cyclodextrin formed an adduct with $[HCo(CN)_5]^{3-}$, and it was this adduct which was responsible for the exclusive formation of the *trans*-alkene.

5.6. Hydrogenations with CO/H_2O mixtures

In the last two decades there have been many studies of the catalysis of the liquid phase water gas shift reaction, WGSR (equation *5.27*) [167].

$$CO(g) + H_2(l) \rightleftharpoons CO_2(g) + H_2(g) \qquad (5.27)$$

This is a thermodynamically downhill reaction, but due to its high activation energy the process does not take place in practice at low or moderate temperatures (below ca. 300°C) without a catalyst. The lower the temperature the more the equilibrium lies in the direction of H_2 production. Thus active, (preferably water)soluble catalysts have been sought.

Currently, however, homogeneous catalysis of the WGSR is not competitive with its heterogeneous counterpart, especially in an industrial context, due to the limited rate, the instability of the catalysts, and the usually high costs. In most cases, the first step of CO conversion is its coordination to the metal ion of the catalyst. When the other ligands are strongly basic, resulting in high electron density at the metal, coordinated CO looses its affinity for nucleophiles, and only powerful nucleophiles such as OH^- or CH_3O^- are able to attack it. Thus the WGSR is most often observed in strongly alkaline solutions. Reaction of {M-CO} and OH^- results in formation of either a metallocarboxylic acid, {M-COOH}, or methanoatometal complexes, {M-O_2CH}, both of which yield metal hydrides by loss of CO_2. Mono- or bimolecular reaction of the latter hydrides leads to elimination of molecular hydrogen (Scheme *5.28*).

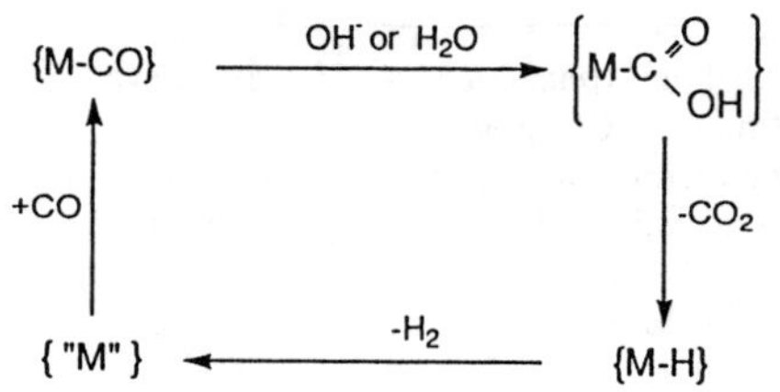

Scheme 5.28.

The intermediate hydrides produced in the WGSR may react with suitable substrates. It should be noted, that by reaction of one CO ligand, a *monohydride* is formed. Many hydrogenation catalysts operate *via* intermediate *dihydrides*, formed by the oxidative addition of dihydrogen to the low valent metal. The reactivity of mono- and dihydrides towards certain group of substrates, such as nitro compounds, may differ so that the same metal complex may catalyse the reduction of different substrates, depending on whether H_2 or CO/H_2O is used as reductant. Moreover, the rate limiting step of most of the WGS reactions is the elimination of molecular hydrogen. Hydrogen transfer to a reducible substrate may be much faster, so that in the absence of a substrate, the WGSR, catalysed by the same complex may often remain unnoticed.

Reduction of nitro compounds is a classic application of WGSR systems (equation *5.29*).

$$PhNO_2 + 3CO + H_2O = PhNH_2 + 3CO_2 \qquad (5.29)$$

In a series of studies on mono and polynuclear metal carbonyls [168] it was established that $[FeH(CO)_4]^-$, produced by the reaction of

$[Fe(CO)_5]$ with OH^-, catalysed the slow reduction of nitrobenzene. To avoid formation of Fe(II) salts, the reaction had to be carried out at 120 bar CO. $[Rh_6(CO)_{16}]$ and $[Ir_4(CO)_{12}]$ showed good catalytic activity at somewhat lower pressures (35 bar CO at room temperature), in aqueous Et_3N solution, at 100–125°C, but the catalytic activity of $[Ru_3(CO)_{12}]$ or $[H_4Ru_4(CO)_{12}]$ was only about 20% of that of the Rh or Ir clusters.

$[Rh_6(CO)_{16}]$ has been studied by several groups. In basic solutions under a CO atmosphere, several carbonylmetallates can be formed [169], for example $[Rh_{12}(CO)_{30}]^{2-}$, $[Rh_5(CO)_{15}]^-$, $[Rh_7(CO)_{16}]^{3-}$, and $[Rh_6(CO)_{14}]^{4-}$, as well as their hydrido- and halo-derivatives. The equilibria amongst these species is sensitive to minor changes in the conditions, such as CO pressure, temperature, base concentration, and nature of the solvent.

In a biphasic system, comprised of a toluene solution of nitrobenzene and an alkaline aqueous solution of any of $[Rh_6(CO)_{16}]$, $[\{RhCl(1,5\text{-hexadiene})\}_2]$ or $RhCl_3.3H_2O$, reduction took place under very mild conditions [170] with a specific rate of 12 mol aniline/mol Rh.h (room temperature, 1 bar CO). The product could be easily isolated from the toluene phase. In homogeneous methanolic solutions there was a slight increase in the rate, but isolation of the aniline became more difficult. Other nitroarenes could also be reduced efficiently, but nitroaliphatics reacted sluggishly.

$[RuCl_2(PPh_3)_3]$ was found a moderately active catalyst for reduction of nitroarenes with CO/OH^- [171] in a phase transfer system. The yield of 4-Me-aniline in the reduction of 4-nitrotoluene increased from 31% to 71%, when CO was replaced by synthesis gas. However, under H_2 alone, no reaction took place. Although employing the same catalyst, this PTC reduction of nitroarenes by CO/H_2O occurs under much milder conditions than hydrogenation in homogeneous solutions.

Aldehydes were reduced by CO in 60:40 water:2-ethoxy-ethanol mixtures under mild conditions (30–80°C, 20 bar CO) with the $[Rh_6(CO)_{16}]$/diamine catalyst system [172] (Scheme *5.30*). When the diamine component was N,N,N′,N′-tetramethyl-1,3-diaminopropane, α,β-unsaturated aldehydes were reduced exclusively to saturated alcohols. On the other hand, in the presence of 4-dimethylaminopyridine only the carbon-carbon double bond was hydrogenated. Reduction of ketones was much slower than that of aldehydes.

Rh-complexes of the mono-sulfonated triphenylphosphine derived from $[Rh(COD)(TPPMS)_2]^+$ catalysed the reduction of 1-phenylbutene-3-one [44] to the saturated ketone (Scheme *5.31*). Under 30 bar CO

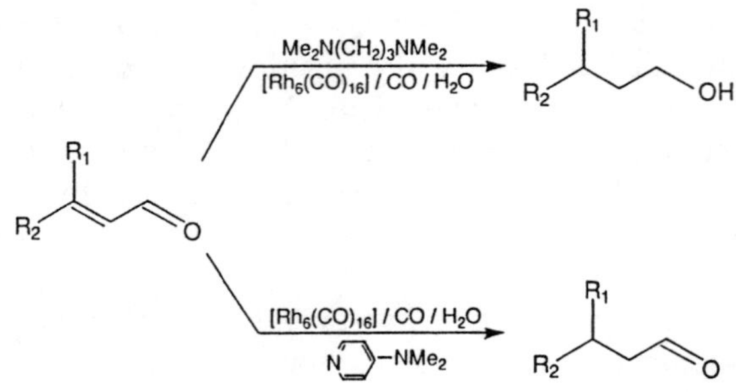

Scheme 5.30.

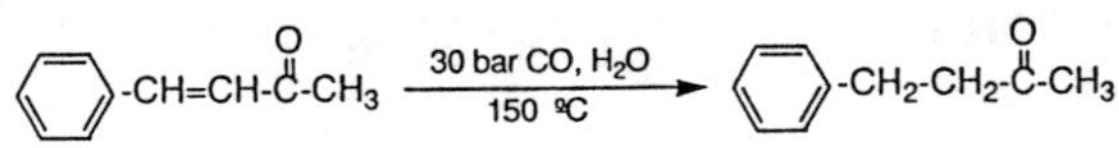

Scheme 5.31.

and at 150°C a reasonable turnover frequency was achieved, *viz.* 170 mol/mol Rhh.

There are only a few systems capable of hydrogenation of arenes and heteroarenes under strictly homogeneous conditions. In aqueous methanol at 150°C and 35 bar CO, $[Fe(CO)_5]$ catalysed the reduction of quinolines [173] exclusively in the heteroaromatic ring. Under identical conditions anthracene was reduced only stoichiometrically, but addition of 1,10-phenanthroline made the reaction catalytic. The reaction may also be carried out under PTC conditions.

Hydrogenation of quinolines and isoquinolines [174] under WGSR conditions (150°C, 56 bar CO at room temperature) can best be carried out using $[Rh_6(CO)_{16}]$ as catalyst. $[Mo(CO)_6]$ and $[Ru_3(CO)_{12}]$ showed negligible activity. All the substrates were reduced only in the heteroaromatic ring, in 81–97% isolated yields.

There are only a few examples of organometallic phase transfer catalysis which operate under acidic conditions. Diarylethenes were hydrogenated [175] in the presence of $[Co_2(CO)_8]$ in a benzene/aqueous $H[BF_4]$ solvent system, using 4-dodecylbenzenesulfonate as the phase transfer catalyst (Scheme *5.32*). Yields were close to quantitative.

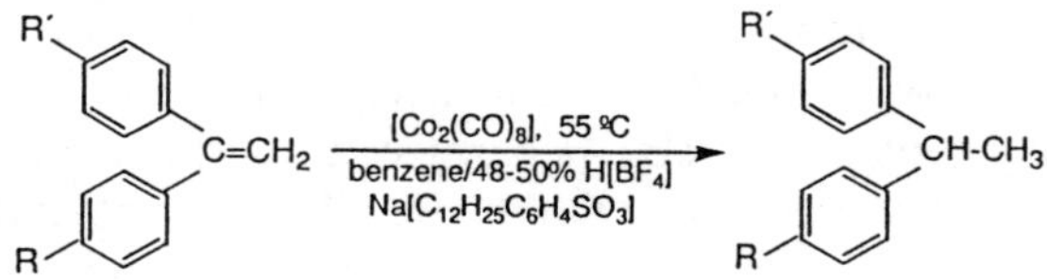

Scheme 5.32.

5.7. The combination of organometallic and enzymatic catalysis

Enzymes are enormously active and highly selective catalysts, but their use in synthesis is often limited by the lack of appropriate cofactors. For example, many enzymatic reactions require dihydronicotinamide nucleotides, NADH and NAD(P)H , and a key step in developing an efficient practical process with an NAD-dependent enzyme is the regeneration of NADH from NAD^+ (equation *5.33*).

$$NAD^+ + 2e^- + H^+ = NADH \qquad (5.33)$$

There are several methods for accomplishing this reduction [177]. However, in many cases the cofactors, NAD^+ and $NAD(P)^+$, are reduced in a non-selective way, i.e. in addition to the 1,4-dihydropyridine derivatives the 1,2- and 1,6-isomers are also formed, which are inactive in the enzymatic reaction. Moreover, the intermediate product of one-electron reduction , the $NAD^{\cdot}$ radical easily forms an inactive dimer. Though the direct hydrogenation [3, 41] of NAD^+ to NADH, catalysed by Rh(I) complexes of **1**, **14** and **20**, and by $[Rh(C_5Me_5)(bipy)]^{2+}$ was found impractically slow, transition metal catalysis has some potential for reactions with other reducing agents. (Reduction with H_2, however, could be achieved using the enzyme hydrogenase as catalyst [178]; see also Chapter 7.) For solubility reasons, these kinds of experiments have to be carried out in aqueous systems.

NADH can be regenerated with several NAD^+-dependent dehydrogenases [177] such as lactate dehydrogenase (LDH). In a double-enzyme system, 2-oxo-propanoate was hydrogenated to 2-hydroxy-propanoate using $[Rh(NBD)L]^+$ complexes (L = **14** and **20**) as catalyst at room temperature and 2.8 bar H_2 pressure. The product racemic 2-OH-propanoate was dehydrogenated with D- and L-LDH, consuming NAD^+ and producing NADH. The latter served as reductant in a horse liver alcohol dehydrogenase (HLADH) catalysed hydrogenation of cyclohexanone and 2-norbornanone (Scheme *5.34*). Thus the net reaction was the hydrogenation of the ketone to alcohol in the rather slow process. The rhodium catalysts were slowly deactivated during the reaction, most probably due to the interaction with the free thiol groups of the enzymes, but conversion of the ketone could be made complete by adding a new batch of the complexes during the run. The most interesting feature of the reaction was that from (±)2-norbornanone 72% *exo*-norbornanol and 28% *endo*-norbornanol was obtained with an enantiomeric purity of 38% and 100%, respectively.

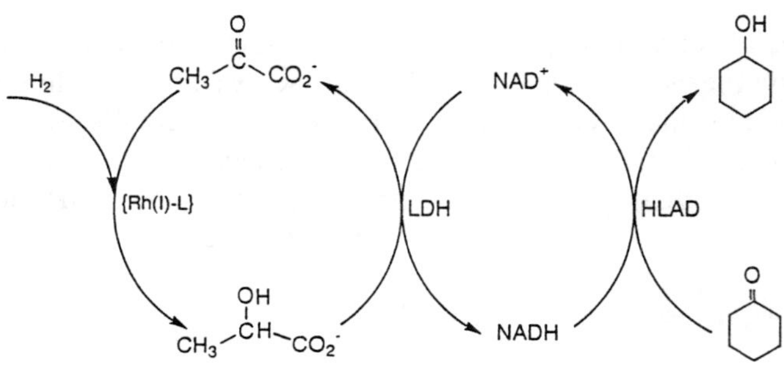

Scheme 5.34.

Both NAD^+ and $NAD(P)^+$ were efficiently reduced by methanoate using $[Rh(C_5Me_5)(bipy)]^{2+}$ as catalyst [41] (equation *5.35*).

$$NAD^+ + HCO_2^- = CO_2 + NADH \qquad (5.35)$$

In sodium phosphate buffer solutions of pH = 6–10, at 25–38°C, turnover frequencies higher than 80 h^{-1} were observed with both substrates. A distinct feature of the system, as opposed to enzymatic reductions, is that the reaction rate is independent of the concentration of the substrate, allowing complete conversion, while the enzymes are inhibited by the products, NADH or NAD(P)H, well before complete reduction. Reduction of NAD^+ by HCO_2^- was catalysed also by $[RhCl(TPPMS)_3]$ [42].

If a suitable catalyst can be found the hydrogen, generated in *photochemical* systems (with a simultaneous consumption of an electron donor molecule), can be transferred to reducible substrates. One such arrangement is described in section 5.4 for the hydrogenation of carbon-carbon double bonds in lipids [27]. Another example is the reduction of NAD^+ which can be achieved by illumination of a solution containing $[Ru(bipy)_3]^{3+}$ as the photosensitizer, ascorbate as the electron donor, and $[RhCl(TPPMS)_3]$ as the catalyst [42].

Model compounds for nicotinamide adenine dinucleotide cofactors are extensively investigated. One such compound is 1-benzylnicotinamide (BNA^+), the reduced form of which being able to reduce carbonyl compounds in the presence of divalent ions such as Mg^{2+} and Zn^{2+}. In the above photochemical system [42] BNA^+ was reduced to BNAH, and the latter reduced diphenylethane dione to 1,2-diphenyl-2-hydroxyethanone. Since BNA^+ is hydrophilic, but BNAH dissolves better in non-polar solvents, a biphasic reaction (Scheme *5.36*) could be devised which allowed easy recovery of the catalyst. Mg^{2+}, which is believed to activate carbonyl compounds acting as a Lewis acid, is transported to the organic phase as associated to BNAH.

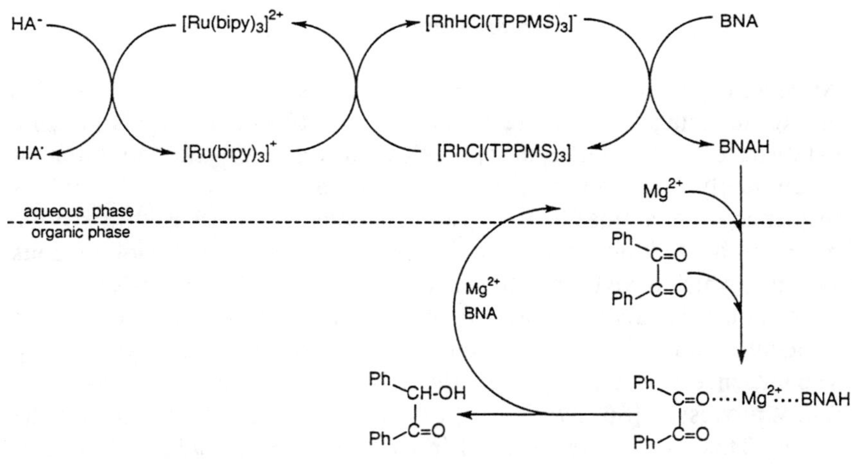

Scheme 5.36.

When triethanolamine was used as the electron donor, $[Ru(bipy)_3]^{2+}$ as the photosensitizer, and $[RhL_3]^{3+}$ as the redox catalyst (L = 2,2′-bipyridine-5-sulfonic acid) cyclohexanone was hydrogenated to cyclohexanol [179] with the NADH-dependent HLADH. The reduced form of the rhodium complex itself does not react directly with the ketone, thus the stereoselectivity depends entirely on the enzymatic reduction. 2-Methylthiacyclohexan-4-one was reduced to the corresponding alcohol having (2**S**,4**R**)- and (2**R**,4**S**)-configuration in 14% and 64% isolated yields, and 95% and 81% optical purity, respectively (Scheme *5.37*). It should be

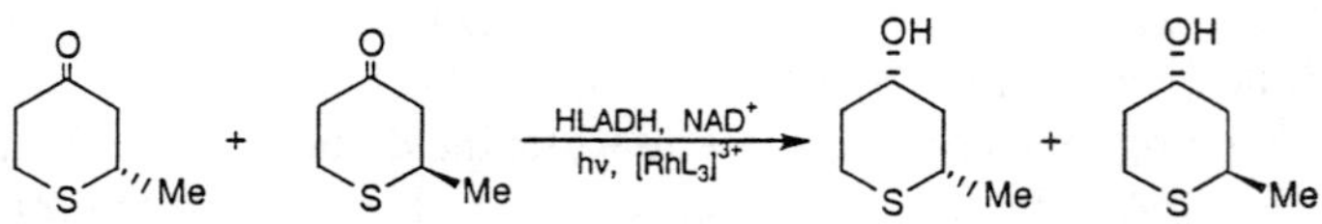

1 : 1 14% 46%

L = bipyridine-5-sulfonic acid

Scheme 5.37.

stressed, that while this photochemical reaction was slow, the system was stable for several days.

5.8. Conclusions

At several points in this Chapter we have discussed some advantages and disadvantages of the use of an aqueous solvent for organometallic catalysis of hydrogenation. The effect of water on the formation of catalytically active complexes was covered in section 5.2.2, and its influence on enantioselective hydrogenations was discussed in section 5.3.3. In this section we wish to discuss a few other features of aqueous systems, which have broad relevance to organometallic catalysis is water.

It should be always borne in mind that water is itself a reactive molecule, and it can be activated by transition metal catalysts [180]. The presence of water may affect the stability of the catalysts by assisting redox processes [36, 37, 55–57] of the metal ion and/or those of the ligand. This was shown in detail in case of the Rh(III)/TPPTS system. Even in a rigorously oxygen-free atmosphere, considerable partial oxidation of the phosphine was noted [57], as a consequence of the oxidative addition of H_2O to Rh(I), followed by reduction of Rh(III) by TPPTS (Scheme *5.38*). In the presence of O_2 the process was catalytic in rhodium, but it could be suppressed by working in strongly acidic solutions, where the intermediate hydroxorhodium complexes were rapidly protonated. Another intermediate in oxydations with O_2, a peroxorhodium complex, $[RhCl(TPPTS)_3O_2]$, could be characterized by ^{31}P NMR spectroscopy, indicating that it is rather stable. Nevertheless, these redox processes are fast enough to deactivate potential catalysts, so that in several processes the presence of a large excess of the phosphine ligand is recommended.

$RhCl_3 . xH_2O \longrightarrow [RhCl_2(OH)(H_2O)_3] + H^+ + Cl^-$ 5-38.1

$RhCl(TPPTS)_3] \rightleftharpoons [RhCl(TPPTS)_2] + TPPTS$ 5-38.2

$[RhCl(TPPTS)_2] \rightleftharpoons 1/2\ [\{RhCl(TPPTS)_2\}_2]$ 5-38.3

$[RhCl_2(OH)(H_2O)_3] + TPPTS \longrightarrow \{Rh(I)\text{-}Cl\} + OTPPTS + H^+ + Cl^-$ 5-38.4

$\{Rh(I)\text{-}Cl\} + [RhCl(TPPTS)_2] \longrightarrow 2\,\{Rh(I)Cl(TPPTS)\}$ 5-38.5

$\{Rh(I)Cl(TPPTS)\} + H_2O \longrightarrow [RhHCl(OH)(TPPTS)]$ 5-38.6

Scheme 5.38.

The hydration energies of small ions (especially that of a proton) are very large, so that water may facilitate reactions producing such ions, which do not generally take place without additives in non-polar organic solvents or in the solid state. An example is the reaction of $[RuCl_2P_n]$ and H_2 (equation *5.39*).

$$[RuCl_2P_n] + H_2 \rightleftharpoons [RuHClP_n] + H^+ + Cl^- \qquad (5.39)$$

In benzene solutions, when P = PPh_3 and n = 3, this process requires a strong (preferably a non coordinating, hindered) base, such as 1,8-*bis*(dimethylamino)naphtalene ("proton sponge") [113]. On the other hand, the hydride is easily formed even in a quite strongly acidic aqueous solution, *e.g.* 0.1 M HCl (L = TPPMS, n = 2) [18].

Another very important example of a water-assisted reaction is that of a phosphine with an activated alkene [36] or alkyne [37], these often being substrates for catalytic hydrogenation (Scheme *5.40*). The phosphonium salts, which are the products of such reaction may influence the course of hydrogenation [19b]. It is very likely, for example, that phosphonium intermediates have to be considered in the deuterium incorporation in the products of hydrogenation of alkenes in D_2O [35].

Being highly polar, water dissolves salts which are insoluble in common organic solvents. These salts may have a dramatic influence on the rate and selectivity of homogeneously catalysed hydrogenations. Examples previously mentioned include hydrogenation of propanal using $[RuCl_2(TPPTS)_3]$ as the catalyst in the presence of various salts [92] (section 5.3.2.1), biphasic hydrogenation of dienes using $[HCo(CN)_5]^{3-}$ in the presence of lanthanide salts [122] (section 5.3.1.2), and reduction of diphenylethane dione by methanoate using $[RhCl(TPPMS)_3]$ as catalyst in the presence of Mg^{2+} [42] (section 5.7). The formation of

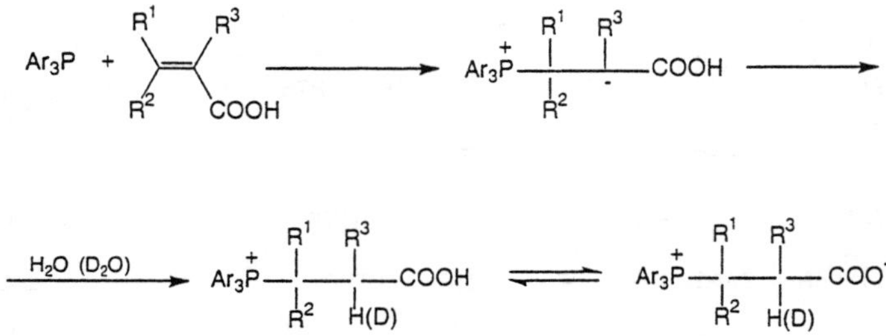

Scheme 5.40.

strong hydrogen bonds between a metal-bound H_2O molecule [183] and the substrate may also have an important effect on the outcome of the reactions. On the other hand, the solubility of H_2 in water is rather small (8×10^{-4} M, at 20°C and 1 bar total pressure [181]), so that competing isomerisation reactions are sometimes important.

In multiphase systems, reactions can be influenced by the distribution of reactants or products between the phases. One example is the biphasic reduction of aldehydes by hydrogen transfer from methanoate [21, 22], catalysed by $[RuCl_2(TPPMS)_2]$ (section 5.5). In that case the neat aldehyde could be used as the organic phase, because the limited solubility of the aldehyde in the aqueous phase kept the substrate/catalyst ratio below the inhibition level. Another interesting example is the biphasic hydrogenation of phenylethanone catalysed by $[RhCl(PPh_3)_3]$ and similar catalysts in the presence of amines. Here, the ammonium chloride formed in the reaction of the amine with $[RhHCl(PPh_3)_3]/PPh_3$ was extracted into the aqueous phase [141a], resulting in a shift of the equilibrium towards the formation of the catalytically active species $[RhH(PPh_3)_n]$ (n = 3 or 4; Scheme *5.41*). In this way a considerable increase in the rate of phenylethanone reduction was observed in biphasic systems (Fig. 5.2), by comparison with homogeneous solutions.

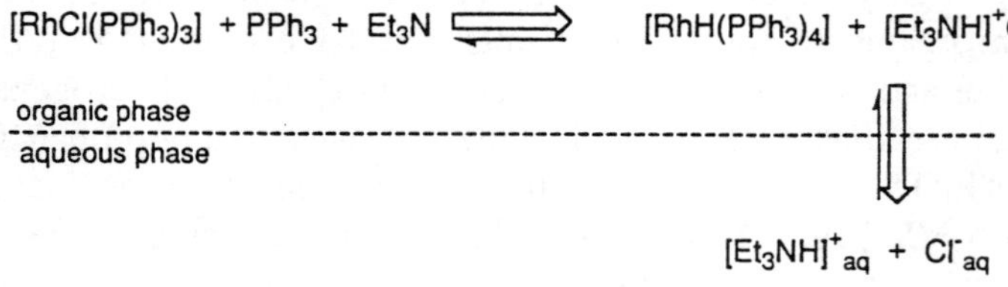

Scheme 5.41.

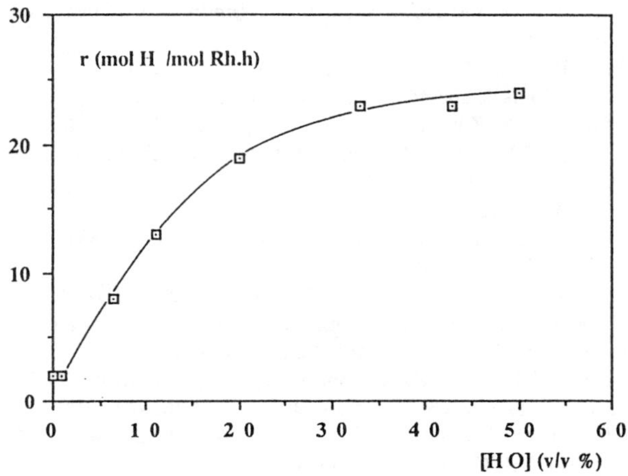

Fig. 5.2. Initial rate of the hydrogenation of phenylethanone as a function of the relative amount of the aqueous phase.
Conditions: 20 mM [$RhCl(PPh_3)_3$], 40 mM Et_3N, 1 bar total pressure, 50 °C.

References

1. E.G. Kuntz, CHEMTECH, **17**, 570 (1987)
2. H. Alper, in *Fundamental Research in Homogeneous Catalysis*, M. Graziani and M. Giongo Eds., Plenum Press, New York, 1984, Vol. 4, p. 79
3. O. Abril and G.M. Whitesides, J. Am. Chem. Soc., **104**, 1552 (1982)
4. P.J. Quinn, F. Joó and L. Vígh, Prog. Biophys. Molec. Biol., **53**, 71 (1989)
5. F. Joó and Z. Tóth, J. Mol. Catal., **8**, 369 (1980)
6. D. Sinou, Bull. Soc. Chim. France, 480 (1987)
7. T.G. Southern, Polyhedron, **8**, 407 (1987)
8. M. Barton and J.D. Atwood, J. Coord. Chem., **24**, 43 (1991)
9. P. Escaffre, A. Thorez and P. Kalck, J. Mol. Catal., **33**, 87 (1985)
10. L. Lecomte, J. Triolet, D. Sinou, J. Bakos and B. Heil, J. Chromatography, **408**, 416 (1987)
11. a) E. Paetzold, A. Kinting and G. Oehme, J. Prakt. Chem., **329**, 725 (1987). b) G. Oehme, in *Coordination Chemistry and Catalysis*, J.J. Ziolkowski Ed., World Scientific, Singapore, 19880, p. 269
12. H. Schindlbauer, Monatsh., **96**, 2051 (1965)
13. T. Bartik, B. Bartik, B.E. Hanson, T. Glass, W. Bebout, Inorg. Chem., **31**, 2667 (1992).
14. S. Ahrland, J. Chatt, N.R. Davies and A.A. Williams, J. Chem. Soc., 264, 276 (1958)
15. a) F. Joó and M.T. Beck, Magyar Kémiai Folyóirat, **79**, 189 (1973). b) F. Joó and M.T. Beck, React. Kinet. Catal. Lett., **2**, 257 (1975)
16. F. Joó, Z. Tóth and M.T. Beck, Inorg. Chim. Acta, **25**, L61 (1977)

17. F. Joó, L. Somsák and M.T. Beck, Proc. *Rhodium in Homogeneous Catalysis*, Veszprém, 1978, p. 59
18. Z. Tóth, F. Joó and M.T.Beck, Inorg. Chim. Acta, **42**, 153 (1980)
19. a) F. Joó, L. Somsák and M.T. Beck, J. Mol. Catal., **24**, 71 (1984). b) A. Bényei, J.N.W. Stafford, Á. Kathó, D.J. Darensbourg and F. Joó, J. Mol. Catal, submitted for publication (1993). c) V.R. Parameswaran and S. Vancheesan, Proc. Indian Acad. Sci. (Chem. Sci.), **103**, 1 (1991)
20. F. Joó and M.T. Beck, J. Mol. Catal., **24**, 135 (1984)
21. F. Joó and A. Bényei, J. Organomet. Chem., **363**, C19 (1989)
22. A. Bényei and F. Joó, J. Mol. Catal., **58**, 151 (1990)
23. E.G. Chepaikin, A.P. Bezruchenko, E.N. Salnikova, F. Joó, Z. Tóth, M.T. Beck and M.L. Khidekel', Izv. Akad. Nauk. SSSR, Ser. Khim., 542 (1989)
24. P.J. Quinn and C.E. Taylor, J. Mol. Catal., **13**, 389 (1981)
25. L. Vígh, F. Joó, P.R. van Hasselt and P.J.C. Kuiper, J. Mol. Catal., **22**, 15 (1983)
26. L. Vígh, F. Joó and Á. Csépl, Eur. J. Biochem., **146**, 241 (1985).
27. F. Joó, É. Csuhai, P.J. Quinn and L. Vígh, J. Mol. Catal., **49**, L1 (1988)
28. A.F. Borowski, D.J. Cole-Hamilton and G. Wilkinson, Nouv. J. Chim., **2**, 137 (1978)
29. Y. Dror and J. Manassen, J. Mol. Catal., **2**, 219 (1976/77)
30. E. G. Kuntz, Ger. Offen. DE 2627354 (1976); DE 2700904 (1976); DE 2733516 (1976), to Rhône-Poulenc
31. J.L. Sabot, European Patent 0104967 (1985), to Rhône-Poulenc
32. R. Gartner, B. Cornils, H. Springer and P. Lappe, Ger. Offen. DE 3235030 (1984), to Ruhrchemie
33. T. Okano, Y. Moriyama, H. Konishi and J. Kiji, Chem. Lett., 1463 (1986)
34. T. Okano, I. Uchida, T. Nakagaki, H. Konishi and J. Kiji, J. Mol. Catal., **54**, 65 (1989)
35. M. Laghmari and D. Sinou, J. Mol. Catal., **66**, L15 (1991)
36. C. Larpent and H. Patin, Tetrahedron, **44**, 6107 (1988)
37. C. Larpent, G. Meignan and H. Patin, Tetrahedron, **46**, 6381 (1990)
38. A.S. Berenblyum, T.V. Turkova and I.I. Moiseev, Izv. Akad. Nauk. SSSR, Ser. Khim., 2153 (1980)
39. B. Fontal, J. Orlewski, C.C. Santini and J.M. Basset, Inorg. Chem., **25**, 4320 (1987)
40. G. Schmid, N. Klein, L. Korste, U. Kreibig and D. Schonauer, Polyhedron, **7**, 605 (1988)
41. a) R. Ruppert, S. Herrmann and E. Steckhan, J. Chem. Soc., Chem. Commun., 1150 (1988). b) E. Steckhan, S. Herrmann, R. Ruppert, E. Dietz, M. Frede and E. Spika, Organometallics, **10**, 1568 (1991)
42. I. Willner, R. Maidan and M. Shapira, J. Chem. Soc., Perkin Trans. 2, 559 (1990)
43. I. Willner and R. Maidan, J. Chem. Soc., Chem. Commun., 876 (1988)
44. J. Kaspar, R. Spogliarich, A. Cernogoraz and M. Graziani, J. Organomet. Chem., **255**, 371 (1983)
45. T. Bartik, B. Bartik, B.E. Hanson, I. Guo, I. Toth, Organometallics, **12**, 164 (1993)
46. J. Jenck and D. Morel, European Patent 0133410 (1984), to Rhône-Poulenc; Chem. Abstr., **103**, 6511j (1985)
47. a) D.J. Darensbourg, C.J. Bischoff and J.H. Reibenspies, Inorg. Chem., **30**, 1144 (1991). b) D.J. Darensbourg and C.J. Bischoff, Inorg. Chem., **32**, 47 (1993)
48. a) J.M. Grosselin, C. Mercier, G. Allmang and F. Grass, Organometallics, **10**, 2126 (1991). b) J.M. Grosselin and C. Mercier, J. Mol. Catal., **63**, L25

(1990). c) J.M. Grosselin and C. Mercier, US Patent 4925990 (1990), to Rhône-Poulenc
49. J.M. Grosselin, European Patent 0362037 (1989), to Rhône-Poulenc; Chem. Abstr., **113**, 77911g (1990)
50. E. Fache, C. Santini, F. Senocq and J.M. Basset, J. Mol. Catal., **72**, 331 (1992)
51. W.A. Herrmann, J. A. Kulpe, J. Kellner, H. Riepl, H. Bahrmann and W. Konkol, Angew. Chem. **102**, 408 (1990); Angew. Chem., Int. Ed. Engl., **29**, 391 (1990)
52. W.A. Herrmann, J.A. Kulpe, W. Konkol and H. Bahrmann, J. Organomet. Chem., **389**, 85 (1990)
53. W. A. Herrmann, J. Kellner and H. Riepl, J. Organomet. Chem., **389**, 103 (1990)
54. C. Larpent, R. Dabard and H. Patin, Tetrahedron Lett., **28**, 2507 (1987)
55. C. Larpent, R. Dabard and H. Patin, Inorg. Chem., **26**, 2922 (1987)
56. C. Larpent and H. Patin, Appl. Organomet. Chem. **1**, 529 (1987)
57. C. Larpent, R. Dabard and H. Patin, New. J. Chem., **12**, 907 (1988)
58. C. Larpent and H. Patin, J. Mol. Catal., **61**, 65 (1990)
59. P. Kalck, in *Organometallics in Organic Synthesis*, A. de Meijere and H. tom Dieck Eds., Springer, Berlin/Heidelberg, 1987, Vol. 2, p. 297
60. A.L. Casalnuovo and J.C. Calabrese, J. Am. Chem. Soc., **112**, 4324 (1990)
61. a) J. Bakos, A. Orosz and D. Sinou, Proc. ISHC-7 (Lyon, 1990) P-43. b) A.S.C. Chan and J.P. Coleman, J. Chem. Soc., Chem. Commun., 535 (1991)
62. M.H. Abraham, P.L. Grellier and R.A. McGill, J. Chem. Soc., Perkin Trans. 2, 339 (1988)
63. I.T. Horvath, R.V. Kastrup, A.A. Oswald and E.J. Mozeleski, Catal. Lett., **2**, 85 (1990)
64. I.T. Horvath, Catal. Lett., **6**, 43 (1990)
65. F. Alario, Y. Amrani, Y. Colleuille, T.P. Dang, J. Jenck, D. Morel and D. Sinou, J. Chem. Soc. Chem. Commun., 202 (1986)
66. L. Lecomte and D. Sinou, J. Mol. Catal., **52**, L21 (1989)
67. L. Lecomte, D. Sinou, J. Bakos, I. Tóth and B. Heil, J. Organomet. Chem., **370**, 277 (1989)
68. Y. Amrani, L. Lecomte, D. Sinou, J. Bakos, I. Tóth and B. Heil, Organometallics, **8**, 542 (1989)
69. L. Lecomte and D. Sinou, Phosphorus, Sulphur and Silicon, **53**, 239 (1990)
70. R.G. Nuzzo, D. Feitler and G.M. Whitesides, J. Am. Chem. Soc., **101**, 3683 (1979)
71. R.G. Nuzzo, S.L. Haynie, M.E. Wilson and G.M. Whitesides, J. Org. Chem., **46**, 2861 (1981)
72. M.E. Wilson and G.M. Whitesides, J. Am. Chem. Soc. **100**, 306 (1978)
73. M.E. Wilson, R.G. Nuzzo and G.M. Whitesides, J. Am. Chem. Soc. **100**, 2269 (1978)
74. T. Jarolim and J. Podlahova, J. Inorg. Nucl. Chem., **38**, 125 (1976)
75. A. Jegorov, J. Podlaha, J. Podlahova and F. Turecek, J. Chem. Soc., Dalton Trans., 2393 (1985)
76. J. Podlahova, B. Kratochvil, V. Langer, J. Silha and J. Podlaha, Coll. Czech. Chem. Commun., **46**, 3063 (1981)
77. M.J.H. Russel and B. Murrer, U.K. Patent Appl. GB 2085874 (1982), to Johnson Matthey
78. M.J.H. Russel, Plat. Met. Rev., **32**, 179 (1988)
79. R.T. Smith and M.C. Baird, Transition Met. Chem., **6**, 197 (1981)

80. a) R.T. Smith and M.C. Baird, Inorg. Chim. Acta **62**, 135 (1982). b) R.T. Smith, R.K. Ungar, L.J. Sanderson and M.C. Baird, Organometallics, **2**, 1138 (1983)
81. a) E. Renaud, R.B. Russel, S. Fortier, S.J. Brown and M.C. Baird, J. Organometal. Chem., **419**, 403 (1991). b) E. Renaud and M.C. Baird, J. Chem. Soc., Dalton Trans., 2905 (1992)
82. A. Andreetta, G. Barberis and G. Gregorio, Chim. Ind. (Milano), **60**, 887 (1978)
83. I. Tóth, B.E. Hanson and M.E. Davis, Organometallics, **9**, 675 (1990)
84. D.J. Daigle, A.B. Pepperman Jr. and S.L. Vail, J. Het. Chem., **11**, 407 (1974)
85. M.Y. Darensbourg and D.J. Daigle, Inorg. Chem., **14**, 1217 (1975)
86. K.J. Fisher, E.C. Alyea and N. Shahnazarian, Phosphorus, Sulphur and Silicon, **48**, 37 (1990)
87. D.J. Darensbourg, F. Joó, M. Kannisto, Á. Kathó and J.H. Reibenspies, Organometallics, **11**, 1990 (1992)
88. a) I. Tóth, B.E. Hanson and M.E. Davis, J. Organomet. Chem., **396**, 363 (1990). b) I. Tóth, B.E. Hanson and M.E. Davis, J. Organomet. Chem., **396**, 109 (1990)
89. a) I. Tóth and B.E. Hanson, Tetrahedron: Asymmetry, **1**, 895 (1990). b) I. Tóth, B.E. Hanson and M.E. Davis, Tetrahedron: Asymmetry, **1**, 913 (1990)
90. I. Tóth, B.E. Hanson and M.E. Davis, Catal. Lett. **5**, 183 (1990)
91. U. Nagel and E. Kinzel, Chem. Ber., **119**, 1731 (1986)
92. a) E. Fache, F. Senocq, C. Santini and J.M. Basset, J. Chem. Soc., Chem. Commun., 1776 (1990). b) E. Fache, C. Santini, F. Senocq and J.M. Basset, J. Mol. Catal., **72**, 337 (1992)
93. M. Meier, *Phosphinkomplexe von Metallen*, Dissertation No. 3988, E.T.H. Zurich, 1967
94. J. Chatt, G.J. Leigh and R.M. Slade, J. Chem. Soc., Dalton Trans., 2021 (1973)
95. a) J.W. Ellis, K.N. Harrison, P.A.T. Hoye, A.G. Orpen, P.G. Pringle and M.B. Smith, Inorg. Chem., **31**, 3026 (1992). b) K.N. Harrison, P.A.T. Hoye, A.G. Orpen, P.G. Pringle and M.B. Smith, J. Chem. Soc., Chem. Commun., 1096 (1989). c) P.A.T. Hoye, P.G. Pringle, M.B. Smith, and K. Worboys, J. Chem. Soc., Dalton Trans., 269 (1993)
96. P.G. Pringle and M.B. Smith, Plat. Met. Rev. **34**, 74 (1990)
97. T. Okano, K. Morimoto, H. Konishi and J. Kiji, Nippon Kagaku Kaishi, 486 (1985)
98. T. Okano, M. Iwahara, H. Konishi and J. Kiji, J. Organomet. Chem. **346**, 267 (1988)
99. T. Okano, M. Iwahara, T. Suzuki, H. Konishi and J. Kiji, Chem. Lett., 1467 (1986)
100. R. Benhamza, Y. Amrani and D. Sinou, J. Organomet. Chem. **288**, C37 (1985)
101. Y. Amrani, D. Lafont and D. Sinou, J. Mol. Catal., **32,** 222 (1985)
102. D. Sinou and Y. Amrani, J. Mol. Catal., **36**, 319 (1986)
103. a) J. Bakos, A. Orosz, B. Heil, M. Laghmari, P. Lhoste and D. Sinou, J. Chem. Soc., Chem. Commun., 1684 (1991). b) C. Lensink and J.G. deVries, Tetrahedron Asymmetry, **2**, 235 (1992)
104. G.P. Schiemenz, Chem. Ber., **99**, 504 (1966)
105. K. Issleib and H. Zimmermann, Z. Anorg. Allg. Chem., **353**, 197 (1967)
106. S.C. Tang, T.E. Paxson and L. Kim, J. Mol. Catal., **9**, 313 (1980)
107. M.E. Ford and J.E. Premecz, J. Mol. Catal., **19**, 99 (1983)
108. B. Salvesen and J. Bjerrum, Acta Chem. Scand., **16**, 735 (1962)
109. J.C. Chang and J. Bjerrum, Acta Chem. Scand., **26**, 815 (1972)
110. J. Bjerrum, R.S. George, C.J. Hawkins and D.C. Olson, in *Proc. Symp. Coord. Chem. (Tihany, 1964)*, M.T. Beck Ed., Akadémiai Kiadó, Budapest, 1965, p. 187

111. R.R. Schrock and J.A. Osborn, J. Am. Chem. Soc., **93**, 2397 (1971)
112. J.M. Brown, P.A. Chaloner, A.G. Kent, B.A. Murrer, P.N. Nicholson, D. Parker and P.J. Sidebottom, J. Organomet. Chem. **216**, 263 (1981)
113. B.R. James, *Homogeneous Hydrogenation*, John Wiley and Sons, New York, 1973
114. B.R. James, Adv. Organomet. Chem., **17**, 319 (1979)
115. a) J.F. Harrod, S. Ciccone and J. Halpern, Canadian J. Chem., **39**, 1372 (1961). b) J. Halpern and B.R. James, Canadian J. Chem., **44**, 671 (1966)
116. J. Halpern, J.F. Harrod and B.R. James, J. Am. Chem. Soc., **88**, 5150 (1966)
117. J. Kwiatek, Catal. Rev., **1**, 37 (1967)
118. D.L. Reger and M.M. Habib, J. Mol. Catal., **7**, 365 (1980)
119. D.L. Reger, M.M. Habib and D.J. Fauth, J. Org. Chem., **45**, 3860 (1980)
120. K. Ohkubo, T. Kawabe, K. Yamashita and S. Sakaki, J. Mol. Catal., **24**, 83 (1984)
121. H.A. Zahalka, K. Januszkiewicz and H. Alper, J. Mol. Catal., **35**, 249 (1986)
122. J.T. Lee and H. Alper, J. Org. Chem., **55**, 1854 (1990)
123. J.T. Lee and H. Alper, Tetrahedron Lett., **31**, 1941 (1990)
124. A.M. Rosan, R.K. Crissey and I.L. Mador, Tetrahedron Lett., **25**, 3787 (1984)
125. J. Kwiatek, I.L. Mador and J.K. Seyler, Adv. Chem. Ser., **37**, 201 (1963)
126. M. Murakami, R. Kawai and K. Suzuki, J. Chem. Soc. Japan, **84**, 669 (1963)
127. I. Amer, T. Bravdo, J. Blum and K.P.C. Vollhardt, Tetrahedron Lett., **28**, 1321 (1987)
128. J. Blum, I. Amer, K.P.C. Vollhardt, H. Schwarz and G. Hohne, J. Org. Chem., **52**, 2804 (1987)
129. A. Laufenberg, A. Behr and W. Keim, Ger. Offen. DE 3841698 (1989), to Henkel
130. E. Bayer and W. Schumann, J. Chem. Soc., Chem. Commun., 949 (1986)
131. E.G. Chepaikin and M.L. Khidekel', J. Mol. Catal. **4**, 103 (1978)
132. Yu.A. Sakharovsky, M.B. Rosenkevich, A.S. Lobach, E.G. Chepaikin and M.L. Khidekel', React. Kinet. Catal. Lett., **2**, 249 (1978)
133. A.V. Bulatov, E.N. Izakovich, L.N. Karklin' and M.L. Khidekel', Izv. Akad. Nauk. SSSR, Ser. Khim., 2032 (1981)
134. A.V. Bulatov, G.P. Voskerchyan, S.N. Dobryakov and A.T. Nikitaev, Bull. Acad. Sci. USSR, Chem. Sci., **35**, 747 (1986)
135. F. Joó, N. Balogh, L.I. Horváth, G. Filep, I. Horváth and L. Vígh, Anal. Biochem., **194**, 34 (1991)
136. B.A. Murrer and J.W. Jenkins, UK Patent GB 2070023 (1981), to Johnson Matthey.
137. T. Hayashi, T. Mise, S. Mitachi, K. Yamamoto and M. Kumada, Tetrahedron Lett., 1133 (1976)
138. K. Achiwa, J. Am. Chem. Soc., **98**, 8265 (1976)
139. H. Brunner and W. Pieronczyk, Angew. Chem., Int. Ed. Engl., **18**, 620 (1979)
140. F.T. Lin and N.J. Leonard, J. Am. Chem. Soc., **101**, 996 (1979)
141. a) F. Joó and E. Trócsányi, J. Organomet. Chem., **231**, 63 (1982). b) V.V. Grushin, A.B. Vymenits, M.E. Vol'pin, J. Organometal. Chem., **382**, 185 (1990)
142. A. Brandstrom, J. Mol. Catal. **20**, 93 (1983)
143. K.R. Januszkiewicz and H. Alper, Organometallics, **2**, 1055 (1983)
144. A. Harada, Y. Hu and S. Takahashi, Chem. Lett., 2083 (1986)
145. E. Paetzold and G. Oehme, Phosphorus, Sulfur and Silicon, **51/52**, 359 (1990)
146. V.V. Grushin and H. Alper, Organometallics, **10**, 1620 (1991)
147. F. Joó and L. Vígh, in *Advances in Catalyst Design*, M. Graziani and C.N.R. Rao, Eds., World Scientific, Singapore, 1991, p. 254

148. L. Vígh, I. Horváth, F. Joó and G. A. Thompson, Jr., Biochim. Biophys. Acta, **921**, 167 (1987)
149. Y. Pak, F. Joó, L. Vígh, Á. Kathó and G.A. Thompson, Jr., Biochim. Biophys. Acta, **1023**, 230 (1990)
150. L. Vígh, Z. Gombos and F. Joó, FEBS Lett., **191**, 200 (1985)
151. I. Horváth, A.R. Mansourian, L. Vígh, P.G. Thomas, F. Joó and P.J. Quinn, Chem. Phys. Lipids, **39**, 251 (1986)
152. M. Schlame, I. Horváth, Zs. Török, L.I. Horváth and L. Vígh, Biochim. Biophys. Acta, **1045**, 1 (1990)
153. D. Chapman and P.J. Quinn, Proc. Natl. Acad. Sci. USA, **73**, 3971 (1976)
154. T.D. Madden, W.E. Peel, P.J. Quinn and D. Chapman, J. Biochem. Biophys. Methods, **2**, 19 (1980)
155. J.S. Conolly, Ed., *Photochemical Conversion and Storage of Solar Energy*, Academic Press, New York, 1981
156. R.S. Coffey, J. Chem. Soc., Chem. Commun., 923 (1967)
157. M.E. Vol'pin, V.P. Kukolev, V.O. Chernyshev and I.S. Kolomnikov, Tetrahedron Lett., 4435 (1971)
158. Y. Watanabe, T. Ohta and Y. Tsuji, Bull. Chem. Soc. Japan, **55**, 2441 (1982)
159. G.A. Chukhadjian, V.P. Kukolev, L.N. Melkonyan, V.A. Matossian and N.A. Balyushina, Armianskii Khim. Zh., **36**, 412 (1983)
160. G.A. Chukhadjian, V.P. Kukolev, V.A. Matossian and N.A. Balyushina, Proc. ISHC-5 (Leningrad, 1984), p. 138
161. R. Bar, L.K. Bar, Y. Sasson and J. Blum, J. Mol. Catal., **33**, 161 (1985)
162. R. Bar and Y. Sasson, Tetrahedron Lett., **22**, 1709 (1981)
163. J. Blum, I. Pri-Bar and H. Alper, J. Mol. Catal. **37**, 359 (1986)
164. S. Oishi, J. Mol. Catal., **39**, 225 (1987)
165. R. Bar, Y. Sasson and J. Blum, J. Mol. Catal., **16**, 175 (1982)
166. J.T. Lee and H. Alper, Tetrahedron Lett., **31**, 4101 (1990)
167. See for example: R.M. Laine and E.J. Crawford, J. Mol. Catal., **44**, 357 (1988)
168. T. Cole, R. Ramage, K. Cann and R. Pettit, J. Am. Chem. Soc., **102**, 6184 (1980)
169. R.S. Dickson, *Organometallic Chemistry of Rhodium and Iridium*, Academic Press, New York, 1983
170. F. Joó and H. Alper, Canadian J. Chem., **63**, 1157 (1985)
171. K. Januszkiewicz and H. Alper, J. Mol. Catal., **19**, 139 (1983)
172. K. Kaneda, M. Yasumura, T. Imanaka and S. Teranishi, J.Chem. Soc., Chem. Commun., 935 (1982)
173. T.J. Lynch, M. Banah, M. McDougall, H. Kaesz and C.R. Porter, J. Mol. Catal., **17**, 10 (1982)
174. S.I. Murahashi, Y. Imada and Y. Hirai, Tetrahedron Lett., **28**, 77 (1987)
175. H. Alper and J. Heveling, J. Chem. Soc., Chem. Commun., 365 (1983)
176. D. Sinou, M. Safi, C. Claver and A. Masdeu, J. Mol. Catal., **68**, L9 (1991)
177. C.H. Wong and G.M. Whitesides, J. Org. Chem., **47**, 2816 (1982)
178. C.H. Wong, L. Daniels, W.H. Orme-Johnson and G.M. Whitesides, J. Am. Chem. Soc., **103**, 6227 (1981)
179. M. Franke and E. Steckhan, Angew. Chem., Int. Ed. Engl., **27**, 265 (1988)
180. T. Yoshida, T. Okano, Y. Ueda and S. Otsuka, J. Am. Chem. Soc., **103**, 3411 (1981)
181. *CRC Handbook of Chemistry and Physics*, R.C. West, Ed., 54th edn., CRC Press, Cleveland, OH, 1974, p. B-94
182. V.V. Grushin and H. Alper, Organometallics, **10**, 831 (1991)

133. P. Leoni, N. Sommovigo, M. Pasquali, S. Midollini, D. Braga and P. Sabatino, Organometallics, **10**, 1038 (1991)
134. F. Farin, H.L.M. van Gaal, S.L. Bonting and F.J.M. Daemen, Biochim. Biophys. Acta, **711**, 336 (1982)
135. S. Ganguly, J.T. Mague and D.M. Roundhill, Inorg. Chem., **31**, 3500 (1992)
136. C. Larpent, H. Patin, N. Thilmont and F.F. Valdor, Syn. Commun., **21**, 495 (1991)
137. W.A. Herrmann and Kohlpaintner, H. Bahrmann and W. Konkol, J. Mol. Catal., **73**, 191 (1992)
138. J.P. Genet, E. Blart and M. Savignac, Synlett, 715 (1992)
139. B. Fell and F. Papadogianakis, J. Mol. Catal., **66**, 143 (1991)
190. T.N. Mitchell and K. Heesche-Wagner, J. Organometal. Chem., **436**, 43 (1992)
191. T. Okano, M. Kaji, S. Isotani and J. Kiji, Tetrahedron Lett., **38**, 5547 (1992)
192. M. Laghmari, D. Sinou, A. Masdeu and C. Claver, J. Organometal. Chem., **438**, 213 (1992)
193. G. Oehme , E. Paetzold and R. Selke, J. Mol. Catal., **71**, L1 (1992)
194. a) T.K. Miyamoto, Y. Suzuki, and H. Ichida, Bull. Chem. Soc. Jpn., **65**, 3386 (1992). b) P. Bernhard, M. Biner, and A. Ludi, Polyhedron, **9**, 1095 (1990). c) G. Laurenczy and A.E. Merbach, J.Chem. Soc., Chem. Commun., 187 (1993). d) T. X. Le, J.S. Merola, Abstr. A.C.S. Mtng., INOR 810 (San Francisco, 1992)
195. a). A. Avey, S.C. Tenhaeff, T.J.R. Weakley, and D.R. Tyler, Organometallics, **10**, 3607 (1991). b) A. Avey and D.R. Tyler, Organometallics, **11**, 3856 (1992). c) A. Avey, D.M. Schut, T.J.R. Weakley, and D.R. Tyler, Inorg. Chem., **32**, 233 (1993)

CHAPTER 6

SUPPORTED METAL COMPLEXES

The subject of supported metal complexes as catalysts was reviewed by Hartley in a book published in 1985 [1], and hydrogenation was extensively treated. Supported metal complexes as hydrogenation catalysts were also reviewed by Yermakov and Arzamaskova in 1986 [2], and enantioselective versions of the reaction by Hetflejš in the same year [3]. Therefore account will focus on work published since 1984.

The success of homogeneous catalysts in all types of hydrogenation reactions led to great interest in attaching the complexes to insoluble supports, thus facilitating catalyst removal and reuse. Much of the literature in this area has therefore been derivative; a homogeneous catalyst is linked to a support, and the same series of reactions undertaken. Over the last five years or so there has been much interest in the field of the use of metal complexes as precursors to supported metal catalysts, especially in the form of metal crystallites and clusters, but this lies beyond the scope of this work [4].

The supports used for metal complexes are very varied. Most early work employed conventional polymers such as polystyrene, bearing functional groups for metal coordination. Recently silica and other metal oxides have been more prominent; they are inexpensive, easily functionalised, and chemically, thermally and mechanically stable.

Whilst there has been no diminution in the number of papers published in this area in the last ten years, many of these describe technological improvements, rather than studies of fundamental chemistry. A number seem to ignore the possibility of metal crystallite formation, although this is known to the common even in cases in which the homogeneous catalyst is stable [5], and there are often few clues as to the nature of the true catalytically active species. The field has been dominated by workers from the former Soviet Union and estern bloc, the People's Republic of China and Japan.

P.A. Chaloner, M.A. Esteruelas, F. Joó and L.A. Oro, Homogeneous Hydrogenation 241–253.

6.1. Catalysts supported on organic polymers

6.1.1. CATALYSTS SUPPORTED ON FUNCTIONALISED POLYSTYRENE

Phosphinated polystyrene was one of the earliest supports to be used for a wide range of metal complexes [6]. Recent years have seen careful studies of the chemical instability of the phosphorus-carbon bonds in these species [7, 8], and the formation of metal crystallites [9].

There are numerous examples of catalysts for reduction of simple alkenes which have been supported on phosphinated polystyrene, including $[RhCl(PPh_3)_3]$, $[RuCl_2(PPh_3)_3]$ and $[PtCl_2(PhCN)_2]/SnCl_2$ [10]. The reaction may usually be run under mild conditions, selectivities towards the reduction of unhindered alkenes are often better that with homogeneous analogues, and the catalyst may frequently be reused. Activity tends to depend on the extact method of catalyst preparation [11]. In an interesting recent example, $[Rh(DPPE)(NBD)][PF_6]$ was entrapped inside cross-linked polystyrene, and in this form proved to be an active catalyst for reduction of 1-hexene in methanol at 3 bar hydrogen pressure [12].

Extensions to other substrates have been relatively few. Cyclododecatriene has been reduced at 100°C and 20 bar hydrogen pressure using $PdCl_2$ on styrene/DVB copolymers modified by various functional groups; catalyst stability depended on the ligand [13]. $PdCl_2$ on phosphinated polystyrene was a selective catalyst for reduction of cyclopentadiene to cyclopentene [14], and 1,4-cyclohexadiene was reduced to cyclohexene using $[Ni(acac)_2]/Al_2Cl_3Et_3$ on the same support [15]. Reduction of nitrogen containing heterocycles over supported $[RhCl(PPh_3)_3]$ was reported to be 10–12 times faster than when the homogeneous catalyst was used, which leads to the suspicion that metal particles might be involved [16]. Phosphide linkages were important in catalysts prepared from **6-1** and $[\{(RhCl(COD)\}_2]$ or $[PtCl_2(COD)]$ for arene reduction.

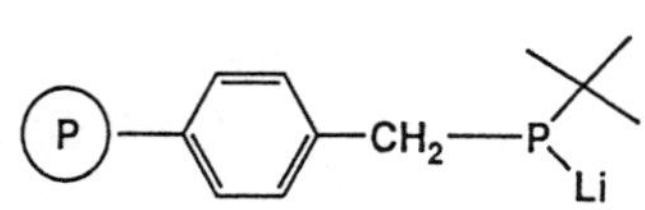

(6-1)

Chloromethylated polystyrene has also been functionalised with nitrogen containing ligands. Metal derivatives of polymer supported

1,2-diaminoethane used in alkene reduction have included complexes of ruthenium [18], rhodium [19] and cobalt [20]. Polystyrene functionalised with bipy units has been complexed with palladium(II) to give a catalyst for alkene reduction [21], and palladium derivatives of supported 2,2′-dipridylmethane have been used to catalyse alkene and alkyne reduction. A supported catalyst prepared by reacting $[PdCl_2(PhCN)_2]$ with bromopyridinomethylated or *N*-diphenylphosphinopyridinomethylated styrene/DVB copolymer was used in the reduction of 4-methylpent-3-ene-2-one to 4-methyl-2-pentanone [22].

An early example of a complex anchored to polystyrene *via* a carbon-metal bond was prepared by reacting $[Pd(PPh_3)_4]$ with cross-linked chloromethylated polystyrene [23]. However, the anchoring of metals *via* an η^5-Cp ring seems to be more generally useful. Polymer anchored titanocene is reasonably stable, and is a catalyst for alkene reduction. This is in contrast to soluble $[TiCp_2]$, which decomposes with titanium-titanium bond formation. Polymer supported $[NbCl_4Cp]$ and $[TaCl_4Cp]$ have been reduced to give catalysts for alkene and alkyne reduction [25].

6.1.2. CATALYSTS SUPPORTED ON POLYVINYL PYRIDINE

Palladium(II) supported on poly-2-vinylpyridine or poly-4-vinylpyridine has been widely used as a catalyst for reduction of unsaturated aldehydes, alcohols, [26] and carboxylic acids [27]. The catalyst is frequently pre-reduced with $Na[BH_4]$, and XPES data implied that both palladium(II) and palladium(0) were present. The derivative of 4-vinylpyridine was the more active catalyst, and there has also been a suggestion that activity may depend of the degree of charge transfer from palladium to the polymer. Related platinum derivatives are active catalysts for alkene reduction, but rates are usually lower than with the palladium analogues. Both palladium and platinum derivatives of 4- or 2-vinylpyridine were used as catalysts for reduction of nitroarenes to anilines; palladium complexes were usually more active, and activity decreased as the molecular weight of the polymer was increased [28].

6.1.3. CATALYSTS SUPPORTED ON ACRYLATE POLYMERS

Rhodium or platinum anchored to polyacrylic acid proved to be good catalysts for reduction of alkenes, dienes, arenes and nitroarenes [29]. Rhodium polyacrylic acid complexes have also been used to catalyse reduction of furan-2-carbaldehyde to 2-hydroxymethyl tetrahydrofuran, and, and after pretreatment with Na[BH_4], phenol to cyclohexanone. The activity of the complex formed from atactic polyacrylic acid and [RhH(PPh_3)$_4$] was improved by reaction with the cross-linked agent 1,6-diaminohexane [30]. Palladium(II) or rhodium(II) salts have been anchored on hydrogel polymers based on 2-hydroxyethylmethacrylate; they were active catalysts for the reduction of alkenes, alkynes and dienes under ambient conditions [31]. Radiation induced copolymerisation of *cis*-[$PdCl_2${$CNMe_2OC(O)CH{=}CH_2$}$_2$] with dimethylacrylamide and methylene bisacrylamide gave a terpolymer which catalysed reduction of phenylethyne, styrene or nitrobenzene under ambient conditions [32].

6.1.4. CATALYSTS SUPPORTED ON FUNCTIONAL POLYALKENES

Alkenes may be hydrogenated in the presence of [RhCl(PPh_3)$_3$] bound on phosphinated polyethene single crystals. The material had a high surface:volume ratio and the catalysts were more active than related bead system [33]. This type of catalyst has been used in a fixed bed reactor and may be separated and reused without loss of activity [34]. The preparation of a soluble version of the catalyst has been described in detail. Although the reaction rates observed for hexene reduction (100°C, 1 atm) were only half those obtained with [RhCl(PPh_3)$_3$], they were an order of magnitude better than those used a rhodium catalyst supported on a phosphinated insoluble cross-linked polystyrene [35]. Palladium supported on a polymer prepared by photochemical grafting of 4-vinylpyridine onto polyethene showed good selectivity for reduction of nitroarenes to anilines [36].

6.1.5. CATALYSTS SUPPORTED ON POLYETHYLENEIMINE

Polyethyleneimine and polyethenylpyrrolidone have been used to support rhodium, palladium or nickel complex catalysts. The rhodium based systems were used to catalyse reduction of phenols, ketones and sugars, and the palladium (after pre-reduction with Na[BH_4]) for alkenes, nitroarenes, imines and alkynes. Nickel derivatives catalysed stereospecific reduction of alkynes to *cis*-alkenes [37]. The palladium based system catalysed hydrogenation of allylphenols to propylphenols, but under similar conditions the rhodium(III) derivative catalysed ring hydrogenation. Pre-reduction of the palladium(II) species with hydrogen has been shown to give a poorer catalyst than that from the $Na[BH_4]$ procedure. An interesting selectivity for the reduction of $ArNO_2$ was observed in the presence of a platinum(II) derivative of an excess of polyethyleneimine reduced by $Na[BH_4]$; the major products are *N*-arylhydroxylamines, obtained in up to 98% yield [38].

6 1.6. CATALYSTS SUPPORTED ON OTHER ORGANIC POLYMERS

The most widely-used of the other polymers used to support metal complex catalysts are amide condensation polymers, formed from, for example, **6-2** and **6-3**. When treated with $PdCl_2$ this polymer gave a catalyst for reduction of alkenes and nitroarenes [39]. A rhodium derivative of the same polymer was also used to catalyse reduction of carbonyl groups. Palladium(II) supported on polyethenylcarbazole was used for nitroarene reduction, and transfer hydrogenation (from 2-propanol) with a related rhodium based system. In the latter case EPR spectroscopic data suggested that Rh(II).O_2^- and Rh(III).O_2^- were present during the reaction [40].

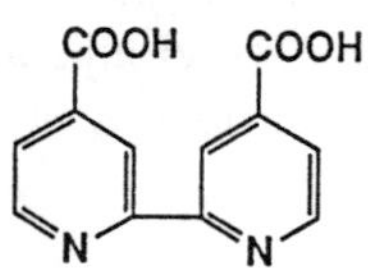

(6-2)

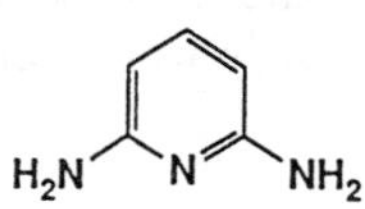

(6-3)

6.2. Catalysts supported on ion exchange resins

The very first supported transition metal catalysts were of this type; their formation involves ion exchange and the forces between catalyst and support are electrostatic in nature. The degree of ion exchange is usually small, and often confined to surface layers. Solvent effects and resin swelling are frequently very important in determining acticity and selectivity.

Rhodium(III)chloride on a 3-carboxy-butenolic acid/DVB cation exchange resin with a macroreticular structure showed selectivity in alkene reduction similar to the homogeneous catalyst derived from linear macromolecules [41]. A range of ion exchange resins has been used to support palladium for the reduction of cottonseed oil. Catalysts with large pore diameters and surface areas showed the best activity, and acidic resins were superior to basic ones [42].

The most common of this type of catalyst, however, involve palladium(II) on anionic exchange resins. Alkenes have been reduced using Pd(II) derivatives of 8-aminoquinoline [43] or $[Pt_{15}(CO)_{30}]^{2-}$ on IRA401 [44]. The latter system is also active for nitroarene and aldehyde reduction. Dienes have been reduced to alkenes with up to 100% selectivity in the best cases. A study of the reduction of nitroarenes over a quaternary ammonium containing polystyrene resin showed that $[PdCl_4]^{2-}$ was the most suitable anion (rather than $[PdBr_4]^{2-}$ or $[PdI_4]^{2-}$). Activity is also affected if a polynuclear palladium containing species is formed [45].

6.3. Catalysts supported on silica or other metal oxides

Catalysts of this type have been popular in recent years; metal oxides as supports have the advantage that they are physically and chemically robust, and essentially inert. Attachment of a ligand to silica is usually accomplished by exchange; surface hydroxyl groups react with species such as **6-4** with the displacement of the alkoxy groups at silicon. Thus the side chain is very well anchored, though problems of metal leaching do remain.

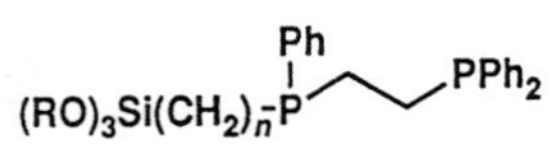

(6-4)

Reduction of simple alkenes over a wide range of silica supported catalysts has been reported. Allyl benzene was reduced to propyl benzene using silica modified with $\{\equiv Si(CH_2)_3NHR\}$ (R = H or Ph) side chains, as rhodium complexes [46], and both alkenes and dienes reacted in the presence of rhodium derivatives of $SiO_2/(EtO)_3Si(CH_2)_nPPh_2$ [47]. Polysiloxane phosphine ligands with $[\{IrCl(COD)\}_2]$ were used in a kinetic study of hexene reduction [48]. Supported Ziegler catalysts were prepared by reduction of nickel(II) 2-ethylhexanoate with alumina-silica treated with Et_3Al or $Na[AlH_2(OCH_2CH_2OCH_3)_2]$, and gave good results for alkene reduction [49].

Palladium and platinum based catalysts have been much studied. A catalyst for reduction of allyl benzene was prepared by complexing palladium(II) with the Schiff bases formed from γ-aminopropyl silica and heterocyclic aldehydes. $Na[BH_4]$ in methanol was used for activation [50]. Silica bearing phosphine side chains such as **6-5** or **6-6** has been used as a support for palladium or platinum for reduction of alkenes and nitroarenes [51]. 4-Vinylpyridine was copolymerised with DVB in the presence of silica gel to give a support for $PdCl_2$ or $Pd(OCOMe)_2$; kinetic data for cyclohexene reduction were obtained under mild conditions [52].

$(RO)_3Si(CH_2)_{11}PPh_2$

(6-5)

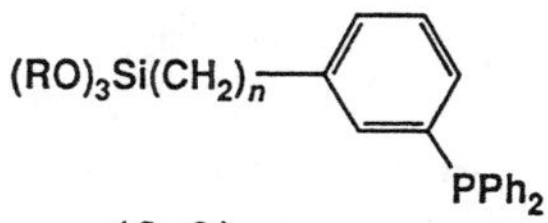

(6-6)

Hydrogenation of alkenes or cycloalkenes has been catalysed by reduced $[TiCl_2Cp_2]$ and its analogue covalently bonded to silica $([\{SiO_2\text{-}(CH_2)_3Cp\}_2TiCl_2])$. The catalyst was readily reusable [53]. $[TiCl_3Cp]/Al_2O_3/SiO_2$ has also been found to be a useful precursor. The well-defined silica supported cluster $[Os_3(CO)_{10}(\mu\text{-H})(\mu\text{-Si}\equiv)]$ catalysed ethene reduction at 90°C, and 1 bar H_2. Kinetic, volumetric and IR spectroscopic studies indicated that the mechanism involved the intact triosmium framework in all the elementary steps [54]. This species has also been used to catalyse reduction of ketones at 130–170°C [55].

Use of **6-4** to immobilise rhodium on silica gave a catalyst for diene reduction [56]. Ruthenium complexes supported on phosphinated silica have been used to catalyse reduction of 3-phenylpropenal to 3-phenyl-prop-2-ene-1-ol or 3-phenylpropanal, depending on the reaction conditions [57]. Imines have been successfully reduced using a palladium complex of silica functionalised with amino groups [58].

Alumina has also provided a useful support for a range of catalysts, though the method of ligand attachment is often less well defined than for silica. Amine containing palladium complexes on alumina have been used in the reduction of alkynes and dienes; they were rather stable to poisoning by sulphur or carbon monoxide [59]. In some cases selective reduction of alkynes to *cis*-alkenes was accomplished. Platinum amine complexes have been used for alkene reduction at room temperature; activity was increased by supporting the catalyst on alumina [60]. Palladium chloride on polyvinylpyrrolidone/alumina was a very active catalyst for reduction of alkenes and α, β-unsaturated aldehydes [61].

6.4. Catalysts supported on clays

Like metal oxides, clays have the advantage as supports that they are chemically and physically robust, and inexpensive. Adjustment of interlayer spacing by the introduction of substituents, pillaring or solvent swelling, and the Lewis acidic nature of the interlamellar regions can also give usefully altered selectivity [62]. Much recent work has involved functionalised montmorillonite, a typical natural smectite (2:1 layered silicate). Palladium(II) complexes of phosphinated montmorillonite have been used to catalyse alkene and alkyne reduction; the result depended on the extent of the swelling. Under favourable conditions alkynes could be reduced to *cis*-alkenes with selectivity better than the Lindlaar catalyst [63]. With non-conjugated diene substrates, terminal double bonds were preferentially reduced, and 1,2-addition to conjugated dienes predominated [64]. Isomerisation is largely supressed. Similar results were obtained when bipyridyl units were used to anchor the palladium [65]. Aromatic nitro compounds have been reduced using a palladium derivative of montmorillonite functionalised with 3-aminopropyl triethoxysilane. Selectivity for reduction of the nitro group was good in the presence of chlorides, aldehydes or ketones, but aryl bromides were preferentially hydrogenolysed [66].

Rhodium phyllosilicate catalyst precursors may be prepared by supporting $[Rh(Me_2CO)_n(NBD)][ClO_4]$ on polygorskite or montmorillonite. These precursors became active for 1-hexene reduction only after a substantial indiction period. Activity then remained constant over several cycles, and no metal leaching was observed [67]. $[Rh(COD)(PPh_3)_2]^+$, intercalated in a synthetic fluorotetrasilicane mica, has been used for reduction of linear and cycloalkenes. With cycloalkene substrates, there is considerable steric influence on the rate from the interlayer separation, which is not noted for unbranched 1-alkenes. Selectivity for 4-ethylcyclohexene from reduction of 4-ethenylcyclohexene is enhanced for the supported catalyst [68]. Hectorite intercalated palladium(II) complex catalysts prepared from $[Pd(NH_3)_4]^{2+}$ or a palladium(II) triazine complex were used for reduction of 1-alkynes in dmso [69].

6.5. Enantioselective reactions

This field was reviewed by Hetflejš in 1986, and only work published since 1985 is detailed here [3]. There has been relatively little substantive progress in this field over the last ten years. As elsewhere in this field, there have been technical improvements, but few breakthroughs. Dehydroamino acids such as **6-7** were reduced in the prsence of a rhodium complex of silica gel derivatised with BPPM. An optical yield of 87% was obtained for the reaction of **6-7**, somewhat better than with the homogeneous analogue [70]. Excellent enantioselectivities were obtained for this reaction using **6-8**, attached to silica, as a catalyst [71]. A catalyst prepared from silica supported $(EtO)_3Si(CH_2)_nPMen_2$ (n = 1, 3 or 5) and $[\{RhCl(C_2H_4)_4\}_2]$ gave a better optical yield in the hydrogenation of **6-9** than homogeneous analogues, but reuse led to metal leaching [72]. The homogeneous catalysts [Rh(COD)(DIOP)][PF_6] and [Rh(COD)(NORPHOS)][PF_6] have been impregnated into Ba[SO_4], celulose, silica, alumina, AgCl and charcoal, and used for the reduction of **6-9**. The best optical yield was 79%, but up to 87% was achieved using acid ion exchange resins as the support [73]. The complexes **6-10**, **6-11** (Ar = 4-dimethylaminophenyl) and **6-12** (Ar′ = 4-trimethy-

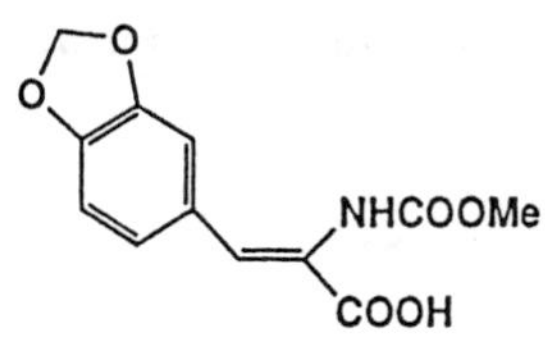

(6-7)

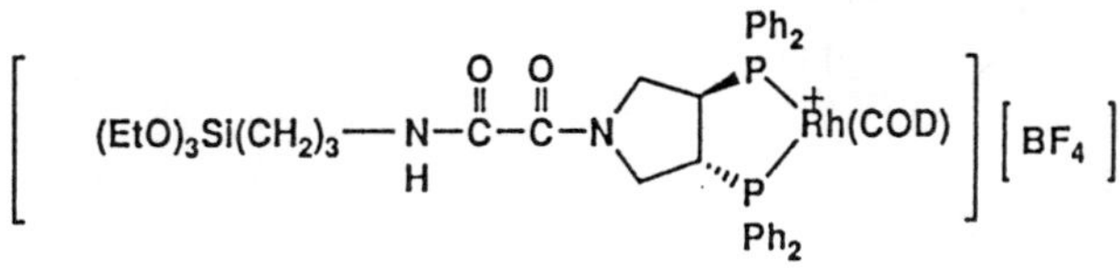

(6-8)

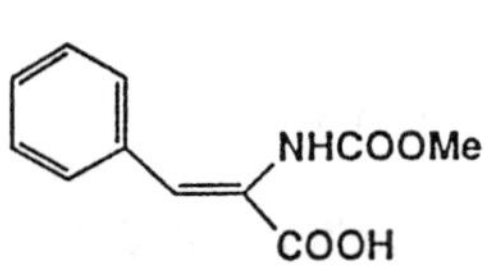

(6-9)

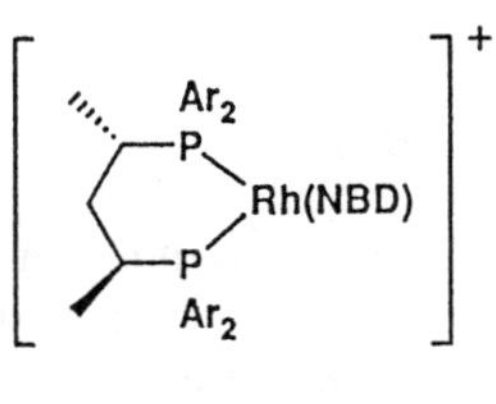

(6-10)

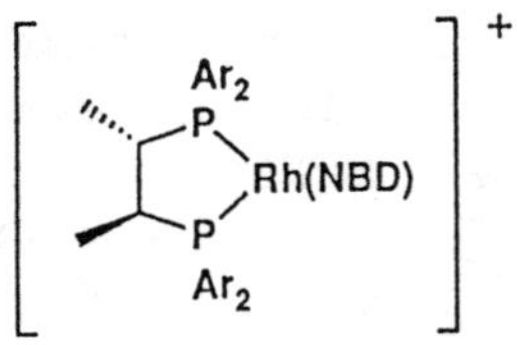

(6-11)

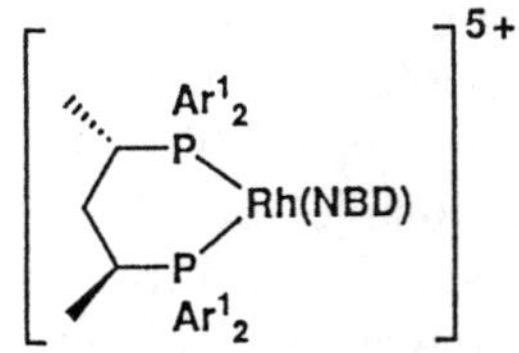

(6-12)

lammoniumphenyl) have been supported on cation exchange resins and used to catalyse reduction of **6-9**. Enantiomer excesses were similar to those in related homogeneous systems but rates were lower. The catalysts could be reused without significant metal loss [74]. Copolymers of **6-13**, **6-14** and **R-6-15** were used to support rhodium(I) to catalyse dehydroamino acid reduction, but optical yields were modest [75].

A rhodium complex, **6-16**, was supported on zeolite or silica. With **6-9** as the substrate up to 98% optical yield was obtained using the zeolite supported material, which is considerably better than with most previous homogeneous complexes of chiral nitrogen ligands [76].

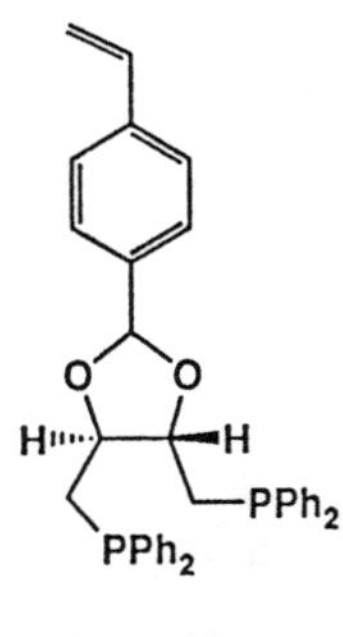

(6-13)

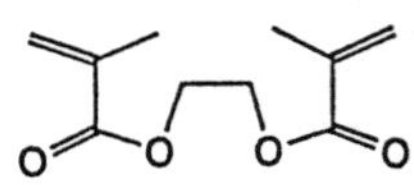

(6-14)

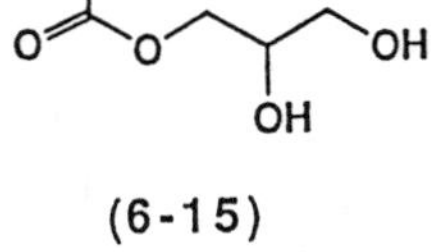

(6-15)

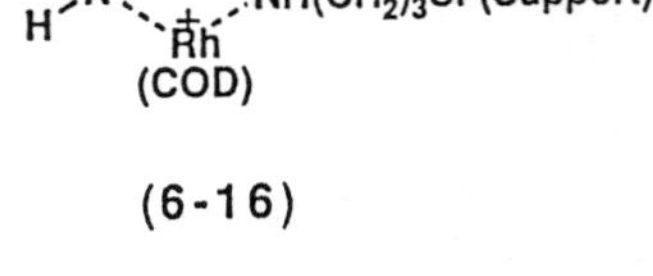

(6-16)

The use of platinum metal modified by cinchona alkaloids as a catalyst for the reduction of ketoesters was first reported by Orito and coworkers [77] and has been considerably developed since [78]. Now an analogue of this, involving a polymer supported catalyst, has been developed.

The catalyst is prepared by reacting chloromethylated cross-linked polystyrene with a chiral tertiary amine such as cinchonine or quinine. The chloride counterion is then exchanged for $[Pt_{15}(CO)_{30}]^{2-}$ and catalytic reduction of $CH_3C(=O)COOMe$ was then accomplished in 80% optical yield using the cinchonine based catalyst. The catalyst is, however, sensitive to minor changes in substrate structure and results were poor with PhC(=O)COOMe or methyl 3-oxobutanaote as substrate [79].

References

1. Supported Metal Complexes, F.R. Hartley, D. Reidel Publishing Company, 1985
2. Yu.I. Yermakov and L.N. Arzamaskova, Stud. Surf. Sci. Catal., **27**, 459 (1986)
3. J. Hetflejš, Stud. Surf. Sci. Catal., **27**, 497 (1986)
4. J. Margitfalvi, S. Szabó and F. Nagy, Stud. Surf. Sci. Catal., **27**, 373 (1986)
5. N. Li and J.M. Fréchet, J. Chem. Soc. Chem. Commun., 1100 (1985)
6. H. Hirai and N. Toshima in Tailored Metal Catalysts, Ed. V. Iwasawa, Reidel Dordrecht 87 (1986)
7. R.A. Dubois and P.E. Garrou, Organometallics, **5**, 466 (1986)
8. P.E. Garrou, Chem. Rev., **85**, 171 (1985)
9. M. Bartholin, C. Graillat and A. Guyot, J. Mol. Catal., **10**, 377 (1980)
10. H.C. Clark, C.A. Fyfe, P.J. Hayes, I. McMahon, J.A. Davies and R.E. Wasylishen, J. Organomet. Chem., **322**, 393 (1987)
11. S. Torroni, G. Innorta, A. Foffani, F. Scagnolari and A. Modelli, J. Mol. Catal., **33**, 37 (1985)
12. A. Patchornik, Y. Ben-David and D. Milstein, J. Chem. Soc., Chem. Commun., 1090 (1990)
13. B. He and W. Li, Yingyong Huaxue, **6**, (2) 52 (1989); Chem. Abstr., **111**, 59944z (1989); B. He. J. Sun and L. Wang, Gaodeng Xuexiao Huaxue Xuebao, **9**, 945 (1988); Chem. Abstr., **110**, 97530y (1989)
14. K. Sato, K. Matuoka, K. Kasano, M. Uemura, K. Kaneda and T. Imanaka, Chem. Express, **3**, 387 (1988); Chem. Abstr., **110**, 114327y (1989)
15. M. Sakai, T. Fujii, T. Kishimoto, K. Suzuki, Y. Sakakibara and N. Uchino, Nippon Kagaku Kaishi, 126 (1986); Chem. Abstr., **106**, 84020z (1987)
16. R.H. Fish, A.D. Thormodsen and H. Heinemann, J. Mol. Catal., **31**, 191 (1985)
17. R.A. Jones and M.H. Seeberger, J. Chem. Soc., Chem. Commun., 373 (1985)
18. J.N. Shah, D. T. Gokak and R.N. Ram, J. Mol. Catal., **60**, 141 (1990)
19. D.T. Gokak, B.V. Kamath and R.N. Ram, React. Polym., **10**, 37 (1989); Chem. Abstr., **112**, 20594f (1990)
20. D.T. Gokak, B.V. Kamath and R.N. Ram, J. Appl. Polym. Sci., **35**, 1523 (1988)
21. G.R. Newkome and A. Yoneda, Makromol. Chem., Rapid Commun., **6**, 451 (1985)
22. Y. Fang. S. Zhang, W. Zhao, Y. Zhang and L. Yang, Lizi Jiaohuan Yu Xifu, **4**, 325 (1988); Chem. Abstr., **111**, 235524y (1989)
23. S.D. Nayak, V. Mahadevan and M. Srinivasan, J. Catal., **92**, 327 (1985)
24. W.D. Bonds, C.H. Brubaker, E.S. Chandrasekaran, C. Gibbons, R.H. Grubbs and L.C. Kroll, J. Am. Chem. Soc., **97**, 2128 (1975)

25. B. Chang, C. Lau, R.H. Grubbs and C.H. Brubaker, J. Organomet. Chem., **281**, 213 (1985)
26. A.K. Zharmagambetova, V.A. Golodov and Y.P. Saltykov, J. Mol. Catal., **55**, 406 (1989)
27. G.R. Newkome and A. Yoneda, Makromol. Chem., Rapid Commun., **6**, 77 (1985)
28. C.R. Saha and S. Bhattacharya, J. Chem. Technol. Biotechnol, **37** (1987) 233; Chem. Abstr., **108**, 55561n (1988)
29. E.A. Karakhanov, A.S. Loktev, V.S. Pshezhetskii and A.G. Dedov, Neftekhimiya, **25** (1985) 176; Chem. Abstr., **103**, 104089x (1985)
30. G. Valentini, G. Sbrana, G. Braca and P. Da Prato, J. Catal., **96**, 41 (1985)
31. J. Mathew and V. Mahadevan, J. Mol. Catal., **60**, 189 (1990)
32. B. Corain, M. Zecca, F.O. Sam, G. Palua and S. Lora, Angew. Chem., Int. Ed. Engl. **29** (1990) 384; Angew. Chem., **102**, 404 (1990)
33. B. Gordon, J.S. Butler and I.R. Harrison, J. Polym. Sci., Part A, Polym. Chem., **25**, 2139 (1987)
34. J.S. Butler, B. Gordon and I.R. Harrison, J. Appl. Polym. Sci., **35**, 1183 (1988)
35. D.E. Bergbreiter and R. Chandran, J. Am. Chem. Soc., **109**, 174 (1987)
36. A.D. Pomogailo and M.V. Klyuev, Bull. Acad. Sci. USSR, Div. Chem. Sci., **34**, 1568 (1985); Izv. Akad. Nauk SSSR, Ser. Khim., 1716 (1985)
37. E. Bayer and W. Schumann, J. Chem. Soc., Chem. Commun., 949 (1986)
38. É.N. Izakovich, A.N. Shupik and Yu.M. Shul'ga, Bull. Acad. Sci. USSR, Div. Chem. Sci., **38**, 1370 (1989); Izv. Akad. Nauk SSSR, Ser. Khim., 1499 (1989)
39. Y.P. Wang and D.C. Neckers, React. Polym., Ion Exch., Sorbents, **3**, 181, 191 (1985); Chem. Abstr., **103**, 71793s., 70994w (1985)
40. V.F. Dovganyuk, V.Z. Sharf, V.K. Belyaeva, I.N. Marov, Zh.L. Dykh, L.I. Lafer and V.I. Yakerson, Bull. Acad. Sci USSR, Div. Chem. Sci., **39**, 217 (1990); Izv. Akad. Nauk SSSR, Ser. Khim., 268 (1990)
41. A. Ueda and S. Nagai, Kagaku to Kogyo (Osaka), **62**, 358 (1988); Chem. Abstr., **110**, 172336f (1989)
42. K. Zheng and M. Zang, Lizi Jiaohuan Yu Xifu, **4**, 425 (1988); Chem. Abstr., **112**, 121016z (1990)
43. B. He and L. Wang, Gaofenzi Xuebao, 404 (1987); Chem. Abstr., **109**, 73046y (1988)
44. S. Bhaduri and K.R. Sharma, J. Chem. Soc., Dalton Trans., 2309 (1984)
45. M.B. Klyuev, T.B. Pogodina and V.D. Kopylova, Russ. J. Phys. Chem., **64**, 428 (1990); Zh. Fiz. Khim., **64**, 809 (1990)
46. V.F. Dovganyuk, L.I. Lafer, V.I. Isaeva, Zh.L. Dykh, V.I. Yakerson and V.Z. Sharf, Bull. Acad. Sci. USSR, Div. Chem. Sci., **36**, 2465 (1987); Izv. Akad. Nauk SSSR, Ser. Khim., 2660 (1987)
47. V. Zbirovský and M. Čapka, Coll. Czech. Chem. Comm., **51**, 836 (1986)
48. I.R. Butler, W.R. Cullen, N.F. Han, F.G. Herring, N.R. Jagannathan and J. Ni, Appl. Organomet. Chem., **2**, 263 (1988); Chem. Abstr., **111**, 154026k (1989)
49. P. Svoboda, J. Hetflejs, W. Schulz and H. Pracejus, React. Kinet. Catal. Lett., **32**, 39 (1986); Chem. Abstr., **107**, 153769k (1987)
50. M.A. Nesterov, L.I. Lafer, Zh.L. Dykh, V.Z. Sharf and V.I. Yakerson, Bull. Acad. Sci. USSR, Div. Chem. Sci., **34**, 21 (1987); Izv. Akad. Nauk SSSR, Ser. Khim., 27 (1987)
51. Y. Chen, C. Xiao, C. Luo and J. Liu, Gaofenzi Tongxun 375 (1986); Chem; Abstr., **106**, 198084d (1987)

52. J.P. Matthew and M. Srinivasan, Chem. Ind. (London) 262 (1990)
53. M. Čapka and A. Reissová, Coll. Czech. Chem. Comm., **54**, 1760 (1989)
54. A. Choplin, B. Besson, L. D'Ornelas, R. Sanchez-Delgado and J.-M. Basset, J. Am. Chem. Soc., **110**, 2783 (1988)
55. J. Kaspar, A. Trovarelli, M. Graziani, C. Dossi, A. Fusi, R. Psaro, R. Ugo, R. Ganzeria and M. Lenarda, J. Mol. Catal., **51**, 181 (1989)
56. V. Zbirovsky, H.J. Kreuzfeld and M. Capka, React. Kinet. Catal. Lett., **29**, 243 (1985); Chem. Abstr., **104**, 231278d (1986)
57. Z. Broucková, M. Czaková and M. Capka, J. Mol. Catal., **30**, 241 (1985)
58. J. Liu, T. Shi, C. Hu and Y. Jiang, Gaofenzi Xuebao, 337 (1988); Chem. Abstr., **113**, 154701t (1990)
59. A.G. Novikova, G.M. Cherkashin, A.I. El'natanova, E.Ya. Mirskaya, O.P. Parenago and V.M. Frolov, Neftekhimiya, **29**, 329 (1980); Chem. Abstr., **112**, 35237r (1990)
60. E.G. Kliger, L.P. Shuikina and V.K. Frolov, Kinet. Katal., **31**, 510 (1990); Chem. Abstr., **113**, 8388 (1990)
61. Y. Xu, S. Zhang, H. Yuan and S. Liao, Yingyong Huaxue, **5**, (5) 71 (1988); Chem. Abstr., **110**, 212049n (1989)
62. T.J. Pinnavaia, in Preparative Chemistry Using Supported Reagents, Ed P. Laszlo, Academic Press 483 (1987)
63. B.M. Choudary, K. Mukkanti and Y.V. Subba Rao, J. Mol. Catal., **48**, 151 (1988)
64. G.V.M. Sharma, E.M. Choudary, M.R. Sarma and K.K. Rao, J. Org. Chem. **54**, 2997 (1989)
65. B.M. Choudary, G.V.M. Sharma and P. Bharathi, Angew. Chem., Int. Ed. Engl., **28** (1980) 465; Angew. Chem., **101**, 506 (1989)
66. Y.V. Subba Rao, K. Mukkanti and B.M. Choudary, J. Mol. Catal., **49**, L47 (1989); K. Mukkanti, Y.V. Subba Rao and B.M. Choudary, Tetrahedron Lett., **30**, 251 (1989)
67. J. Herrero, C. Blanco, M.A. Esteruelas and L.A. Oro, Appl. Organomet. Chem., **4**, 157 (1990); Chem. Abstr., **113**, 190694b (1990)
68. T. Miyazaki, A. Tsuboi, H. Urata, H. Suzuki, Y. Morikawa, Y. Moro-oka and T. Ikawa, Chem. Lett., 793 (1985)
69. S. Shimazu, W. Teramoto, T. Iba, M. Miura and T. Uematsu, Catal. Today, **6**, 141 (1989); Chem, Abstr., **112**, 121037g (1990)
70. N. Ishizuka, M. Togashi, M. Inoue and S Enomoto, Chem. Pharm. Bull., **35**, 1686 (1987)
71. U. Nagel and E Kinzel, J. Chem. Soc., Chem. Commun., 1098 (1986)
72. A. Kinting, H. Krause and M. Čapka, J. Mol. Catal., **33**, 215 (1985)
73. H. Brunner, E. Bielmeier and J. Wiehl, J. Organomet. Chem., **384**, 223 (1990)
74. I. Tóth, B.E. Hanson and M.E. Davies, J. Organomet, Chem., **397**, 109 (1990)
75. R. Deschenaux and J.K. Stille, J. Org. Chem., **50**, 2299 (1985)
76. A. Corma, M. Iglesias, C. del Pino and F. Sánchez, J. Chem. Soc., Chem. Commun., 1253 (1990)
77. Y. Orito, S. Imai and S. Niwa, Nippon Kagaku Kaishi, **8**, 1118 (1979)
78. I.M. Sutherland, A. Ibbotson, R.B. Mayes and P.B. Wells, J. Catal., **125**, 77 (1990); P.A. Maheux, A. Ibbotson and P.B. Wells, J. Catal., **128**, 387 (1991)
79. S. Bhadura, V.S. Darshane, K. Sharma and D. Mukesh, J. Chem. Soc., Chem. Commun., 1738 (1992)

CHAPTER 7

HYDROGEN ACTIVATION IN BIOLOGICAL SYSTEMS

7.1. Introduction

It has been known for almost a century that certain microorganisms consume or produce hydrogen. Activation of molecular hydrogen in biological systems is catalysed by the enzyme hydrogenase. By definition, hydrogenase enzymes (EC class 1.12) are able to catalyse both the forward and reverse reactions of equation *7.1*.

$$H_2 \rightleftarrows 2H^+ + 2e^- \qquad (7.1)$$

Though not a process of physiological importance, the enzyme promotes both the *para/ortho* H_2-conversion, and H/D (or T) exchange reactions, the study of which has proved invaluable in establishing some features of the mechanism of the enzyme catalysed reaction.

In contrast with other enzymes (*e.g.* nitrogenase) the hydrogenases found in various organisms differ widely with respect to their protein structure and the nature of their electron carriers. There is a substantial research effort in achieving their isolation, purification and characterization. In addition, studies of the mechanism of H_2-activation by hydrogenase date back for more than 40 years. The research work gained a new impetus in 1980 when it was first shown that certain hydrogenases contain nickel at their active center. More recent work suggests that this is a rule rather than an exception, although there are also Ni-free hydrogenases.

Independent of their origin, the purified enzymes catalyse both the forward and the reverse reactions. However, their physiological role usually requires catalysis of either the production or the consumption of H_2. Thus the terms "uptake", "hydrogen-producing" and "reversible" or "bidirectional" hydrogenases are widely used.

The purpose of this Chapter is to give a short introduction to the

P.A. Chaloner, M.A. Esteruelas, F. Joó and L.A. Oro, Homogeneous Hydrogenation 255–270.

chemistry and biochemistry of hydrogenase. However, since this enzyme can be found in so many different microorganisms, assisting in many diverse functions of the cell, it is beyond the scope of this account to give a detailed description of all the important aspects of research in this field. Instead, a few specific examples have been chosen. Fortunately, there are several general and some more specialized reviews which can be consulted for further details [1–4].

7.2. Physiological function of hydrogenase

In a very broad sense, hydrogenase provides reducing power for the proper functioning of the cell by oxidizing H_2, or eliminates unneeded reducing power by reduction of H^+. Electrons produced in H_2-oxidation or required for H^+-reduction are transported between the enzyme and the rest of the relevant cell components by the primary electron carriers, such as quinones, cytochrome c_3, or ferredoxin, rubredoxin, etc. Reduction of nicotinamide pyridine nucleotides (NAD) with hydrogen is an interesting reaction catalysed by a special class of hydrogenases which contain flavin covalently bound to the protein. The reducing power may eventually be used for reduction of such terminal acceptors as CO_2, E-2-butene dioate, or inorganic ions, such as $[SO_4]^{2-}$, and $[NO_3]^-$. Hydrogen in the reduced product(s) is derived from the aqueous solvent. The transport of electrons may be mediated by a variety of secondary electron carriers and other enzymes, but we do not discuss hydrogen usage in living organisms in any detail here. A selected list of microorganisms, together with the normal substrates for their hydrogenase and the reactions catalysed is collected in Table *7.1*.

It is appropriate to mention here that hydrogenases catalyse the reduction of such artificial electron acceptors as viologens and methylene

TABLE 7.1
Selected reductions catalysed by hydrogenases from different sources (after [1])

Reaction	Source of hydrogenase
$[SO_4]^{2-} + 4H_2 = S^{2-} + 4H_2O$	*Desulfovibrio*
$CO_2 + 4H_2 = CH_4 + 2H_2O$	*Methanobacterium*
$2CO_2 + 4H_2 = CH_3COOH + 2H_2O$	*Acetobacterium*
$S + H_2 = H_2S$	*Campylobacter*
$2\ [NO_3]^- + 5H_2 + 2H^+ = N_2 + 6H_2O$	*Paracoccus denitrificans*

blue. This serves as basis for methods of assay for the enzymes and also for systems for photoproduction of H_2 (see later).

Hydrogenase enzymes may be employed to assist in the maintenance of the energy balance of the cell. Thus energy is supplied to *Methanobacteria* by reduction of CO_2 to methane, and a related example is provided by reduction of O_2 to water in the so called "knall-gas" bacteria (aerobic H_2-oxidizing bacteria). Some species are able to use CO_2 as the sole carbon-source and H_2 as the sole electron donor, and in these cases hydrogenases evidently play a very important role in metabolism and growth of the cells.

An interesting example of efficiency is provided by aerobic nitrogen fixing bacteria. It is known that nitrogenase, either in the living root nodule bacteria, or in cell-free preparations, generates H_2 as well as reducing N_2, and the overall process can be formulated by equation 7.2.

$$N_2 + 8e^- + 8H^+ + 16ATP \longrightarrow 2NH_3 + 16ADP + 16P_i + H_2 \quad (7.2)$$

As written, this equation accounts for a 25% waste of ATP as molecular hydrogen, but measured losses are usually higher than 30%. In the absence of a suitable substrate (N_2 or ethyne) all the ATP in the cell free extracts could be expended in generation of hydrogen. However, H_2 inhibits nitrogenase, and though this may be beneficial under N_2-free conditions, nitrogen fixation is hindered by the evolution of H_2. The presence of an "uptake" hydrogenase in certain strains of *Azotobacter* and *Rhizobium* leads to an efficient recycling of H_2, generated by nitrogenase. The energy obtained in the oxyhydrogen ("knall-gas") reaction is used for ATP synthesis, and this reaction also effectively removes O_2 from the immediate vicinity of the very oxygen-sensitive nitrogenase. Comparison of crops of leguminous plants infected with certain *Rhizobium* strains having and lacking, respectively, hydrogenase activity, showed that hydrogenase had a marked beneficial effect, an observation of clear importance in agriculture.

The above example also relates to the physiological role of H_2-production. It was found that, under illumination with visible light, the purple non-sulphur bacterium *Rhodospirillium rubrum* evolved H_2 even in a pure H_2-atmosphere, or under noble gases, but not under N_2. It follows that the ATP produced in excess by photophosphorylation is removed *via* H_2-formation in the absence of the N_2 substrate. Although in this example H_2 was produced by nitrogenase, hydrogenase may play the same role in maintaining the internal redox balance of the cells.

Maintaining a balance requires the accessibility of both forward and reverse reactions. However, instead of having one "reversible" hydrogenase, several species contain two "unidirectional" hydrogenases for the production and uptake of H_2, respectively. The reason may derive from the fact that cells rely on production or consumption of hydrogen under different conditions (*e.g.* of pH), and the two enzymes may have different optimum pH's for their operation. The "uptake" and "hydrogen producing" hydrogenases found in the same organism are spatially separated on the opposite sides of the cell membrane.

Most hydrogenases are membrane bound enzymes. If coupled to a vectorially organized electron transport chain, they can create a transmembrane electrochemical potential and proton gradient [5].

7.3. Properties of hydrogenase

All hydrogenases are iron-sulfur proteins, containing iron in form of $\{Fe_4S_4\}$ clusters, bound to the protein by S-coordination of cysteine residues. Some of them, such as the soluble hydrogenase of *Alcaligenes eutrophus* also contain $\{Fe_2S_2\}$ clusters. A number of Ni-containing hydrogenases show, in the oxidized state, an EPR signal due to $\{Fe_3S_4\}$ clusters [6]. Selenium was also detected in some Ni-hydrogenases.

Most of the hydrogenases are built up of subunits, in many cases with a larger unit of about 60 kDa molar mass, and a smaller one of approx. 30 kDa molar mass. However, there are many exceptions, with regard both to the number of subunits, and to the mass of the enzyme, which can be as high as 900 kDa. Although Ni is generally found in the *active centre* of the enzyme, in the case of the aerobic H_2-oxidizing bacterium, *Nocardia opaca*, some of the nickel plays an important role in binding the four subunits together [7]. In the absence of Ni the hydrogenase complex dissociates into two major parts each consisting of a larger and a smaller subunit.

The case of the hydrogenase of *N. opaca* provides a good illustration of the difficulties of working with hydrogenase. This enzyme is located in the cytoplasm and (rather exceptionally) directly reduces NAD [6]. If, however, under Ni-free conditions the enzyme dissociates, none of the resulting, still dimeric, units retains any activity for NAD-reduction. On the other hand, the smaller dimer, containing 2 Ni atoms and 1 $\{Fe_4S_4\}$ cluster per dimer, remains highly active in the usual hydrogenase assay, in the reduction of artificial electron carriers, such as viologens. Most importantly, however, mixing of the two dimers reconstitutes the full NAD-reducing activity. These observations suggest the possibility of specific binding sites for natural electron acceptors being on different subunits of the enzyme. It may well happen that during isolation and purification such subunits, which themselves are inactive in hydrogenase assays, are lost. Another severe problem when working with hydrogenases is in that they are generally very sensitive to oxygen, and complete or partial impairment of catalytic activity, together with protein structural changes [8] and loss of subunits may ensue if purification is not done under stringent anaerobic conditions.

The isoelectric point of hydrogenases may vary widely between 4.2 and 6.6. In the usual assays (hydrogen uptake in the catalysed reduction of methyl viologen or methylene blue, or hydrogen evolution from dithionite-reduced methyl viologen) there is a rather sharp maximum in the variation of the catalytic activity with pH. However, in addition to changing the $[H^+]$ concentration which serves either as substrate or product of hydrogenase activity, changes in pH may also result in protonation/deprotonation of the protein and a change in the redox potential of the dithionite. Therefore it is not easy to establish the relationship between activity and pH, and the microscopic events of the mechanism of H_2-activation at the active site of the enzyme. In support of this view one may cite the observation that several hydrogenases have their optimum pH for H_2-*uptake* in acidic media [9].

Purified hydrogenases are rather stable to heat, and some of them retain activity at temperatures up to 80 °C. Stability, including thermal stability, can be improved by addition of reagents such as glycerol or bovine serum albumin, or by immobilization of the enzyme in solid matrices.

7.4. The active site of hydrogenase

Most of our knowledge of the active site of hydrogenase is derived from studies of core-extrusion [10] (isolation of the iron-sulfur clusters in form of $[Fe_nS_n(SR)_n]$, by replacing cysteinyl residues of the protein by thiols), and investigation of spectroscopic properties of the enzyme, principally by EPR, Mössbauer, and increasingly by EXAFS, ENDOR, and electron spin echo envelope modulation (ESEEM) spectroscopy [11–13]. Interestingly, the first hint of the possible presence of Ni(III) in hydrogenase, or in coupled cofactors came also from EPR measurements on soluble and membrane fractions of *Methanobacterium bryantii* [14].

One of the best characterized groups of hydrogenases are those of the enzymes found in the sulfate reducing *Desulfovibrio sp.* [15]. This group also exemplifies the complexity encountered in the biochemistry of hydrogenases. Three different types of the enzyme are known within this group.

1. Fe hydrogenases. The enzyme purified from *Desulfovibrio vulgaris* contains only iron-sulfur centers, namely two $\{Fe_4S_4\}$ clusters, and a third one, considered atypical, which has been proposed as the site of H_2 oxidation and H_2 production.

2. NiFe hydrogenases. The enzyme purified, *inter alia*, from *D. gigas*, contains one nickel center and various iron-sulfur centers, generally two $\{Fe_4S_4\}$ and one $\{Fe_3S_x\}$ clusters.

3. NiFeSe hydrogenases. The enzyme, purified, *inter alia*, from *D. desulfuricans* (Norway strain) showed the presence of equivalent amounts of Ni and Se in addition to the two $\{Fe_4S_4\}$ clusters.

From spectroscopic studies it has been deduced that Ni-containing hydrogenases in their oxidized form contain a low-spin Ni(III) center, surrounded by at least two sulfur atoms and one nitrogen atom in rhombically distorted octahedral coordination. In NiFeSe hydrogenases selenium replaces one of the sulfur atoms.

Hydrogenases, especially those obtained from strictly anaerobic microorganisms (*e.g.* the sulfate reducing bacteria) are sensitive to oxygen, and the enzymes are not catalytically active in their oxidized state. In the *D. gigas* hydrogenase [2], as isolated, the Ni center is characterized by EPR signals at g = 2.31, 2.23 and 2.02 (Ni-A signal) and another series at g = 2.32, 2.16 and 2.02 (Ni-B signal). On prolonged treatment under hydrogen, these signals are replaced by a new series with values of g = 2.19, 2.14 and 2.02 (Ni-C signal). In conjunction with Mössbauer spectroscopic studies the EPR signal at g = 2.02 is assigned to an $\{Fe_3S_x\}^{3+}$ cluster. The Ni-C form of the enzyme is fully active, while Ni-A is inactive. The Ni-B state can be reached from Ni-C with mild oxidants, and converted back to the fully active form (as measured in the viologen assay) with only a short induction period on treatment with reductants, including also H_2. The three forms were termed the unready (Ni-A), ready (Ni-B) and active (Ni-C) states of the enzyme [16]. Both the unready and ready states appear to involve an Ni(III) paramagnetic centre. (It is also possible that the formation of the unready state of *D. gigas* hydrogenase is an artefact, due to the isolation procedure in air.) The "ready" enzyme has an activated protein configuration but with a masked H_2-binding site, since it does not readily react with H_2, without prior reduction to the active state by strong reductants, such as reduced viologens. Additionally, the "ready" form is unable to catalyse the D_2/H^+ exchange. The oxidation state of Ni in the active form has been subject of controversy, and both Ni(I) and Ni(III) were considered. It is important to note that the midpoint redox potential of the Ni-A signal is pH-dependent, changing by –60 mV/pH unit. This indicates that reduction of the Ni-centre is associated with protonation of the enzyme. In several cases, however, the active enzyme was found to be EPR-silent, and this was interpreted in terms of a Ni(II) oxidation state at the active site, or of strong magnetic coupling of Ni(III) to an $\{Fe_4S_4\}$ cluster.

Interestingly, the "Fe-only" hydrogenases, such as those purified from *D. vulgaris* and the anaerobic N_2-fixing bacterium *Clostridium pasteurianum* are amongst the most active hydrogenases known. For example, the hydrogenase from *D. vulgaris* catalysed hydrogen evolution from reduced methyl viologen with an activity [17] of 10,400 μmol H_2/min/mg, whilst the activity of the enzyme from *D. gigas* was 440 μmol H_2/min/mg protein [18]. Based on the results of EPR, Mössbauer, magnetic circular dichroism, and ^{57}Fe and ^{1}H ENDOR spectroscopy, it has been suggested that the enzyme from *D. vulgaris* may contain an $\{Fe_6\}$ center in addition to the usual iron-sulfur clusters. It is believed that the $\{Fe_6\}$ cluster plays the same role as nickel in Ni-containing hydrogenases. They are the sites for hydrogen activation, whilst the conventional $\{Fe_4S_4\}$ clusters function as electron carriers.

As mentioned earlier in connection with *Nocardia opaca*, hydrogenases have specific binding sites for primary electron acceptors (carriers). Such a "ping-pong mechanism" was suggested [19] for the hydrogenase from *Rhizobium japonicum*; according to this H_2 is reversibly activated at one site, whilst an electron carrier interacts at a second site. Thus carbon monoxide is a competitive inhibitor of H_2-oxidation, though it does not necessarily compete with reduced electron carriers. As with the hydrogenase from *Nocardia*, that from *Alcaligenes eutrophus* contains a "built-in" primary acceptor, *viz.* 1 flavin mononucleotide per 2 Ni atoms. Most interestingly, this enzyme is insensitive to CO. This suggests that the H-binding site in this enzyme, unlike that in other hydrogenases, is not able to bind CO.

7.5. Hydrogen activation by hydrogenases

Most of our understanding of the mechanism of H_2-activation at the active site of hydrogenases comes from studies of either *para-* to *ortho-*hydrogen conversion or H_2/D^+ (or D_2/H^+) exchange reactions.

Krasna and Rittenberg [20] have shown that a partially purified enzyme preparation from *Proteus vulgaris* actively catalysed the *para/ortho* conversion when the solvent was H_2O, but the reaction was almost halted in D_2O. The exchange reaction between H_2 and D_2O was catalysed approximately twice as efficiently as *para/ortho* conversion. The kinetics of the two processes was found to be essentially similar. They concluded, that their data could be rationalized in terms of a heterolytic hydrogen activation (equation *7.3*) at variances with earlier suggestions of a mechanism involving homolysis of dihydrogen.

$$E + H_2 \rightleftharpoons EH^- + H^+ \qquad (7.3)$$

The reaction of EH^- with H^+ explains the *para/ortho* conversion, whilst that with D^+ leads to exchange. However, there are more subtle experimental observations which need clarification.

If one assumes a fast equilibration of H^+, produced in the reaction, with D^+ in the solvent, then during the initial phase of the reaction exclusive formation of HD is expected, followed in the later stages by formation of D_2 *via* exchange of HD with D^+. However, this is not generally the case, and formation of some D_2 is observed, even when the concentration of the enzyme approaches zero (or by extrapolation). A possible explanation to this finding is, that the enzyme also has a binding site for the proton released in the heterolytic step, and both the bound H^- and H^+ can be separately exchanged, albeit at different rates, before recombination to give molecular hydrogen. This conclusion was also derived from experiments on D_2/H^+ exchange with enzymes from three different species of *Desulfovibrio* [21], using membrane inlet mass spectrometry for precise measurement of the dissolved gases. Interestingly, in one case, the pH optima for the evolution of HD and H_2 were significantly different (3 and 4.5, respectively).

The assumption of the existence of two binding sites also explains the reduced rate of *para/ortho* conversion in D_2O. An efficient conversion in D_2O would require that the conversion rate in H_2O be faster than H_2/D^+ exchange, or hydrogen would be replaced by deuterium before recombination. This was not the case in the above experiments with *Proteus vulgaris* hydrogenase, but with a purified enzyme from *Desulfovibrio vulgaris* significant *para/ortho* conversion was observed both in H_2O and D_2O.

The activation energy for oxidation of H_2 to $2H^+$ has been determined from Arrhenius plots of enzyme activity for various hydrogenases; E_a falls in the range 25–90 kJ mol^{-1} [1, 22].

From the data available in the literature, it is impossible to establish definitively the nature and sequence of the elementary steps of hydrogen activation at the active site. However, recent developments in the coordination chemistry of H_2 may shed some light on these elementary processes. As discussed in Chapter 2, η^2-coordination of dihydrogen is much more widespread as was believed even ten years ago. It is also well documented that these H_2 complexes are rather acidic and can dissociate when the process is assisted by an appropriate base or solvent. It was suggested by Crabtree [23], that nickel containing hydrogenase may operate *via* formation of dihydrogen complexes, followed by deprotonation as shown in Scheme *7.4*. This suggestion is supported by the observation that there are known dihydrogen complexes capable of catalyzing H/D exchange between H_2 and D_2O. Moreover, this kind of H_2-activation does not require oxidation of the metal (in contrast to oxidative additions) and therefore may be particularly appropriate for metals in high oxidation state, such as Ni(III).

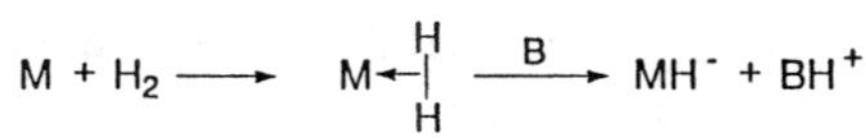

Scheme 7.4.

7.6. Chemical models for hydrogenase

Synthetic models of enzymes can be evaluated from various viewpoints, with regard to their composition and structure, and more importantly, similarities to particular characteristics of the enzymatic function. Considering the diverse nature of the hydrogenases, no model will possess all the important features, but individual models may reflect certain typical properties.

It has recently been reported by Crabtree and co-workers [24], that the Ni(II) complex, $[NiL_2]Cl_2$, where L = 2-HOC_6H_4CH = N-$NHCSNH_2$ could be reduced either electrochemically or with $NaBH_4$. The ESR spectrum of the reduced complex (g = 2.25, 2.12 and 2.06) is consistent with an axially distorted Ni(I) species. This is the first example of sulfur-coordinated Ni(I), and the compound was found to be rather stable to air. Most interestingly, $[NiL_2]Cl_2$ catalysed the D_2/H exchange with EtOH at 25 °C and 1 bar pressure of D_2, especially in the presence of promoters such as HI or $H[BF_4]$. It was established that in solution the phenolic hydroxide is not coordinated to the metal. This observation led to the suggestion that hydrogen bonding between the coordinated

H_2 and the free OH-groups in the ligand promotes the activation of hydrogen. The structure of the proposed catalytic intermediate is shown on Scheme *7.5*.

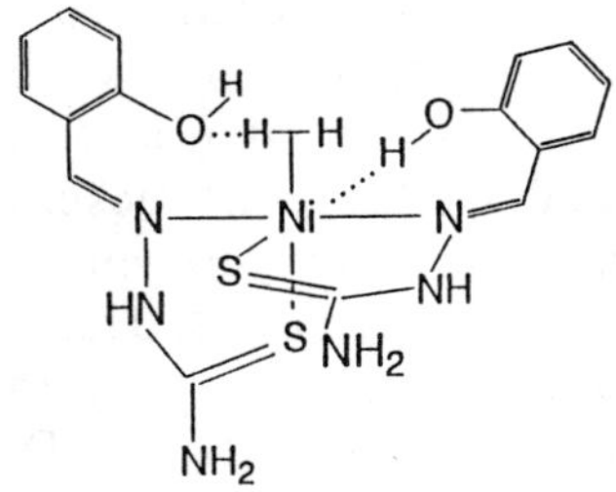

Scheme 7.5.

A different approach was taken by Vizi-Orosz and Markó, who investigated the reduction of aromatic nitro compounds by hydrogen transfer from hydrazine, catalysed by *Ni(III)* complexes [25] of the formula $[Bu_4N][Ni(dt)_2]$, dt = toluene-3,4-dithiolate or benzene-1,2-dithiolate. In refluxing tetrahydrofuran solutions the substrates were effectively reduced to anilines, although in some cases the intermediate hydroxylamines could be isolated (or even made the major product) (Table *7.2*). The catalyst mimics a possible composition of the enzyme active site as sulfur-coordinated Ni(III), but the conditions under which it operates are very far from being biocompatible.

Khidekel' and co-workers attempted the construction of hydrogenase models [26, 27] by using ligands with extended π-conjugation, such as indigosulfonates or Alizarin Red in complexes of Rh, Pt or Pd. In aqueous solutions and at ambient temperature these complexes readily promote hydrogenation of nitro compounds, catalyse D_2/H^+ exchange, and their reactivity is inhibited by CO. Although the detailed mechanistic chemistry leading to these observations is not always clear, it could be established that during activation of the catalyst the ligands are partially reduced to give radical species, and the electrons from H_2-oxidation are transferrred to the substrates *via* the reduced ligands (see also section 5.3.2.3). Thus the ligands mimic the role of electron carriers in hydrogenase mediated reductions.

It has long been known that in aqueous hydrochloric acid solutions the complexes $[RuCl_2(H_2O)_{6-n}]^{(n-3)-}$ activate H_2 heterolytically, and as well as reduction also catalyse D_2/H^+ exchange [28]. Interestingly, in this exchange reaction, the initial H_2/HD ratio is non-zero (0.3) (*cf.* section

TABLE 7.2
Hydrogenation of aromatic nitro compounds[a] by hydrogen transfer from $N_2H_4.H_2O$, catalysed by Ni(III) complexes, **1** and **2** [25]

Substrate (S)	Cat.	Reaction time (h)	Conv. (%)	Product (yield, %)	
nitrobenzene	**1**	5	42	aniline	(42)
4-nitrotoluene	**1**	5	0.5	4-Me-aniline	(0.5)
4-nitrochlorobenzene	**1**	5	20	4-chloroaniline	(20)
1,3-dinitrobenzene	**1**	5	100	3-nitroaniline	(100)
1,3-dinitrobenzene	**2**	4	100	3-nitroaniline	(100)
1,3-dinitrobenzene[b]	**1**	15	100	3-nitroaniline	(2.5)
				N-(3-nitrophenyl)HA	(97.5)
1,4-dinitrobenzene[c]	**1**	3	100	4-nitroaniline	(4)
				N-(4-nitrophenyl)HA	(96)
1,4-dinitrobenzene[c]	**1**	5	100	4-nitroaniline	(95)
				N-(4-nitrophenyl)HA	(5)

HA = hydroxylamine
1 = $[Bu_4N][Ni(tdt)_2]$; tdt=toluene-3,4-dithiolate
2 = $[Bu_4N][Ni(tdt)_2]$; bdt=toluene-1,2-dithiolate
[a] Conditions: 0.1 mmol catalyst, 10 ml THF, 65 °C, [S]/[Cat.] = 5
[b] 40 °C
[c] [S]/[Cat.] = 50

7.5). This system, therefore, has a great deal of similarity to hydrogenase, although the reaction conditions are rather severe (see also section 5.3.1.1 and Scheme *5.9*).

The ruthenium-octaethylporphyrin complex, $[Ru(OEP)(THF)_2]$, heterolytically activated H_2 in the presence of a base like KOH [38]. The resultant anionic monohydride, K[HRu(OEP)(THF)] reduced the NAD^+ analogue, [1-benzyl-*N,N*-diethylnicotinamide]$[PF_6]$ to yield the corresponding 1,6-dihydronicotinamide. This hydrogen transfer had to be induced by adding small amounts of pyridine to the reaction mixture in THF, *i.e.* by replacing the axial THF with a ligand of larger trans effect. Studies of protonation of the hydride revealed that under the particular experimental conditions bimolecular H_2-elimination played no important role, and that in the presence of D_2O there was some H/D exchange, but it was considerably slower than hydrogen elimination.

Detailed kinetic investigation [29] of the complex [Pd(II)(SALEN)], where SALEN = N,N′-ethylenebis(salicylidene-imine), suggested close similarity between catalytic activity of this complex and the action of hydrogenase. Since the complex is poorly soluble in alcohols, some of its properties were investigated in alcoholic suspensions. In such systems it catalysed H/D exchange both in EtOD with H_2 (some D_2 was observed together with the major product HD), and in EtOH with D_2. The rate of 1-hexene hydrogenation passed through a maximum with increasing concentration of NaOH in EtOH. These observations, together with kinetic measurements in homogeneous N,N′-dimethylmethanamide solutions led to the conclusion that H_2 was activated heterolytically and that the activation step was assisted by protonation of one of the oxygen atoms of the ligand (Scheme *7.6*).

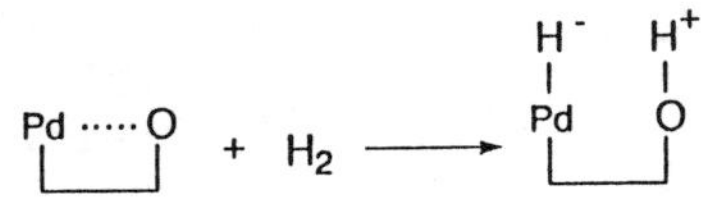

Scheme 7.6.

As was reported recently, the catalytically active species may not be the monomeric square-planar [Pd(SALEN)] complex, as was originally thought, but rather an oligomeric derivative of unknown structure [30].

7.7. Practical applications of hydrogenase

In principle, hydrogenase could be used for various purposes, such as hydrogen generation, as a component of fuel cells, or in biotechnological processes for reduction of a range of substrates [31]. However, despite extensive research, none of these possibilities has thus far matured to a level exceeding that of small scale demonstrations. This is partly due to the relative instability of hydrogenase itself, or of other components of the reaction mixture, and partly to economic considerations given the more acceptable costs of alternative processes.

Much effort has been expended in studying the use of hydrogenases as catalysts for H_2 evolution in photochemical systems [3]. Thus the hydrogenases from *Desulfovibrio vulgaris* (strain Miyazaki), *D. desulfuricans* (strain Norway 4), *D. baculatus* 9974 were used in conjuction with various photosensitizers (*e.g.* $[Ru(bipy)]^{2+}$, Zn-TPPS$_4$, Zn-PcTS, and semiconductors, such as TiO_2) and with various electron donors (EDTA, ascorbate, *tris*(2-hydroxyethyl)amine, cysteine, thiols, methanol, etc.) [32, 33]. As a non-sacrificial electron donor, NAD(P)H, continuously regenerated by photoreduction at the grana from green plants, was also studied. This latter case is an example of using H_2O as the ultimate source of hydrogen, through the photochemically assisted splitting of water by the action of chloroplasts. The reader is referred to the many reviews in this field [1, 3, 4] for further details.

All these variations of attempted H_2-production have several drawbacks. Many of the photosensitizers, particularly chloroplasts, are prone to degradation, though the latter showed increased stability (but decreased activity) when immobilized in various matrices. When H_2O is split then O_2- and H_2-evolution must be strictly compartmentalized to avoid recombination, and because the enzymes involved in H_2-generation are highly oxygen sensitive. The evolution of molecular oxygen can be circumvented by using sacrificial electron donors, but the latter should be extremely cheap to render the process economically viable (*e.g.* available as wastes). In this respect the use of intact photosynthetic bacteria, such as *Rhodospirillium rubrum*, grown on lactic acid-containing wastes show considerable promise [4]. Ironically, however, with regard to the topic of this book, in this latter case hydrogen is evolved by the nitrogenase of the microorganism.

The catalytic reduction of NAD with hydrogen by certain hydrogenases (such as the soluble hydrogenase from *Alcaligenes eutrophus*) allows the use of some NAD-dependent enzymes for organic transformations in a coupled system. Combining H_2, hydrogenase, NAD, and alcohol dehydrogenase led to the reduction of ketones [34] (2-butanone, 2-pentanone and cyclohexanone) to alcohols with small but constant activity. After 65 h at 30 °C, the turnover number of NAD in 2-butanone hydrogenation was 64, corresponding to the reduction of 87% of the ketone present (Scheme 7.7). Similar reactions, but utilizing synthetic catalysts in certain steps of hydrogen activation or electron transport are treated in section 5.7.

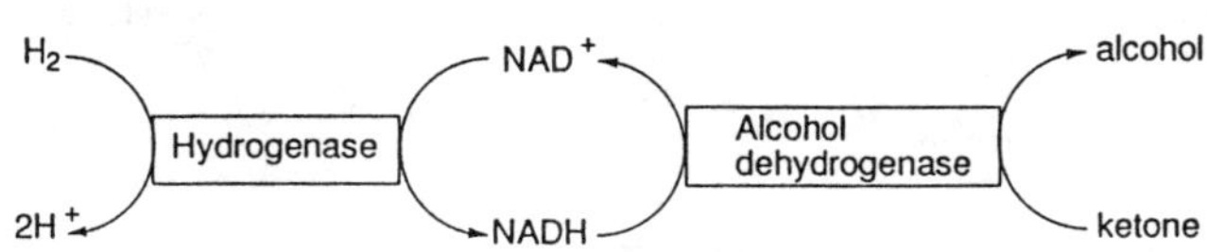

Scheme 7.7.

The reduction of ketones was investigated using a combination of whole cells [35] of the *Halobacterium halobium* strain MMT_{22}, and *Escherichia coli*. Photoactivation of the bacteriorhodopsin present within the purple membrane of *H. halobium* results in proton transfer from the interior of the cells to the external space. These protons are used for efficient hydrogen generation by the hydrogenases of *E. coli* in the absence of a reducible substrate. However, in the presence of cyclohexanone, reduction of the ketone takes place, first to cyclohexanol and later to cyclohexane. Obviously, other enzymes, such as alcohol dehydrogenase should be involved in the process, which is rather efficient, giving 75 μmol alcohol per mg wet weight of the bacteria per minute. Unfortunately, the product cyclohexane is toxic to the cells.

Biotechnological production of fine chemicals [36] has been attempted using a reverse micellar medium to separate the aqueous solution of enzymes and mediators from the substrate. In such a system (Scheme 7.8) progresterone was reduced to 20β-hydroxy-pregn-4-en-3-one. In the reaction mixture, containing hydrogenase from *A. eutrophus*, NAD was reduced directly by the hydrogenase, and used for reduction of progresterone by the enzyme 20β-hydroxysteroid dehydrogenase (HSDH) from *Streptomyces hydrogenans*. In long term experiments turnover numbers for NADH regeneration as high as 500 could be achieved. In the case of hydrogenases incapable of direct NAD reduction(such as the one from *D. vulgaris*) methylviologen was used as the primary electron acceptor, and reduction of NAD was mediated by pigheart lipoamide dehydrogenase (LipDH). However, these more complicated systems gave lower conversion of the steroid and showed lower long-term stability.

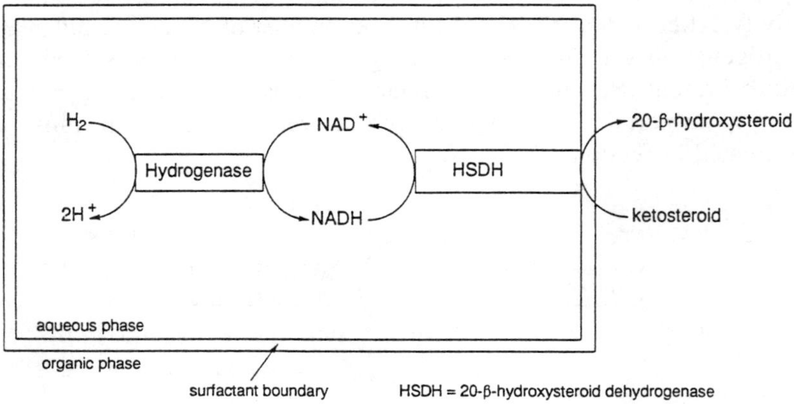

Scheme 7.8.

Although attempts to construct cell-free systems for using hydrogenase will undoubtedly result in efficient catalytic transformations, at present, in most cases, enzyme chemistry is better "done" by Nature. This is also true for the biotransformation of steroids [37].

References

1. M.W.W. Adams, L.E. Mortenson and J.S. Chen, Biochim. Biophys. Acta, **594**, 105 (1981)
2. J.J.G. Moura, I. Moura, M. Teixeira, A.V. Xavier, G.D. Fauque and J. LeGall, in *Metal Ions in Biological Systems*, H. Sigel, Ed., Marcel Dekker, New York and Basel, 1988, Vol. 4, p. 285
3. R. Cammack, D.O. Hall and K.K. Rao, in *Microbial Gas Metabolism*, R.K. Poole and C.S. Dow, Eds., Academic Press, London, 1985, p. 75
4. a) P.M. Vignais, A. Colbeau, J.C. Willison and Y. Jouanneau, in *Adv. Microbial Physiol.*, Academic Press, London, 1985, Vol. 26, p. 155. b) R. Cammack, V.M. Fernandez and K. Schneider, in *The Bioinorganic Chemistry of Nickel*, J.R. Lancaster, Ed., VCH Publishers, Deerfield Beach, Florida, 1988, p. 167
5. a) K.L. Kovács, Z. Szőkefalvi-Nagy, I. Demeter, and Cs. Bagyinka, Proc. 3rd Int. Conf. Molec. Biol. Hydrogenases (Tróia, Portugal, 1991), p. 196. b)K.L. Kovács and Cs. Bagyinka, FEMS Microbiol. Rev., *87*, 407 (1990), and references therein
6. J.J. Moura, M. Teixeira, I. Moura, A.V. Xavier and J. LeGall, J. Mol. Catal., **23**, 303 (1984)
7. S. Hornhardt, K. Schneider and H.G. Schlegel, Biochimie, **68**, 15 (1986)
8. G. Tigyi, Cs. Bagyinka and K.L. Kovács, Biochimie, **68**, 69 (1986)
9. P. A. Lespinat, Y. Berlier, G. Fauque, M. Czechowski, B. Dimon and J. LeGall, Biochimie, **68**, 55 (1986)
10. W.O. Gillum, L.E. Mortenson, J.S. Chen and R.H. Holm, J. Am. Chem. Soc., **99**, 584 (1977)
11. R.A. Scott, S.A. Wallin, M. Czechowski, D.V. DerVartanian, J. LeGall, H.D. Peck, Jr. and I. Moura, J. Am. Chem. Soc., **106**, 6864 (1984)
12. P.A. Lindahl, N. Kojima, R.P. Hausinger, J.A. Fox, B.K. Teo, C.T. Walsh and W.H. Orme-Johnson, J. Am. Chem. Soc., **106**, 3062 (1984)
13. S.L. Tan, J.A. Fox, N. Kojima, C.T. Walsh and W.H. Orme-Johnson, J. Am. Chem. Soc., **106**, 3064 (1984)
14. J.R. Lancaster, Jr., FEBS Lett., **115**, 285 (1980)
15. G. Fauque, H.D. Peck, J.J. Moura, B.H. Huynh, Y. Berlier, D.V. DerVartanian, M. Teixeira, A.E. Przybyla, P. A. Lespinat, I. Moura and J. LeGall, FEMS Microbiol. Rev., **54**, 299 (1988)
16. R. Cammack, V.M. Fernandez and K. Schneider, Biochimie, **68**, 85 (1986)
17. M.W.W. Adams, M.K. Johnson, I.C. Zambrano and L.E. Mortenson, Biochimie, **68**, 35 (1986)
18. R. Cammack, D.S. Patil, A.C. Hatchikian and V.M. Fernandez, Biochim. Biophys. Acta, **912**, 98 (1987)
19. D.J. Arp and R.H. Burris, Biochemistry, **20**, 2234 (1981)
20. A.I. Krasna and D. Rittenberg, J. Am. Chem. Soc., **76**, 3015 (1954)
21. D. Lloyd and R.I. Scott, in *Microbial Gas Metabolism*, R.K. Poole and C.S. Dow, Eds., Academic Press, London, 1985, p. 247
22. V.M. Fernandez, E.C. Hatchikian and R. Cammack, Biochim. Biophys. Acta, **832**, 69 (1985)
23. R.H. Crabtree, Inorg. Chim. Acta, **125**, L7 (1986)
24. M. Zimmer, G. Schulte, X.L. Luo and R.H. Crabtree, Angew. Chem., Int. Ed. Engl., **30**, 193 (1991); Angew. Chem., **103**, 205 (1990)
25. A. Vizi-Orosz and L. Markó, Transition Met. Chem., **13**, 221 (1988)

26. E.G. Chepaikin and M.L. Khidekel', J. Mol. Catal., **4**, 103 (1978)
27. Yu.A. Sakharovskii, M.B. Rozenkevich, A.S. Lobach, E.G. Chepaikin and M.L. Khidekel', React. Kinet. Catal. Lett., **8**, 249 (1978)
28. B.R. James, in *Comprehensive Organometallic Chemistry*, G. Wilkinson, F.G.A. Stone and E.W. Abel, Eds., Pergamon Press, Oxford, 1982, Vol. 8, p. 330
29. G. Henrici-Olivé and S. Olivé J. Mol. Catal., **1**, 121 (1975/76)
30. J.M. Kerr, C.J. Suckling and P. Bamfield, J. Chem. Soc., Perkin Trans. 1, 887 (1990)
31. a) M. Klibanov, Process Biochem., **8**, 13 (1983). b) I.N. Gogotov, Biochimie, **68**, 181 (1986)
32. I. Okura, Biochimie, **68**, 189 (1986)
33. P. Cuendet, K.K. Rao, M. Gratzel and D.O. Hall, Biochimie, **68**, 217 (1986)
34. K. Otsuka, S. Aono, I. Okura and H. Hasumi, J. Mol. Catal., **51**, 35 (1989)
35. M.M. Taqui Khan and P.J. Bhatt, J. Mol. Catal., **63**, L15 (1990)
36. A. van Berkel-Arts, M. Dekker, C. van Dijk, H.J. Grande, W.R. Hagen, R. Hilhorst, M. Kruse-Wolters, C. Laane and C. Veeger, Biochimie, **68**, 201 (1986)
37. A. Szentirmai, D.Sc. Thesis, Hungarian Academy of Sciences, Budapest, 1980
38. J.P. Collman, P.S. Wagenknecht, and N.S. Lewis, J. Am. Chem. Soc., **114**, 5665 (1992)

LIST OF THE ABBREVIATIONS

acacH	=	pentane-2,4-dione
ADP	=	adenosine diphosphate
AlaH	=	alanine, $CH_3CH(NH_2)COOH$
Aliquat-336	=	trioctylmethylammonium chloride
Alizarin Red	=	sodium 1,2-dihydroxy-9,10-anthraquinone-3-sulfonate
AMPHOS	=	2-diphenylphosphinoethylammonium ion
An	=	9-methylanthracene
o-An	=	2-methoxyphenyl
Ar	=	aryl
ATP	=	adenosine triphosphate
BCPM	=	2**S**,4**S**-*N*-*t*-butoxycarbonyl-4-dicyclohexylphosphino-2-diphenylphosphinomethylpyrrolidine
BDPP	=	2,4-bis(diphenylphosphino)pentane
BDPPODP	=	1**R**,3**R**-bis(diphenylphosphinoxy)-1,3-diphenyl propane
BICHEP	=	2,2′-bis(dicyclohexylphosphino)-6,6′-dimethyl-1-1′-biphenyl
bim	=	2,2′-biimidazolate
R-BINAP	=	**R**-2,2-bis(diphenylphosphino)-1,1′-binaphthyl
bipy	=	2,2′-bipyridine
bipym	=	2,2′-bipyrimidine
Bn	=	benzyl, CH_2Ph
BNA	=	1-benzylnicotinamide
Boc	=	*tert*-butoxycarbonyl
S,**R**-BPPFA	=	**S**-*N*,*N*-dimethyl-1-{**R**-1′,2-bis(diphenylphosphino)-ferrocenyl}ethylamine
S,**R**-BPPFOH	=	**S**-1-{**R**-1′,2-bis(diphenylphosphino)-ferrocenyl}-ethanol

BPPM	=	2**S**,4**S**-*N*-*t*-butoxycarbonyl-4-diphenylphosphino-2-diphenylphosphinomethylpyrrolidine
Brij 35	=	polyoxyethylated-*n*-dodecyl alcohol
Bu	=	butyl
Bz	=	benzoyl, –C(=O)Ph
bzn	=	benzonitrile
CAMP	=	cyclohexyl-2-methoxyphenylmethylphosphine
$C_5Me_5^-$	=	5-pentamethylcyclopentadienyl
CD	=	cyclodextrin
S,**S**-CHIRAPHOS	=	**S**,**S**-2,3-bis(diphenylphosphino)butane
COD	=	1,5-cyclooctadiene
Cp	=	η^5-cyclopentadienyl
Cp*	=	η^5-pentamethylcyclopentadienyl
CTAB	=	hexadecyltrimethylammonium bromide
Cy	=	cyclohexyl
DANTE	=	Delays alternating with nutations for tailored excitation
dba	=	1,5-diphenyl-1,4-pentadiene-3-one
de	=	diastereoisomer excess
R,**R**-DIOCP	=	**R**,**R**-*trans*-4-diphenylphosphinomethyl-5-dicyclohexylphosphinomethyl-2,2-dimethyl-1,3-dioxolan
R,**R**-DIOP	=	**R**,**R**-*trans*-4,5-bis(diphenylphosphinomethyl)-2,2-dimethyl-1,3-dioxolan
DIOXOP	=	2**R**,4**R**-bis(dimethylphosphinomethyl)-1,3-dioxolan
R,**R**-DIPAMP	=	**R**,**R**-1,2-ethanediylbis(2-methoxyphenylphenylphosphine)
DMF	=	*N*,*N*-dimethyl methanamide
DMG	=	dimethylglyoxime
dmso	=	dimethylsulfoxide
L-DOPA	=	**S**-3,4-dihydroxyphenylalanine
DOPC	=	dioleoylphosphatidylcholine
DOPE	=	dioleoylphosphatidylethanolamine
DPPB	=	1,4-bis(diphenylphosphino)butane
DPPE	=	1,2-bis(diphenylphosphino)ethane
DPPM	=	bis(diphenylphosphino)methane
DPPP	=	1,3-bis(diphenylphosphino)propane
DVB	=	1,4-bis(ethenyl)benzene

EDTA = ethylenediaminetetraacetic acid
e.e. = enantiomeric excess
en = 1,2-diaminoethane
EPR = electron paramagnetic resonance
Et = ethyl

h_0 = hexane
h_1 = 1-hexene
HD = 1,5-hexadiene
HLADH = horse liver alcohol dehydrogenase
HSDH = 20β-hydroxysteroid dehydrogenase

LipDH = pigheart lipoamide dehydrogenase
LDH = lactate dehydrogenase

M = a metal atom or ion
MBZ = methyl benzoate
Me = methyl
Men = menthyl

NADH = nicotinamide adenine dinucleotide
NBD = bicyclo[2.2.1]hepta-2,5-diene
NMDPP = (+)-neomenthyldiphenylphosphine
NMR = nuclear magnetic resonance
NORPHOS = **R**,**R**-5-*endo*-6-*exo*-bis-(diphenylphosphino)-bicyclo[2.2.1]hept-2-ene

P = monodentate tertiary phosphine
PAMP = phenyl-2-methoxyphenylmethylphosphine
P-P = bidentate tertiary phosphine
PcTS = tetrasodium phthalocyaninetetrasulfonate
PCy_3 = tricyclohexylphosphine
PEG 400 = polyethylene glycol, average molecular mass: 400
PEI = polyethyleneimine
Ph = phenyl
PhCAPP = 2**S**,4**S**-*N*-phenylamidocarbonyl-4-diphenylphosphino-2-diphenylphosphinomethylpyrrolidine
PHELLANPHOS = 5,6-bis(diphenylphosphino)-8-*iso*-propyl-2-methyl-bicyclo[2.2.2]oct-2-ene
phen = 1,10-phenanthroline

P_i = inorganic phosphate
PMHS = polymethylhydrosiloxane
Pr = propyl
R-PROPHOS = **R**-1,2-bis(diphenylphosphino)propane
PTA = 1,3,5-triaza-7-phosphaadamantane
PTC = phase transfer catalytic
PVP = poly-*N*-vinylpyrrolidinone
py = pyridine
pz = pyrazolate

Q^+ = quaternary ammonium or phosphonium ion
quin = quinoline

R = alkyl
r. t. = room temperature

SALEN = N,N′-ethylenebis(salicylideneimine)
SDS = sodium dodecylsulfonate

TEOC = 2(trimethylsilyl)ethoxycarbonyl
TFB = tetrafluorobenzobicyclo[2.2.2]octadiene
TfOH = trifluoromethyl sulphonic acid
THF = tetrahydrofuran
TPPDS = disodium [(phenyl)-3-sulfophenylphosphino]benzene-3-sulfonate (disulfonated triphenylphosphine)
TPPMS = sodium diphenylphosphinobenzene-3-sulfonate-(monosulfonated triphenylphosphine)
$TPPS_4$ = tetrasodium tetraphenylporphinetetrasulfonate
TPPTS = trisodium tris(3-sulfophenyl)phosphine (trisulfonated triphenylphosphine)
tpt = 2,4,6-tris(2-pyridyl)-*s*-triazine

WGSR = water gas shift reaction

X = one electron ligand, usually a halogen
XPES = X-ray photoelectron spectroscopy

INDEX

Catalysis by Metal Complexes

1.* F. J. McQuillin: *Homogeneous Hydrogenation in Organic Chemistry*. 1976
ISBN 90-277-0646-8

2. P. M. Henry: *Palladium Catalyzed Oxidation of Hydrocarbons*. 1980
ISBN 90-277-0986-6

3. R. A. Sheldon: *Chemicals from Synthesis Gas*. Catalytic Reactions of CO and H_2. 1983
ISBN 90-277-1489-4

4. W. Keim (ed.): *Catalysis in C_1 Chemistry*. 1983 ISBN 90-277-1527-0

5. A. E. Shilov: *Activation of Saturated Hydrocarbons by Transition Metal Complexes*. 1984 ISBN 90-277-1628-5

6. F. R. Hartley: *Supported Metal Complexes*. A New Generation of Catalysts. 1985
ISBN 90-277-1855-5

7. Y. Iwasawa (ed.): *Tailored Metal Catalysts*. 1986 ISBN 90-277-1866-0

8. R. S. Dickson: *Homogeneous Catalysis with Compounds of Rhodium and Iridium*. 1985
ISBN 90-277-1880-6

9. G. Strukul (ed.): *Catalytic Oxidations with Hydrogen Peroxide as Oxidant*. 1993
ISBN 0-7923-1771-8

10. A. Mortreux and F. Petit (eds.): *Industrial Applications of Homogeneous Catalysis*. 1988 ISBN 90-277-2520-9

11. N. Farrell: *Transition Metal Complexes as Drugs and Chemotherapeutic Agents*. 1989
ISBN 90-277-2828-3

12. A. F. Noels, M. Graziani and A. J. Hubert (eds.): *Metal Promoted Selectivity in Organic Synthesis*. 1991 ISBN 0-7923-1184-1

13. L. I. Simándi: *Catalytic Activation of Dioxygen by Metal Complexes*. 1992
ISBN 0-7923-1896-X

14. K. Kalyanasundaram and M. Grätzel (eds.), *Photosensitization and Photocatalysis Using Inorganic and Organometallic Compounds*. 1993 ISBN 0-7923-2261-4

15. P. A. Chaloner, M. A. Esteruelas, F. Joó and L. A. Oro: *Homogeneous Hydrogenation*. 1994 ISBN 0-7923-2474-9

16. G. Braca (ed.): *Oxygenates by Homologation or CO Hydrogenation with Metal Complexes*. 1994 ISBN 0-7923-2628-8

Kluwer Academic Publishers – Dordrecht / Boston / London

**Volume 1 is previously published under the Series Title: Homogeneous Catalysis in Organic and Inorganic Chemistry.*